Nonlinear Phenomena in Complex Systems: From Nano to Macro Scale

NATO Science for Peace and Security Series

This Series presents the results of scientific meetings supported under the NATO Programme: Science for Peace and Security (SPS).

The NATO SPS Programme supports meetings in the following Key Priority areas: (1) Defence Against Terrorism; (2) Countering other Threats to Security and (3) NATO, Partner and Mediterranean Dialogue Country Priorities. The types of meeting supported are generally "Advanced Study Institutes" and "Advanced Research Workshops". The NATO SPS Series collects together the results of these meetings. The meetings are co-organized by scientists from NATO countries and scientists from NATO's "Partner" or "Mediterranean Dialogue" countries. The observations and recommendations made at the meetings, as well as the contents of the volumes in the Series, reflect those of participants and contributors only; they should not necessarily be regarded as reflecting NATO views or policy.

Advanced Study Institutes (ASI) are high-level tutorial courses to convey the latest developments in a subject to an advanced-level audience

Advanced Research Workshops (ARW) are expert meetings where an intense but informal exchange of views at the frontiers of a subject aims at identifying directions for future action

Following a transformation of the programme in 2006 the Series has been re-named and re-organised. Recent volumes on topics not related to security, which result from meetings supported under the programme earlier, may be found in the NATO Science Series.

The Series is published by IOS Press, Amsterdam, and Springer, Dordrecht, in conjunction with the NATO Emerging Security Challenges Division.

Sub-Series

A.	Chemistry and Biology	Springer
B.	Physics and Biophysics	Springer
C.	Environmental Security	Springer
D.	Information and Communication Security	IOS Press
E.	Human and Societal Dynamics	IOS Press

http://www.nato.int/science
http://www.springer.com
http://www.iospress.nl

Series C: Environmental Security

Nonlinear Phenomena in Complex Systems: From Nano to Macro Scale

edited by

Davron Matrasulov
Turin Polytechnic University in Tashkent
Tashkent, Uzbekistan

and

H. Eugene Stanley
Boston University, Department of Physics
Boston, MA, USA

Springer

Published in Cooperation with NATO Emerging Security Challenges Division

Proceedings of the NATO Advanced Research Workshop on
New Challenges in Complex Systems: Disaster Forecasting,
Crisis Modeling and Sustainable Development
Samarkand, Uzbekistan
20–24 May, 2013

Library of Congress Control Number: 2014936100

ISBN 978-94-017-8707-9 (PB)
ISBN 978-94-017-8703-1 (HB)
ISBN 978-94-017-8704-8 (e-book)
DOI 10.1007/978-94-017-8704-8

Published by Springer,
P.O. Box 17, 3300 AA Dordrecht, The Netherlands.

www.springer.com

Printed on acid-free paper

Preface

This book contains the papers presented at the NATO Advanced Research Workshop, "New Challenges in Complex Systems: Disaster forecasting, crisis modeling and sustainable development" held in Samarkand, Uzbekistan, from May 20 to 24, 2013.

The focus of the Workshop was to discuss problems and prospects in the area of complex system physics and its applications to a broad spectrum of problems from physics, biology, natural disasters and social sciences with a special focus on nonlinear evolution, extreme events, crisis and critical phenomena.

In the past decade, the physics of complex systems has become one of the most interdisciplinary subjects of contemporary science. In particular, this field has penetrated into various areas of natural and even social sciences for the modeling of complex dynamics, data analysis, prediction and monitoring purposes.

Talks presented at the Workshop covered such topics as econophysics, sociophysics, earthquake dynamics, crisis and conflict modeling, complex networks and complex nanoscale systems.

A total of 58 talks were presented by invited and plenary speakers, with 35- and 20-min durations, respectively. Panel discussions attracted a broad audience of experts working on physics, mathematics, seismology, economics and sociology.

The conference was successfully organized due to the efforts of the local organizing committee with K. Sharipov, U. Tashkenbaev, K. Muminov, B. Eschanov, J. Yusupov and A. Saidov. Our special thanks to Samarkand State University for its great contribution to the local organizing activity.

Finally, we would like to thank NATO Science for Peace and Security Program for their funding of the Workshop. Additional support was provided by Physical Society of Uzbekistan and Turin Polytechnic University in Tashkent.

Tashkent, Uzbekistan
November, 2013

H. Eugene Stanley
Davron Matrasulov

Contributors

Elena Agliari Dipartimento di Fisica e Scienza della Terra and INFN, Universita' di Parma, Parma, Italy

Yoji Aizawa Department of Applied Physics, Advanced School of Science and Engineering, Waseda University, Shinjuku, Tokyo, Japan

Adnan Aydemir Kimya Mühendisliği Bölümü, Yeditepe Üniversitesi, İstanbul, Turkey

Alexander Blumen Physikalisches Institut, Universität Freiburg, Freiburg, Germany

Evgeny N. Bulgakov Institute of Physics, Krasnoyarsk, Russia

Raffaella Burioni Dipartimento di Fisica e Scienza della Terra and INFN, Universita' di Parma, Parma, Italy

Guido Caldarelli IMT Alti Studi Lucca, Lucca, Italy

Davide Cassi Dipartimento di Fisica e Scienza della Terra and INFN, Universita' di Parma, Parma, Italy

Claudio Castellano Istituto dei Sistemi Complessi (ISC-CNR), Roma, Italy

Dipartimento di Fisica, "Sapienza" Università di Roma, Roma, Italy

Taksu Cheon Kochi University of Technology, Kami, Kochi Prefecture, Japan

Alessandro Chessa IMT Alti Studi Lucca, Lucca, Italy

L.Y. Chew Division of Physics and Applied Physics, Nanyang Technological University, Singapore

N.N. Chung Temasek Laboratories, National University of Singapore, Singapore

Chester Curme Center for Polymer Studies, Department of Physics, Boston University, Boston, MA, USA

Gregorio D'Agostino ENEA – CR Casaccia, Roma, Italy

Alfonso Damiano Dipartimento di Ingegneria Elettrica ed Elettronica, Univ. di Cagliari, Cagliari, Italy

Angelo De Santis Istituto Nazionale di Geofisica e Vulcanologia (INGV), Rome, Italy

Pavel Exner Doppler Institute for Mathematical Physics and Applied Mathematics, Czech Technical University, Prague, Czech Republic

Nuclear Physics Institute, ASCR, Řež near Prague, Czech Republic

Serge Galam CEVIPOF – Centre for Political Research, CNRS and Sciences Po, Paris, France

Güngör Gündüz Kimya Mühendisliği Bölümü, Orta Doğu Teknik Üniversitesi, Ankara, Turkey

Oksana B. Ismailova Institute Ion-Plasma and Laser Technology, Uzbekistan Academy of Sciences, Tashkent, Uzbekistan

Choy Heng Lai Department of Physics and Yale-NUS College, National University of Singapore, Singapore

Haojiong Liu New Jersey Institute of Technology, Newark, NJ, USA

Dmitrii N. Maksimov Institute of Physics, Krasnoyarsk, Russia

Oksana Manzhura New Jersey Institute of Technology, Newark, NJ, USA

M.B. Marasulov Institute of Polymer Chemistry and Physics, Tashkent, Uzbekistan

Davron Matrasulov Turin Polytechnic University in Tashkent, Tashkent, Uzbekistan

Helen Susannah Moat Behavioural Science, Warwick Business School, University of Warwick, Coventry, UK

Cesar A. Mujica-Martinez I. Institut für Theoretische Physik, Universität Hamburg, Hamburg, Germany

Oliver Mülken Physikalisches Institut, Universität Freiburg, Freiburg, Germany

Mario Mureddu Linkalab, Complex Systems Computational Laboratory, Cagliari, Italy

K. Nakamura Turin Polytechnic University in Tashkent, Tashkent, Uzbekistan

Peter Nalbach I. Institut für Theoretische Physik, Universität Hamburg, Hamburg, Germany

Edip Niver New Jersey Institute of Technology, Newark, NJ, USA

B.L. Oksengendler Institute of Polymer Chemistry and Physics, Tashkent, Uzbekistan

Sakshi Pahwa Department of Electrical and Computer Engineering, College of Engineering, Kansas State University, Manhattan, KS, USA

Konstantin N. Pichugin Institute of Physics, Krasnoyarsk, Russia

Tobias Preis Behavioural Science, Warwick Business School, University of Warwick, Coventry, UK

Walter Quattrociocchi London Institute of Mathematical Sciences, Mayfair, London, UK

K.K. Sabirov Turin Polytechnic University in Tashkent, Tashkent, Uzbekistan

Almas F. Sadreev Institute of Physics, Krasnoyarsk, Russia

A.A. Saidov Turin Polytechnic University in Tashkent, Tashkent, Uzbekistan

Antonio Scala ISC-CNR Physics Department, Univ. La Sapienza, Roma, Italy

Robert Schrader Institut für Theoretische Physik, Freie Universität Berlin, Berlin, Germany

Dieter Schuch Institut für Theoretische Physik, J.W. Goethe-Universität Frankfurt am Main, Frankfurt am Main, Germany

Caterina Scoglio Department of Electrical and Computer Engineering, College of Engineering, Kansas State University, Manhattan, KS, USA

Zarif A. Sobirov Tashkent State Financial Institute, Tashkent, Uzbekistan

H. Eugene Stanley Boston University, Department of Physics, Boston, MA, USA

Ibrahim Tekin New Jersey Institute of Technology, Newark, NJ, USA

Michael Thorwart I. Institut für Theoretische Physik, Universität Hamburg, Hamburg, Germany

Satoru Tsugawa Department of Applied Physics, Advanced School of Science and Engineering, Waseda University, Shinjuku, Tokyo, Japan

N.N. Turaeva Biological Department, Webster University, Louis, MO, USA

J.R. Yusupov Turin Polytechnic University in Tashkent, Tashkent, Uzbekistan

Part I
Complexity in Earthquake Dynamics

Chapter 1
Geosystemics, Entropy and Criticality of Earthquakes: A Vision of Our Planet and a Key of Access

Angelo De Santis

Abstract Earth is a system of interconnected systems, whose complexity is far from being fully understood by a reductionist approach alone. In this chapter we introduce the concept of geosystemics and the use of the entropy to characterize some aspects of the phenomena under study. We will show how entropy and criticality of the system are central to better understand the most important general features of earthquakes. We will analyze two recent seismic sequences culminated with a main-shock (2009 L'Aquila and 2012 Emilia, both in Italy) to show the potential of this approach and to understand some important characteristics of the seismicity under scrutiny.

1.1 Introduction

If we define a rare event as some process that occurs much less frequently than the times it does not occur, earthquakes are rare events: even in the most known seismic fault, like S. Andrea's fault (USA), the times with no earthquakes are more frequent than the times when there is some seismic activity. In addition, the more energetic the earthquake, the more rare its occurrence ([17]; see below for more details). However, as in many other physical processes, it is not only the frequency of occurrence to be important, but also the energy released by the seismic phenomenon, and, consequently, its impact to society. When a rare event has a large impact, especially in terms of catastrophic consequences, it is said to be an extreme event. One of the most recent frontier of science is represented by the study of extreme events, meaning with these, many natural environmental events

A. De Santis (✉)
Istituto Nazionale di Geofisica e Vulcanologia (INGV), Rome, Italy
e-mail: angelo.desantis@ingv.it

D. Matrasulov and H.E. Stanley (eds.), *Nonlinear Phenomena in Complex Systems: From Nano to Macro Scale*, NATO Science for Peace and Security Series C: Environmental Security, DOI 10.1007/978-94-017-8704-8_1,

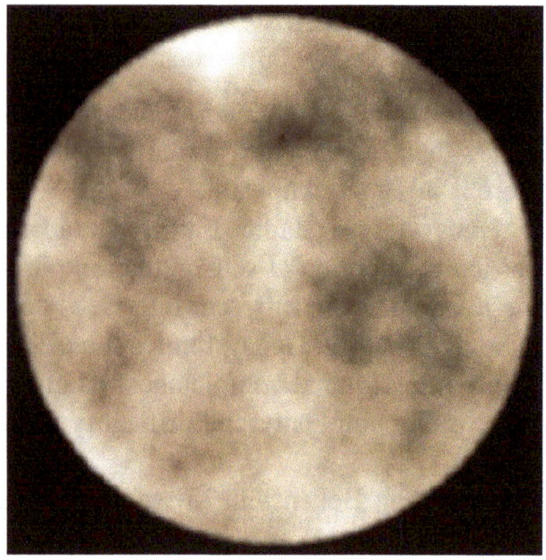

Fig. 1.1 The core-mantle boundary of Earth (2,900 km deep from the surface) might be fractal (De Santis and Barraclough [8]). This figure shows a simulation of CMB with $D = 2.2$

This measure is powerful because with a simple real number (i.e. D) usually less than the embedding topological dimension E (see also below), it characterizes the repeatability of the physical phenomenon or distribution at different scales: the closer D to the embedding dimension, the more the process tends to cover all the space at disposal.

Figure 1.1 shows an example of fractal interpretation with $D = 2.2$ that has been given for the core-mantle boundary (CMB) of Earth, from the study of the geomagnetic field power spectra over the last 400 years projected to the CMB [8].

The concept of fractal dimension is also important in characterizing the dynamical evolution of complex systems (see next section).

1.3.2 Phase Space

The phase space of a dynamical system is the ideal space where each state of the system can be represented by a single point. The minimum number E of phase space axes, which contain all orbits of the dynamics, defines the degrees of freedom of the system, i.e. the number of variables that are required to describe that system. E is also said embedding dimension. The theorem for reconstructing a pseudo phase space from one dimensional signal [37] is fundamental to find the possible chaos of a dynamical system: generally a chaotic process has an orbit in its phase space that falls in a restricted region with a fractal dimension (strange attractor).

A reconstructed phase space of the geomagnetic field is shown in Fig. 1.2. It has been obtained by placing along the three Cartesian axes the magnetic field variations taken at three distant geomagnetic observatories: Hartland (UK), Alibag (India) and Irkutsk (Russia). The reconstructed orbit resembles the picture paradigm of chaos: the Lorenz's butterfly, which is the attractor of a toy-model for the atmosphere dynamics [19].

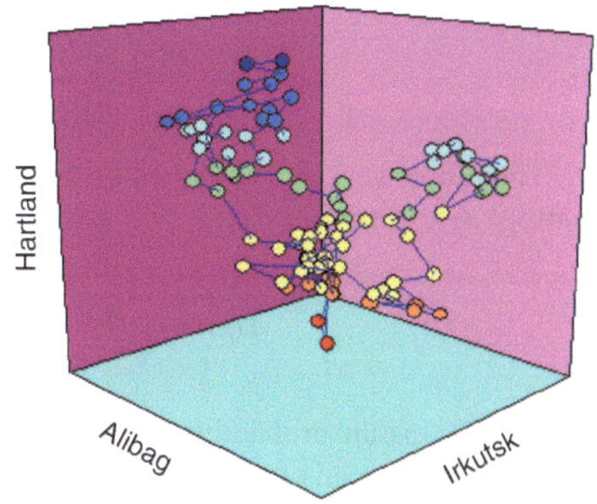

Fig. 1.2 Reconstructed phase space of the geomagnetic field for the last 150 years (De Santis and Barraclough [11])

1.3.3 Shannon Entropy

Shannon Entropy [33] of a system characterized by N independent states and a probability distribution $p_i(t)$ is defined as follows:

$$H(t) = -\sum_{i=1}^{N} p_i(t) \cdot \log p_i(t) \tag{1.2}$$

(for convenience, we impose $\log p_i = 0$ for $p_i = 0$). The base of the logarithm can be arbitrarily chosen. Usually it is taken in a way to be comparable with the number of possible states, N. Some physical interpretations of this important quantity are possible, some of them overlapping and non exclusive:

1. A generalized measure of disorder;
2. Measure of the average missing information content to the knowledge about the state of the system;
3. Measure of unpredictability of the state of the system among many alternatives;
4. Measure of the degree of dispersion of an observable among the system's parts.

From the above variety of meanings one can realize the importance of studying entropy for grasping something about the space-time evolution of a complex system.

1.3.4 Shannon Information and Entropy over a Sphere

If $B(t)$ is a physical quantity defined over a sphere, we can express it by a linear combination of orthonormal spherical harmonics Ψ_n^m with maximum degree N (which defines the smallest detail of the representation) through a set of coefficients c_n^m, i.e.:

$$B(t) = \sum_{i=1}^{N} \sum_{m=0}^{n} c_n^m(t) \Psi_n^m \tag{1.3}$$

Therefore we can consequently define the information content $I(t)$ (and its entropy $H(t)$) as:

$$I(t) = -H = \sum_{n=1}^{N} p_n(t) \cdot \log p_n(t) \tag{1.4}$$

where $p_n(t)$ is the probability to have a certain n-degree spherical harmonic power contribution instead of another at different time t [12], i.e.:

$$p_n = \frac{\langle B_n^2 \rangle}{\langle B^2 \rangle} = \frac{\sum_{m=0}^{n} (c_n^m)^2}{\sum_{n'=0}^{N} \sum_{m=0}^{n'} (c_{n'}^m)^2}$$

where $\langle x^2 \rangle$ is the mean square value of the quantity x.

The information content has been estimated for the geomagnetic field of the last 100 years to suggest a possible imminent magnetic polarity transition [9, 12] and extended to the past 7,000 years [10]. In principle, the definitions (1.3) and (1.4) can be applied to any physical quantity over the Earth, supposed a spherical planet in first approximation.

1.4 Entropy of Earthquakes

Any approach to understanding earthquake physics is worth making. Our geosystemic approach is holistic, although it does not intend to be against the reductionist approach, but it wants to be in parallel/complementary. This approach looks at

- The earth as a whole, and
- The specific phenomenon under study in its most important macroscopic features.

Here we will see some cases of seismicity. Before doing that, we will recall some important empirical statistical laws which characterize the typical seismicity over the globe, and/or over a certain region.

1.4.1 Some Empirical Statistical Laws

Some statistical laws have been established in seismology, based on a great amount of empirical studies on earthquakes. The most important is the Gutenberg-Richter (G-R) Law (1944) which is valid for all earthquakes in a given time and region. Other important laws are valid for aftershocks(i.e. the earthquakes occurred after

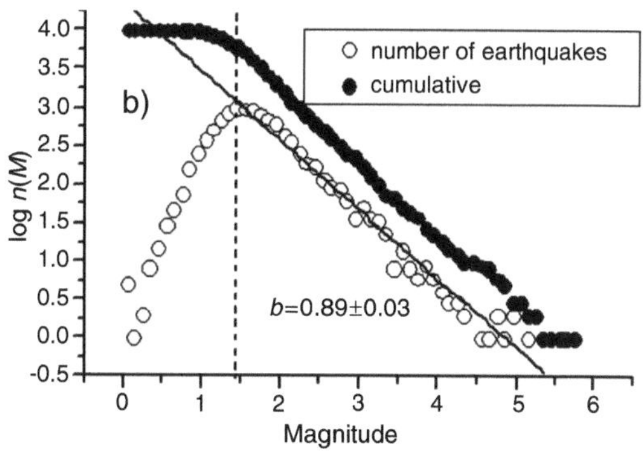

Fig. 1.3 Gutenberg-Richter (G-R) diagram for the 2009 L'Aquila seismic sequence (De Santis et al. [13])

a main-shock): Omori [26], Båth [2] and Felzer and Brodsky [16] laws. In the following we will describe these laws with some more details.

(i) **G-R Law** (1944). *The rate of earthquakes occurrence (number of earthquakes n in a certain time interval) in a given region follows an exponential law of the magnitude M* (i.e., small earthquakes are many more than larger ones). Usually this law is given in terms of the cumulative frequency distribution of earthquakes (n earthquakes equal to or above a given magnitude M):

$$\log n = a - bM \quad (b \cong 1) \tag{1.5}$$

here log is the decimal logarithm, a and b are constant positive parameters to be estimated in the given region; the latter parameter, the b-value, is the slope of the cumulative earthquake distribution, and is an important quantity in seismology assuming almost everywhere a value close to 1.

Because there is an exponential relation between energy and magnitude, Eq. (1.5) becomes a power law relation between n and energy. In this framework, the G-R law is fractal and $b = D/2$, where D is the fractal dimension [38]. For a typical $b = 1$, then $D = 2$: this result can also be found from the statistical distribution of the power spectra vertical to horizontal seismic components ratio [34].

For Italy (Fig. 1.3 shows the case of 2009 L'Aquila seismic sequence where $b = 0.89$; De Santis et al. [13]), we have the following frequencies for different magnitudes:

Case of M5 (magnitude 5 and above): 1/year Case of M6: 0.1/year, i.e. 1/10 yrs.
Case of M7: 0.01/year, i.e. 1/100 yrs.

As we can understand, the above values are purely statistical and cannot provide any deterministic prediction on when and where a large earthquake will occur exactly in Italy, but only some rough spatial and temporal bounds which are the base for any map of seismic hazard.

On the other hand, there are other three statistical laws in seismology which are valid for aftershocks:

(ii) **Omori Law** (1894; slightly modified by Utsu in 1961 [40]; we give here the modified version). It is an inverse power law of the rate n(t) of aftershocks occurrence:

$$n(t) = K/(c+t)^p \quad \text{with } p \cong 1 \tag{1.6}$$

Here K, c and p are appropriate constant parameters, while t is the time after the main-shock.

(iii) **Båth Law** (1965). This rule predicts that the largest aftershock occurs with a given magnitude $\max[M_{after}]$ around a unit less than the main-shock magnitude, M_{main}. That is:

$$\Delta M = M_{main} - \max[M_{after}] \cong 1.2 \pm 0.2 \tag{1.7}$$

However, among all the laws, this is the most approximated rule. For example, for the L'Aquila seismic sequence, culminated with a $M6.2$ main-shock on 6 April 2009, condition (1.7) is satisfied because $M = 1.0$; while for the more recent 2012 Emilia seismic sequence culminated with a $M5.9$ main-shock on 20 May, 2012, this rule does not fit the real situation, because another major earthquake ($M5.8$) occurred 9 days later, so M would be 0.1. However, if we consider the two major earthquakes as a unique physical entity, their subsequent largest aftershock was around 1 unit of magnitude less, thus substantially confirming Båth law.

(iv) **Felzer & Brodsky** (2006). These authors have proposed an inverse power law for the probability P of having an aftershock at distance r from the main-shock epicenter (valid at least up to 100 km):

$$P(r) = K'/r^s \tag{1.8}$$

where K' and s are appropriate constant parameters; the latter exponent has a range of variability between 1 and 2 (see also [21]). This law would be related to the attenuation with distance of the energy released by an earthquake.

1.4.2 Entropy and b-Value of Earthquakes

De Santis et al. [13] have applied the concept of entropy H to earthquakes and then, for the first time, related this quantity to the b-value of the G-R law, such as:

$$H(t) = k' - \log b \tag{1.9}$$

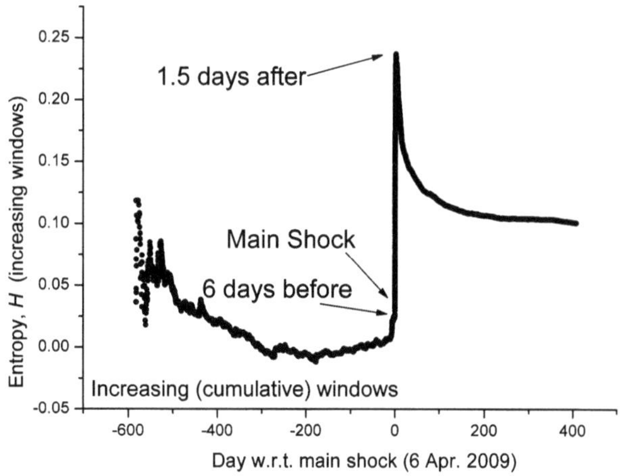

Fig. 1.4 Entropy of the 2009 L'Aquila seismic sequence (De Santis et al. [13])

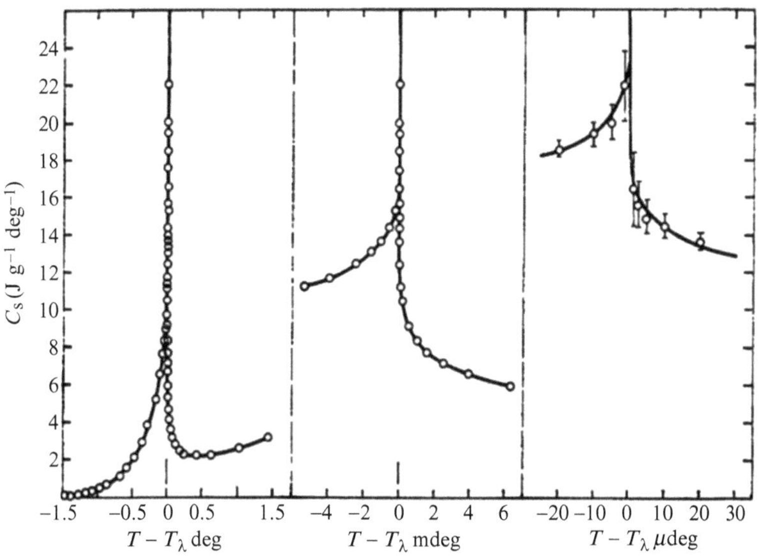

Fig. 1.5 Specific heat of ^{4}He as a function of $T - T_\lambda$ in Kelvin at different resolutions. T_λ is the temperature at which the critical system has a transition (critical point) (Reproduced from Stanley [36])

where $k' = \log(e \log e)$. Equation (1.9) can also be approximated as:

$$b \approx \frac{1.2}{10^H} \tag{1.10}$$

These equations provide a new insight into the b-value meaning [13]. Figure 1.4 represents the entropy for the 2009 L'Aquila seismic sequence when considering an increasing (cumulative) time windows.

It is interesting to notice that this picture (when we translate entropy to specific heat and time to temperature) resembles the plots of Fig. 1.5 (which is a reproduction of Fig. 1.16 by Stanley [36]) for the phase transitions of some critical system, where the main-shock of Fig. 1.4 belongs to the main singularity of the curve. This feature would support the idea that the sequence preceding a main-shock is the manifestation of a critical system moving toward a critical point (the main-shock). This idea is nothing new and has been already proposed by several authors (e.g. [4, 18]).

In general, from the behaviour of $H(t)$ in Fig. 1.4 some major features of the earthquakes evolution can be noticed:

- A preparation phase as a generalized increase (months before main-shock);
- A concentration (or singularity) phase as a sudden jump from days to hours before main-shock;
- A diffusive phase, as a general decrease of $H(t)$ a few hours or days after the main-shock.

Although the main-shock is not perfectly the entropic singularity, it belongs to the population of events characterising the singular behaviour of H. This aspect should be better exploited to evaluate the possibility of some main-shock predictability.

1.4.3 Magnetic Transfer Function Entropy

The entropy concept can also be applied to other quantities which are not seismic, although still related to earthquakes. For instance we show here and in the next section how it can be important when the entropy is used to characterize the geomagnetic field variations recorded by two different platforms of observation (i.e. ground observatory and satellite) above the epicentral region during a given seismic sequence. In the frequency domain the time variations of the components X, Y, Z of the geomagnetic field observed at Earth surface are each other coupled (e.g., [15]):

$$Z(\omega) = A(\omega)X(\omega) + B(\omega)Y(\omega) \tag{1.11}$$

$A(\omega)$ and $B(\omega)$ are the Magnetic Transfer Functions which are related with the conductivity at a certain depth inversely proportional to the square root of frequency ω.

The (normalised) entropy contribution E_i of the harmonic ω_i is given by:

$$E_i(t) = -\frac{p(\omega_i, t) \cdot \log p(\omega_i, t)}{\log N} \quad \text{where} \quad p(\omega_i, t) = \frac{K_r^2(\omega_i, t)}{\sum_{i=1}^{N} K_r^2(\omega_i, t)}$$

K_r = real parts of A or B.

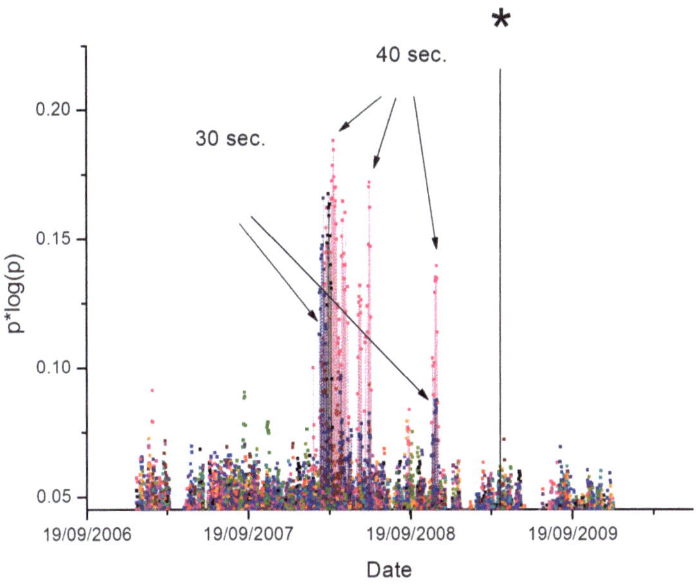

Fig. 1.6 Contributions to the Transfer Function Entropy from magnetic field records of L'Aquila Geomagnetic Observatory (Central Italy), in the period 2007–2009 (the *star* represents the occurrence of the main-shock). The most emerging peaks are the characteristic periods of 30–40 s. (penetrating till 15–20 km depth) (Reproduced from Cianchini et al. [6])

The results shown in Fig. 1.6 in terms of the $p \log p$ contributions to $E_i(t)$ for the two most emerging characteristic periods (30 and 40 s) which correspond to magnetic field variations that can penetrate till 15–20 km in this area, can be suitably interpreted as due to some upward migration of fluids from below to the hypocentral zone (10 km) that activated the 2009 L'Aquila seismic sequence [6].

1.4.4 Wavelet Entropy of Satellite Magnetic Signal

From the Fourier analysis of a certain signal characterised by a power P_i at a given frequency f_i, it is possible to estimate its spectral entropy S [28], i.e.:

$$S = -\sum_i P_i \log P_i \quad \text{where} \quad P_i = \frac{|f_i|^2}{\sum_{i'} |f_{i'}|^2}$$

This entropic quantity measures how energy *concentrates* (low spectral entropy) or *spreads* (high spectral entropy) in frequency.

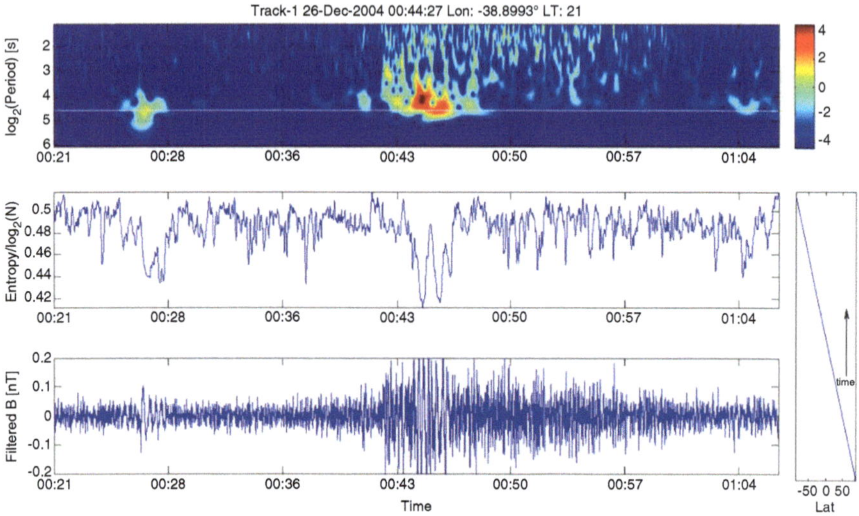

Fig. 1.7 Wavelet spectrum (*top*); wavelet entropy (*middle*) and magnetic satellite signal behavior (*bottom*) of almost 1 h around the occurrence of M9 Sumatra mega-earthquake (26 December, 2004). The lateral plot on the *right* shows the latitudinal variations of the CHAMP satellite orbit (Reprinted after Cianchini et al. [5])

On the other hand, if we consider a *wavelet analysis* that decomposes a signal $f(t)$ both in time and scale (or frequency), transforming it to $W(s, \tau)$, analogously to the Fourier analysis we can define a wavelet entropy such as:

$$S_W = -\sum p_{s,\tau} \log_2 p_{s,\tau} \tag{1.12}$$

where the probability is expressed in term of proportion of energy E with respect to the total:

$$p_{s,\tau} = \frac{E(s,\tau)}{\sum_s E(s,\tau)} \quad \text{with} \quad E(s,\tau) = |W(s,\tau)|^2$$

Please note that, for convenience, we preferred here to use the logarithm in base 2, just because the scale frequency is usually given in powers of 2.

We applied (1.12) to the magnetic signal recorded by CHAMP satellite (e.g. [31]), that flew from 2000 to 2010 in a quasi-polar orbit at 350–450 km of altitude with a couple of vector and scalar magnetometers aboard. Figure 1.7 shows (from top to bottom): wavelet spectrum; corresponding wavelet entropy and signal behavior of almost 1 h around the occurrence of M9 Sumatra mega-earthquake (26 December, 2004). There is an anomalous behavior that appears in all graphs and anticipates the main-shock by a few minutes [5]. If this anomaly is really related with the mega-earthquake, this will represent the result of a solid earth-ionosphere coupling, a scheme that is often advocated to justify such kinds of anomalies in the ionosphere (e.g. [29]).

1.5 Criticality and Multi-scale Space-Time Focalization

This section concerns with the peculiar behavior of seismicity in time and space that often occurs before a large earthquake. As didactical example we will take the lessons learnt from the 2009 L'Aquila seismic sequence to apply them to the most recent 2012 Emilia seismic sequence.

Two recent papers by De Santis et al. [10, 13] analyzed the former sequence and found some interesting results. The L'Aquila seismic sequence showed a sort of space-time focalization, in the sense that seismicity accelerated in time around a year before the main-shock, and in space most of the seismic events migrated from an almost circular periphery around the impending large earthquake toward the epicentral region. The latter aspect was already discovered in Japan by Mogi [24], with the so-called *Mogi doughnut* of seismicity. In the case of L'Aquila sequence this doughnut was characterized by a seismicity gap in the fault till sometime before activation of the epicentral area.

The acceleration in time was characterized by a power law behavior of the cumulative Benioff strain s(t) (e.g. [4]) such as the following one:

$$s(t) = A + B(t_f - t)^m \tag{1.13}$$

with the vertical temporal derivative at the main-shock (i.e. at $t = t_f$). $A > 0$, $B < 0$ and $0 < m < 1$ are appropriate constant parameters; t_f is also called the time of failure of the region under study, because it is the predicted time of occurrence of the main-shock.

This behavior has been interpreted as that of the critical process before the large earthquake moving toward the critical point, i.e. the main-shock, although some recent works criticize this interpretation (e.g. [23]). Studying the acceleration of L'Aquila seismicity before the 2009 main-shock, De Santis et al. [10] also found that the process of preparation behaved as a chaotic process, with characteristic time of around 10 days.

Figure 1.8 shows the seismicity distribution in Emilia of historical earthquakes for the last 1,000 years (left) and the more recent distribution of the $M \geq 4$ earthquakes for the last 7 years. Comparing the two sides of Fig. 1.8 it is evident that the recent 2012 seismicity filled the gap at the central of a partial Mogi doughnut [24], that was characterized by some kind of long quiescence (no significant seismicity).

Regarding the acceleration in time, also the 2012 Emilia sequence shows a similar behavior in time (Fig. 1.9), although the acceleration appears weaker than in the case of 2009 L'Aquila seismic sequence.

To better assess the evolution of the associated seismic events with the distances from the epicenter, we considered the same dataset of $M \geq 4$ events occurring in a circular region of 200 km from the epicenter of the first major earthquake on 20 May 2012 (Fig. 1.10 left). This distribution of the considered distances with time can be fit with an anti-diffusion equation of the following type:

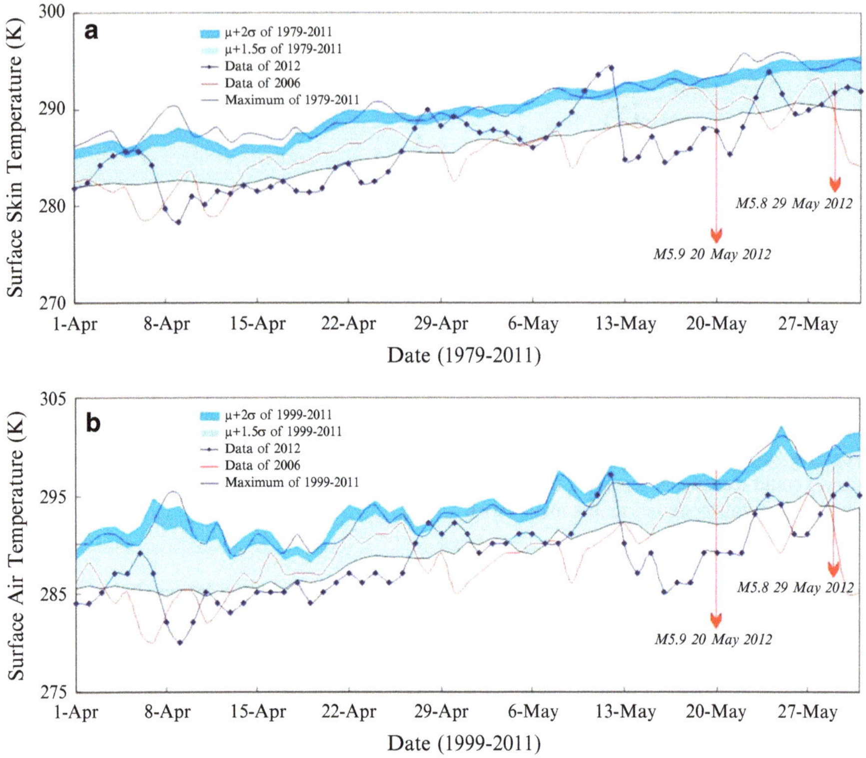

Fig. 1.11 Surface temperatures from satellite (*top*) and from land (*bottom*) a month and half before the two major earthquakes in Emilia (Qin et al. [30])

couple of major earthquakes is compatible with a partial **Mogi doughnut** model characterized by a precursory space-time focalization process. This time evolution, as also confirmed by a power law acceleration of the cumulative Benioff strain, has many characteristics of a critical point process both in space and in time.

1.6 Conclusions

We defined a new systemic approach (the *vision*) to Earth system study called **geosystemics** where multi-platform/parameter/scale observations are fundamental to take a whole picture of our planet, in general, or of some important phenomenon on it, in particular. Fundamental tools (the *keys*) have been proposed, mainly based on **Entropy (and/or Information) and criticality** analysing the whole and the relationships among components. We showed an important application to seismology (*disclosing the relationship between b-value and entropy*) and then something more about criticality and space-time focalisation of a seismic process (cases of 2009 L'Aquila and 2012 Emilia earthquakes). Future can provide other cases of application in seismology and in other fields of Earth sciences.

Acknowledgements Some work was made in the frame of SAGA-4-EPR (Italian Foreign Office funded project) and during a visit of ADS at North Eastern University (China) invited by Professors Liu and Wu. Their hospitality has been appreciated very much. I thank also some of my collaborators (Gianfranco Cianchini and Enkelejda Qamili) for their assistance in making some analyses or graphs, and/or for inspiring some concepts or analyses. I finally thank the organizers of the NATO workshop on New Challenges in Complex System Physics, held on 20–24 May 2013 in Samarkand, Uzbekistan. I thank in particular Dr. Davron Matrasulov and his collaborators, for providing the right atmosphere of cooperation and stimulating discussion during the workshop.

References

1. Baranger M (2001) Chaos, complexity and entropy: a physics talk for non-physicists. Wesleyan University Physics Dept. Colloquium
2. Båth M (1965) Lateral inhomogeneities in the upper mantle. Tectonophysics 2:483–514
3. Bekenstein JD (2003) Information in the holographic universe. Sci Am 289(2):61
4. Bowman DD, Ouillon G, Sammis CG, Sornette A, Sornette D (1998) An observational test of the critical earthquake concept. J Geophys Res 103:24359–24372
5. Cianchini G, De Santis A, Balasis G, Mandea M, Qamili E (2009) Entropy based analysis of satellite magnetic data for searching possible electromagnetic signatures due to big earthquakes. In: Proceedings of the 3rd IASME/WSEAS international conference on GEology and Seismology (GES'09), Cambridge, pp 29–35
6. Cianchini G, De Santis A, Barraclough DR, Wu LX, Qin K (2012) Magnetic transfer function entropy and the 2009 Mw6.3 L'Aquila earthquake (Central Italy). Nonlinear Process Geophys 19:401–409
7. De Santis A (2009) Geosystemics. In: Proceedings of the 3rd IASME/WSEAS International conference on Geology and Seismology (GES'09), Cambridge, Feb 2009, pp 36–40
8. De Santis A, Barraclough DR (1997) A fractal interpretation of the topography of the scalar geomagnetic scalar potential at the core-mantle boundary. Pure Appl Geophys 149(4):747–760
9. De Santis A, Qamili E (2008) Are we going toward a global planetary magnetic change? In: Proceedings of 1st WSEAS international conference on environmental and geological science and engineering (EG'08), Malta, Sept 2008, pp 149–152
10. De Santis A, Qamili E (2010) Shannon information of the geomagnetic field for the past 7000 years. Nonlinear Process Geophys 17:77–84
11. De Santis A, Barraclough DR, Tozzi R (2002) Nonlinear variability in the geomagnetic secular variation of the last 150 years. Fractals 10(3):297–304
12. De Santis A, Tozzi R, Gaya-Piqué L (2004) Information content and K-entropy of the present geomagnetic field. Earth Planet Sci Lett 218(3):269–275
13. De Santis A, Cianchini G, Favali P, Beranzoli L, Boschi E (2011) The Gutenberg-Richter law and entropy of earthquakes: two case studies in Central Italy. Bull Seism Soc Am 101(3):1386–1395
14. De Santis A, Cianchini G, Wu LX, Qin K (2013) Multiscale space-time focalization as precursory pattern of seismicity of the Emilia (Italy) major earthquakes in May 2012 (in preparation)
15. Di Mauro D, Armadillo E, Bozzo E, Cerv V, De Santis A, Gambetta M, Meloni A (1998) GDS (geomagnetic depth sounding) in Italy: applications and perspectives. Ann Geophys 41(3):477–490

16. Felzer KR, Brodsky EE (2006) Decay of aftershock density with distance indicates triggering by dynamic stress. Nature 441:735–737
17. Gutenberg B, Richter CF (1944) Frequency of earthquakes in California. Bull Seism Soc Am 34:185–188
18. Jaumé SC, Sykes LR (1999) Evolving towards a critical point: a review of accelerating seismic moment – energy release prior to large and great earthquakes. Pure Appl Geophys 155:279–306
19. Lorenz EN (1962) Deterministic nonperiodic flow. J Atmos Sci 20:130–141
20. Lovelock J (2000) Gaia: a new look at life on earth. Oxford University Press, New York
21. Marsan D, Lengline O (2010) A new estimation of decays of aftershock density with distance to the mainshock. J Geophys Res 115:B09302. doi:10.1029/2009JB007119
22. Meyers RA (2009) Extreme environmental events. Springer, New York, p 1250
23. Mignan A (2012) Seismicity precursors to large earthquakes unified in a stress accumulation framework. Geophys Res Lett 39:L21308. doi:10.1029/2012GL053946
24. Mogi K (1969) Some features of recent seismic activity in and near Japan (2), Activity before and after great earthquakes. Bull Earthq Res Inst Univ Tokyo 47:395–417
25. Nott J (2006) Extreme events: a physical reconstruction and risk assessment. Cambridge University Press, Cambridge, p 297
26. Omori F (1894) On the aftershocks of earthquakes. J Coll Sci Imper Univ Tokyo 7:111–200
27. Poincaré H (1905) Science and hypothesis. The William Scott, New York, p 244
28. Powell G, Percival I (1979) A spectral entropy method for distinguishing regular and irregular motion of Hamiltonian systems. J Phys A 12:2053–2071
29. Pulinets S, Boyarchuk K (2004) Ionospheric precursors in ionosphere. Springer, Berlin, p 315
30. Qin K, Wu LX, De Santis A, Cianchini G (2012) Preliminary analysis of surface temperature anomalies preceding the two major 2012 Emilia (Italy) earthquakes. Ann Geophys 55(4):823–828
31. Reigber Ch, Luehr H, Schwintzer P (2002) CHAMP mission status. Adv Space Res 30(2):129–134
32. Ruhl HA et al (2011) Societal need for improved understanding of climate change, anthropogenic impacts, and geo-hazard warning drive development of ocean observatories in European seas. Prog Oceanogr 91:1–33
33. Shannon C (1948) A mathematical theory of communication. Bell Syst Tech J 27:379–423; 623–656
34. Signanini P, De Santis A (2012) Power-law frequency distribution of H/V spectral ratio of the seismic signals: evidence for a critical crust. Earth Planet Space 64:49–54
35. Skinner BJ, Porter SC (1995) The blue planet: an introduction to earth system science. Wiley VCH, New York
36. Stanley HE (1971) Introduction to phase transitions and critical phenomena. Oxford University Press, New York
37. Takens F (1981) Detecting strange attractors in turbulence. In: Rand DA, Young LS (eds) Detecting strange attractors in turbulence. Dynamical systems and turbulence. Lecture notes in mathematics, vol 898. Springer, Berlin, pp 366–381
38. Turcotte DL (1997) Fractal and chaos in geology and geophysics, 2nd edn. Cambridge University Press, Cambridge/New York, p 348
39. Turcotte DL, Schubert G (2002) Plate tectonics. In: Geodynamics, 2nd edn. Cambridge University Press, Cambridge/New York, pp 1–21
40. Utsu T (1961) A statistical study of the occurrence of aftershocks. Geophys Mag 30:521–605

Chapter 2
Aftershock Cascade of the 3.11 Earthquake (2011) in Fukushima-Miyagi Area

Yoji Aizawa and Satoru Tsugawa

Abstract Details of the aftershock cascade in [35°–40°N, 140°–145°E] are reported from the viewpoint of three empirical laws; the Omori law, the Gutenberg-Richter law and the Weibull law for the interoccurrence times, and the universal relationship among those three empirical laws is theoretically derived under the quasi-stationary condition. The generalization of the Omori law enables us to derive the extrapolation formula of the GR law, and the multi-fractal relation confirmed universally in moving ensembles combines the magnitude distribution and the interoccurrence time distribution. Furthermore, the generalized Omori formula is interpreted in terms of the quasi-stationary interoccurrence time distribution.

2.1 Introduction

Here we give a brief sketch and comments about the empirical laws in seismic statistics, which will be used in the latter analysis.

2.1.1 Aftershock Frequency: Omori Law

The rate of aftershocks is first formulated by Omori in 1894, where the aftershock frequency dN/dt was very well adjusted by,

$$\frac{dN}{dt} \propto t^{-p}, \tag{2.1}$$

Y. Aizawa (✉) • S. Tsugawa
Department of Applied Physics, Advanced School of Science and Engineering, Waseda University, Shinjuku, Tokyo 169-8555, Japan
e-mail: aizawa@waseda.jp; s.tsugawa@aoni.waseda.jp

D. Matrasulov and H.E. Stanley (eds.), *Nonlinear Phenomena in Complex Systems: From Nano to Macro Scale*, NATO Science for Peace and Security Series C: Environmental Security, DOI 10.1007/978-94-017-8704-8_2,

Here N stands for the total number of aftershocks and t is the time measured from the main shock. As the p-value is close to unity, the aftershock frequency leads to a logarithmic scaling of N in relatively long time behavior,

$$N(t) \approx \ln(t + c), \tag{2.2}$$

This formula is often called the Omori law [8]. Some generalizations of the Omori formula were pursued to obtain the better prediction of aftershocks. Enya (1901) discussed by the following form [3],

$$\frac{dN}{dt} \propto \ln\left(1 + \frac{1}{t + c}\right), \tag{2.3}$$

and another generalization by Utsu [9] is,

$$\frac{dN}{dt} \propto (c + t)^{-p}, \tag{2.4}$$

where c means the characteristic time in each formula. In these generalizations, it should be noted that the stationary activity, which will be realized at $t \to \infty$, is discarded in practical treatment. In the latter part of this paper, some refined aspects of aftershocks will be reported based on the generalized form of the Omori law.

Aftershocks are obviously nonstationary phenomena and reveal remarkable clustering where a huge number of aftershocks are directly induced by the main shock. Moreover, the aftereffect of the big main shock remains for long time, for instance, in the case that Omori reported in 1894, the Omori formula is well justified for very long time more than 80 years since the main shock. From these facts we are obliged to be skeptic whether we can admit any stationary statistical laws in the sequence of earthquakes or not. In the present paper, however, we define the stationary regime from a practical point of view, that is to say, where the seismic activity is not in high level but is relatively static one. We assume that the ensemble which describes the stationary regime could be obtained if we consider a very long time series of shocks happened in the definite area. Indeed, some important laws are known in the stationary ensembles, which will be briefly introduced in what follows.

2.1.2 Intensity Distribution: Gutenberg-Richter Law

Gutenberg and Richter [4] suggested, by use of the magnitude m introduced by Richter, that the cumulative number of earthquake $n(m)$ (for the magnitude larger than m) obeys the GR formula,

$$\ln n(m) \approx a - bm, \tag{2.5}$$

where a and b are assumed to be constant. As the GR law was approximately confirmed in many cases of worldwide data, the magnitude became a useful measure which characterizes the intensity of earthquake in stationary regime.

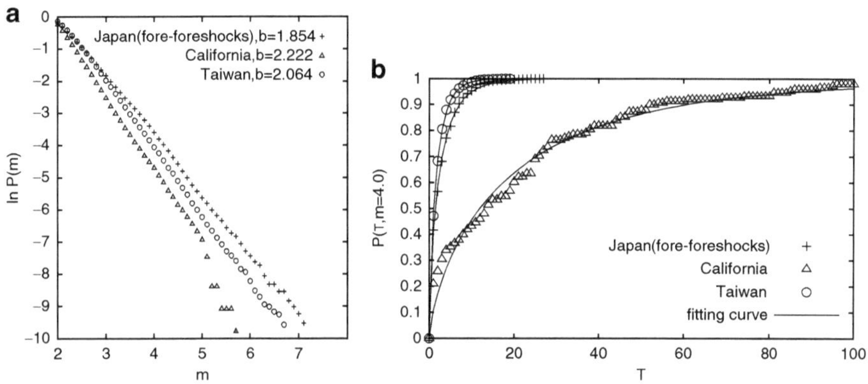

Fig. 2.1 The Gutenberg-Richter law (**a**) and the Weibull law (**b**) in stationary regime. The fore-foreshock region for 10 years (2001–2010) in Fukushima-Miyagi area (+) is compared with other results for worldwide data in South California (△) and in Taiwan (○)

Figure 2.1 shows the GR law realized clearly. To compare with other worldwide data from 2001 to 2010, we showed the GR law for the earthquakes in Taiwan (21°–26°N, 119°–123°E) and South California (32°–37°N, 114°–122°E). Though the mean magnitudes are different, the GR law seems to be well satisfied in each case.

In the nonstationary regime, the GR formula reveals peculiar deviations from the exponential one as shown in latter sections, where we give the extrapolation formula of the GR law in aftershocks.

2.1.3 *Interoccurrence Time Distribution: Weibull Law*

Interoccurrence times play the most important role in the prediction theory of earthquakes. When the cutoff magnitude m increases, the corresponding interoccurrence time τ is prolonged in statistical sense. So the interoccurrence time distribution is parameterized by the magnitude m, i.e., $P(\tau; m)$. If we fix the threshold value m, the sequence of interoccurrence times defines a renewal process. The purpose of the interoccurrence time statistics is to determine the functional form of the cumulative probability $P(\tau; m)$ and to find out the universal nature hidden behind those statistical distributions.

Recently, we have shown that the interoccurrence time distribution $P(\tau; m)$ is very well adjusted by the superposition of the Weibull distribution $P_w(\tau; m)$ and the log-Weibull one $P_{lw}(\tau; m)$ for many natural earthquakes in stationary regime [2, 5–7].

$$P(\tau; m) = pP_w(\tau) + (1 - p)P_{lw}(\tau) \tag{2.6}$$

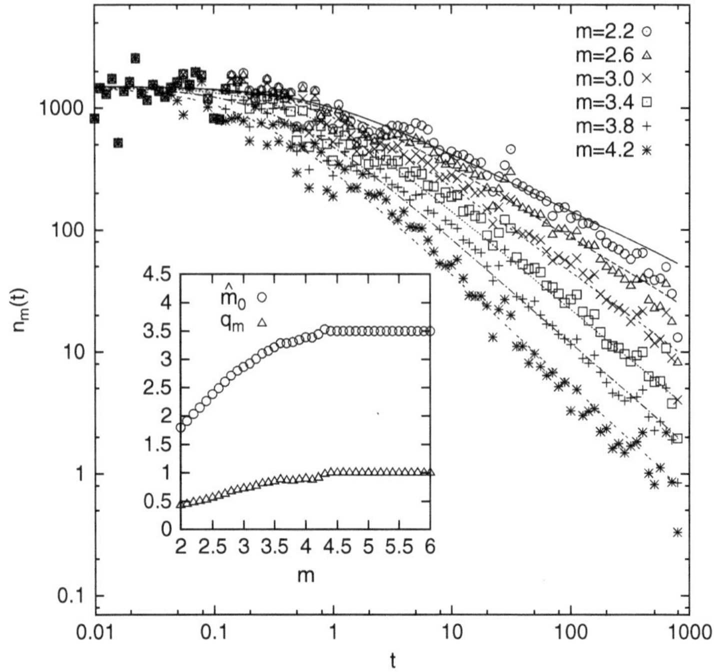

Fig. 2.4 Time courses of the shock frequency $n_m(t)$

In this section, the detailed structure in the Omori law is elucidated, and the statistical aspects in the nonstationary shock sequence are mainly studied by using the moving ensembles, where the interval of each ensemble is defined by the span $[t - \Delta/2, t + \Delta/2]$ at $\Delta = 100$ days (fixed). More details are seen in Ref. [2].

2.2.1 *Refined Formula of the Omori Law*

Figure 2.4 shows the aftershock frequency (per 1 day) $n_m(t)(= dN_m(t)/dt)$, where $N_m(t)$ is the cumulative number of aftershocks in $[0, t]$ (for the magnitude larger than m). Each curve is well fitted by the following forms,

$$n_m(t) = d(1 + t/c_m)^{-q_m}, c_m = e^{-\hat{b}(m-\hat{m}_0)}. \tag{2.11}$$

$N_m(t)$ is given by the q_m-extension; $N_m(t) = dc_m(1 - q_m)^{-1}((1 + t/c_m)^{1-q_m} - 1)$, and $\hat{m}_0$ and q_m are monotonically increasing, but they are almost constant for $m > m_0(\simeq 4.0)$, $q_m \simeq 1$ and $\hat{m}_0 \simeq 3.5$.

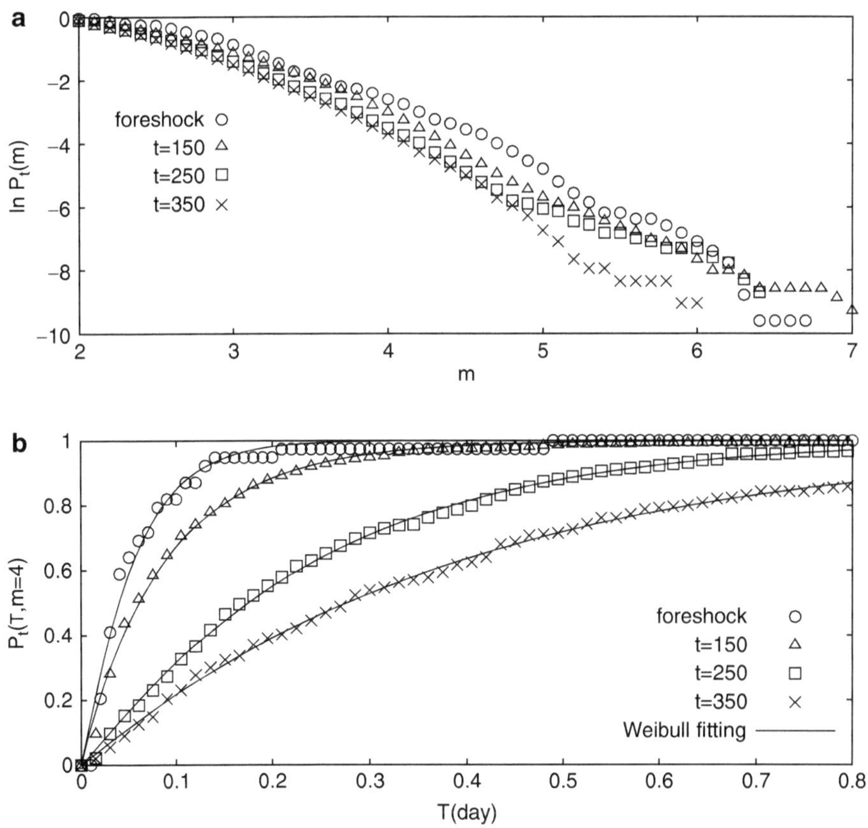

Fig. 2.5 Temporal change of the aftershock statistics in moving ensembles; (**a**) the Gutenberg-Richter law $P_t(m)$, (**b**) the interoccurrence time distribution $P_t(\tau, m)$ ($t = 150$ (△), 250 (□) and 350 (×))

2.2.2 Generalization of Multi-fractal Relation in Moving Ensembles

Figure 2.5 displays the magnitude distribution $P_t(m)$ and the interoccurrence time distribution $P_t(\tau : m)$ for three moving ensembles, which enable us to derive the multi-fractal relation in each time span. One of the remarkable points is that the Weibull law is well satisfied, and that the GR law is a convex function of which fitting curves are given in the next section. The universal aspect discussed in the previous section appears even in the aftershock region, and the rescaled multi-fractal universality is given in Fig. 2.6. Only difference from the stationary case is that the earthquake constant k_{EQ} is not a constant, but is a certain function

When t goes to large enough, this ensures the previous universal relation in the stationary ensemble, and that the earthquake constant $k_{EQ}(t)$ is connected to the coefficient d and other parameters $(\hat{b}, \hat{m}_0)$.

In the case of moving ensembles $[t - \frac{\Delta}{2}, t + \frac{\Delta}{2}]$, the Weibull parameters ($\alpha_{m,\Delta}(t)$ and $\beta_{m,\Delta}(t)$) satisfy the following time-dependent relation (in re-scaled form) for $m > \hat{m}_0$,

$$\left(\frac{\beta_{m,\Delta}(t)}{\beta_{m_c,\Delta}(t)}\right) \Gamma\left(1 + \frac{1}{\alpha_{m,\Delta}(t)}\right) \frac{\ln(1 + \frac{\Delta}{t + c_m - \Delta/2})}{\ln(1 + \frac{\Delta}{t + c_{m_c} - \Delta/2})} e^{-\hat{b}(m - m_c)} = 1. \tag{2.17}$$

In this paper we consider only the Weibull distribution for $P^W_{t,m}$, but the more consistent way to derive the interoccurrence time distribution is given by the quasi-stationary assumption. Here we discuss only the idea toward the theoretical unification between the Omori law and the generalized interoccurrence time statistics. Consider the interoccurrence time distribution density $p_{t,m}(\tau)(=\frac{dP_{t,m}(\tau)}{d\tau})$, then the expectation value of the renewal event in $[t, t + \tau]$ is determined by using the convolution of the density $p_{t,m}(\tau)$, i.e., $\sum_{r=1}^{\infty} P^{r*}_{t,m}$, where $P^{r*}_{t,m}$ denotes the r-th convolution. The number of events in $[t, t + \tau]$ is surmised to obey the generalized formula Eq. (2.11) if the aftershock sequence is quasi-stationary,

$$N_m(t + \tau) - N_m(t) = \sum_r P^{r*}_{t,m}(\tau). \tag{2.18}$$

This implies that the interoccurrence time distribution can be derived only from the aftershock statistics $N_m(t)$. The distribution density is given by the inverse transformation $\mathscr{L}_\tau p_{t,m}(\tau) = \hat{p}_{t,m}(s)$, and the interoccurrence time distribution is subordinate (for small τ) to the Omori law,

$$\hat{P}_{t,m}(s) = 1 - \frac{1}{1 + ac_m\hat{n}_m(as)}, \tag{2.19}$$

where $\hat{n}_m(s)$ is the Laplace transformation of $n_m(\tau)$ and $a = 1 + t/c_m$.

2.3.3 Birth and Death Cascade in Aftershocks

We have not yet succeeded to derive the theoretical multi-fractal relation (Fig. 2.6b) in nonstationary regime, but the data-analysis for the 3.11 EQ (2011) shows that there exist clear hierarchical time-dependent structures among different magnitude-scales. Here, the regularity hidden behind the nonstationary aftershock sequence will be formulated and the birth and death cascade in aftershocks is discussed based on the generalized Omori law at $q_m \simeq 1$ for the simplicity sake.

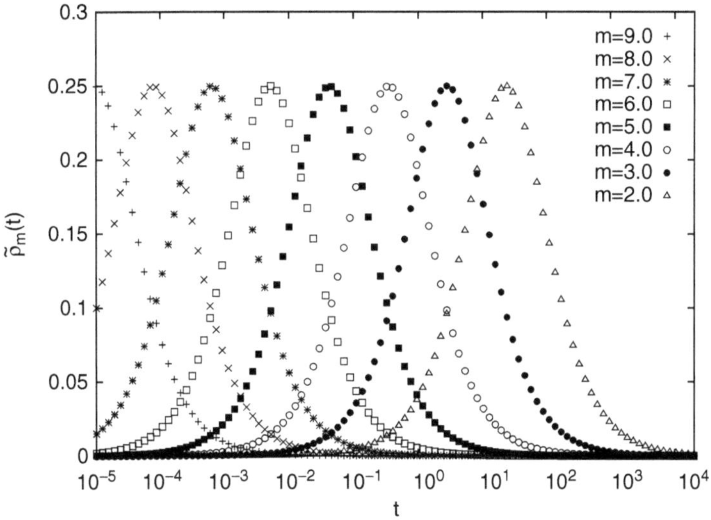

Fig. 2.7 The solitary wave propagation in the space of magnitude m; $\tilde{\rho}_m(t) = (d\tilde{b})^{-1}\rho_m(t)$

We define the magnitude density $\rho_m(t)$ that describes the number of shocks with the magnitude m,

$$\begin{aligned} \rho_m(t) &= \left| \frac{dn_m(t)}{dm} \right| \\ &= d\hat{b}(t/c_m)(1+t/c_m)^{-2}, \qquad (\ln c_m = \hat{b}(\hat{m}_0 - m)) \end{aligned} \tag{2.20}$$

By introducing a scaled variable $z = t/c_m$, it is known that the magnitude density obeys a universal behavior ($\rho_m(t) = \rho(z)$), in all magnitude classes. This indicates that the magnitude scale controls the time scale of shocks, and vice versa. The control mechanism is formulated in the following by using new scaled variables, $\tilde{z}(= \ln z = \tau + \hat{b}m)$ and $\tau(= \ln t - \hat{b}m_0)$, i.e.,

$$\begin{aligned} n_m(t) &= \frac{d}{1+e^{\tilde{z}}}, \text{ and} \\ \rho_m(t) &= \frac{d\hat{b}e^{\tilde{z}}}{(1+\tilde{z})^2} = -\hat{b}\frac{dn_m(t)}{d\tilde{z}} \\ &= -\hat{b}n_m(1-n_m/d) \end{aligned} \tag{2.21}$$

One can see that these solutions represent the typical nonlinear wave (kink and soliton) in m-space, of which traveling velocity is determined by $\hat{b}^{-1}(=dm/d\tau)$. Figure 2.7 shows the cascade process of $\rho_m(t)$ in (m, τ) space, which corresponds to the soliton-propagation in m-space.

It is difficult to derive the nonlinear wave equation uniquely for the aftershock cascade only from the above special solution, but we can surmise the essential mechanism leading to the solitary wave mentioned above. From Eq. (2.11), $\rho_m(\tau)$ obeys,

$$\frac{\partial \rho_m}{\partial \tau} = \rho_m \left\{ n_m/d - (1 - n_m/d) \right\}, \tag{2.22}$$

in other words, by using the relations $n_m = \int_m^{m_M} \rho_m dm$, and $\int_{m_\mu}^{m_M} \rho_m dm = d$,

$$\frac{\partial \rho_m}{\partial \tau} = \frac{1}{d}\rho_m \left\{ \int_m^{m_M} \rho'_m dm' - \int_{m_\mu}^{m} \rho'_m dm' \right\}, \tag{2.23}$$

where m_M and m_μ indicate the maximum and the minimum magnitude in practical analysis respectively, and $m_M = \infty$ and $m_\mu = -\infty$ are assumed in the present treatment. The 1st and the 2nd terms of Eq. (2.23) show the growth and decay effects of $\rho_m(\tau)$. One can check easily that Eq. (2.21) is the particular solution of Eq. (2.23).

The interaction between two shocks is not known clearly, but the interaction obtained in Eq. (2.23) seems to give us a hint, which may enlighten on the hidden coupling mechanism among many shocks with different magnitude-scales. As an approximation, let us consider the birth and death model described by the generalized transition probability $W_{m,m'}$,

$$\frac{d\rho_m}{d\tau} = \int_{m_\mu}^{m_M} W_{m,m'} \rho'_m dm' - \int_{m_\mu}^{m_M} W_{m,m'} \rho_m dm'. \tag{2.24}$$

If we assume $W_{m,m'} \propto \rho_m (m' \geq m)$ and $W_{m,m'} = 0 (m' < m)$, Eq. (2.23) is recovered and the birth and death cascade shown in Fig. 2.7 is obtained again as a particular solution; as a matter of course, there are many other possible solutions in Eq. (2.24). The details studied in this section will be reported in the next paper by using much longer data of the aftershocks.

2.4 Discussions and Prospects

The aftershock cascade reveals a very clear regularity not only in the birth and death process but also in the statistical aspects. In this paper, some parts of the regularity are confirmed even in the short time region $t \simeq 10^{-2}$ (days). The more precise studies immediately after the big shock seem to be important to elucidate the precursive mechanism leading to the main shock. The onset time $t = 0$ is the critical point in statistical seismology, but some analytical continuations beyond the singular point must be pursued to obtain the information in the prestage of big shocks ($t \leq 0$). The birth and death model may give the hint for this end, where some latent variables, for instance, the stress accumulated in the plate interface, should be taken into account. These subjects are still open and in our future challenge.

References

1. Aizawa Y (2011) Foundations of earthquake statistics in view of non-stationary chaos theory. BusseiKenkyu (Kyoto University) 97(3):309
2. Aizawa Y, Hasumi T, Tsugawa S (2013) Seismic statistics: universality and interim report on the 3. 11 Earthquake (2011) in Fukushima-Miyagi Area. Int J Nonlin Phen Comp Sys 16(2):116–130
3. Enya O (1901) Comments on the aftershocks of earthquakes. Rep Imp Earthq Investig Comm 35:35. (in Japanese)
4. Gutenberg B, Richter CF (1956) Magnitude and energy of earthquakes. Ann Geofis 9:1
5. Hasumi T, Chen C, Akimoto T, Aizawa Y (2010) The Weibull/Log Weibull transition of interoccurrence time for synthetic and natural earthquakes. Tectonophys 485:9
6. Hasumi T, Chen C, Akimoto T, Aizawa Y (2012) The Weibull/Log Weibull transition of interoccurrence time of earthquake. In: D'Amico S (ed) Earthquake Research and Analysis. InTech, Croatia, Chap 1, pp 3–24
7. Hasumi T, Chen C, Akimoto T, Aizawa Y (2013) Statistical seismicity in view of complex systems. In: Konstantinou K (ed) Earthquakes: triggers, environmental impact and potential hazards. Nova Science Publisher, New York, Chap 5, pp 109–163
8. Omori F (1894) On the aftershocks of earthquakes. J Coll Sci Imp Univ Tokyo 7:111
9. Utsu T (1961) Statistical study on the occurrence of aftershocks. Geophys Mag 30:521

Part II
Socio-, Econo- and Biophysics

Chapter 3
Is It Necessary to Lie to Win a Controversial Public Debate? An Answer from Sociophysics

Serge Galam

Abstract Controversial public debates driven by incomplete scientific data where nobody can claim absolute certainty, due to the current state of scientific knowledge, are studied. To adopt a cautious balanced attitude based on clear but inconclusive data appears to be a lose-out strategy. In contrast overstating arguments with incorrect claims which cannot be scientifically refuted appears to be necessary but not sufficient to eventually win a public debate. The underlying key mechanisms of these puzzling and unfortunate conclusions are identified using the Galam Unifying Frame (GUF) of opinion dynamics. It reveals that the existence of inflexible agents and their respective proportions are the instrumental parameters to determine the faith of incomplete scientific data in public debates. Acting on one's own inflexible proportion modifies the topology of the flow diagram, which in turn can make irrelevant the value of initial support. On the contrary focusing on open-minded agents may be useless given some topologies. Accordingly, the inflexibles rather than the data are found to drive the opinion of the population. The results shed a new but disturbing light on designing adequate strategies to win a public debate. The cases of global warming is briefly discussed.

3.1 The Central Role of Public Opinion

Public opinion is today a key factor to be accounted for by any policymaker in order to make decision whatever the scale of the associated issue is. This reality holds true in democratic countries but not only as seen recently in a few countries in the arab

S. Galam (✉)
CEVIPOF – Centre for Political Research, CNRS and Sciences Po, 98, rue de l'Université, 75007 Paris, France
e-mail: serge.galam@sciencespo.fr

D. Matrasulov and H.E. Stanley (eds.), *Nonlinear Phenomena in Complex Systems: From Nano to Macro Scale*, NATO Science for Peace and Security Series C: Environmental Security, DOI 10.1007/978-94-017-8704-8_3,

world. Polls are heavily used to measure the current stand of a public opinion about a sensitive issue. And although their accuracy is sometimes questioned, most of the times, they provide rather good estimates of the reality.

Such a novel situation is praised by all democrats on the "obvious assumption" that the public opinion is the direct expression of everyone voice aggregated equally to express the will of the majority. Put under those terms, the role of public opinion deserves to be the major instrument to be used for any major and even minor collective decision.

But this paradigmatic view of the public opinion is an a priori which ignores what makes the public opinion. Its "obvious democratic nature" is never questioned. And yet, no one would dismiss the fact that people do discuss together in small informal groups about a given issue and that those local discussions eventually trigger some shifts of individual opinions. Accordingly, one should ask what are the mechanisms by which a public opinion is formed before any judgment is stated about the political weight to be given to the public opinion.

The identification of those rules should be a prerequisite to the assessment of the democratic character of the dynamics of public opinion, such a feature being never disputed though its internal making is far from being understood. In particular, it happens that contrary to the expected "democratic" perception of a free public debate, several works of sociophysics [1] have hint to the possibility that, although hardly conceivable, the holding of a free and open public debate may give rise to a natural attraction towards initial tiny minority views making a huge majority of individuals to change naturally their minds along the initial minority view [2–5, 29].

It is worth to recall that while sociophysics [6] was adamantly rejected by physicists at its first steps during the earlier 1980s [7] it is today flourishing field of research among physicists all over the world [8]. Among a large series of topics of investigation, opinion dynamics is very active with numerous papers published every year [9–20]. Indeed a unifying frame denoted Galam Unifying Frame (GUF) has been shown to embed most of the discrete models [21].

Voting has also attracted a good deal of works [22–26]. Coexistence of opposite sates were found [27, 28]. Economical applications have been done like for instance for the wine market [29] and innovations [30]. Several review are available [31–35].

In this paper using GUF for opinion dynamics [21] it is shown how a few percents of liars about an issue may overcome a majority of 80 % by just holding on the same opinion whatever arguments are given to them. They do not win by more powerful arguments, they simply behave as inflexibles within the local group of discussions, just sticking to their opinion [20, 36]. We use here the term "liar" to emphasize the fact that given a controversy based on partial scientific results, those agents overstate their conviction by asserting that their view has been scientifically proved. Such an attitude is emblematic of the debate about the anthropic cause of the recently observed global warming [2, 3].

3.2 The Ideal Model of Opinion Dynamics

We start from a societal issue for which two opposite views A and B compete to enforce an eventual new policy about some observed problems. Let denote the real problem as RR and rr the leading explanation about RR. Opinion A claims that rr = RR while opinion B disclaims the equality stating that at the moment rr $\neq$ RR. Usually, the mere available data and facts are consistent with the equality rr=RR but neither prove it nor forbid an other explanation. Let us denote $p_A(t)$ and $p_B(t) = 1 - p_A(t)$ the respective proportions of people supporting respectively opinion A and B at time t =, each person having an opinion.

As time passes by, starting from the initial configuration $p_A(t)$, people keep on discussing the issue. the debate is not formally set and gatherings are shaped by social life. Those informal meeting convene a few agents not large groups.

No advantage being given to either A or B, we assume that while discussing within a group, the agents end up adopting the opinion which had the initial majority support. In case of a tie, each agents keeps its initial opinion. However during a public debate, people keep on discussing for a long period of time moving from one group to another. The associate dynamics is monitored by a series of successive random distributions of the agents in various size groups, all being reshuffled once local majority rule have been applied to each group. An illustration is given in Fig. 3.1.

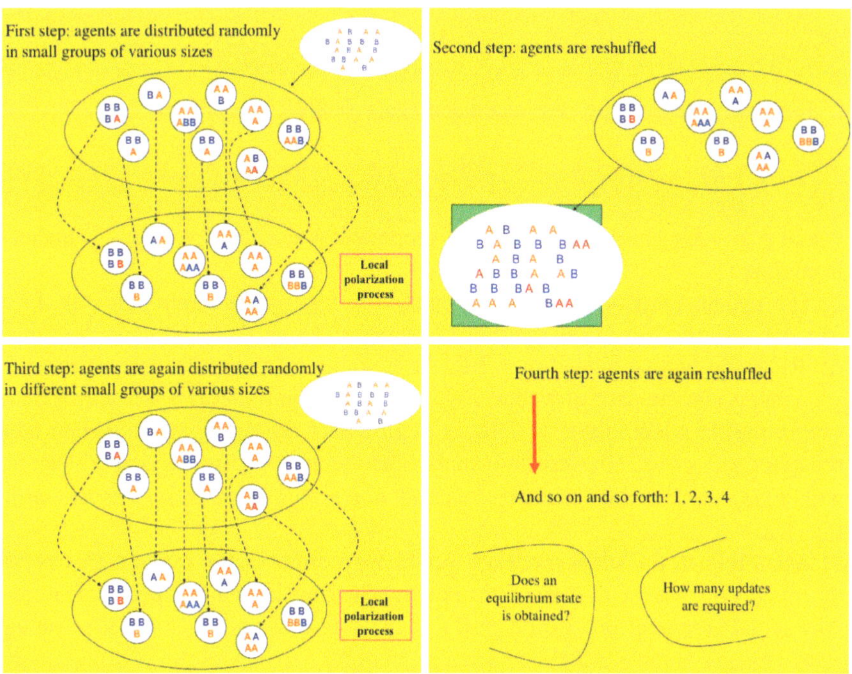

Fig. 3.1 Illustration of the various steps making two successive local updates

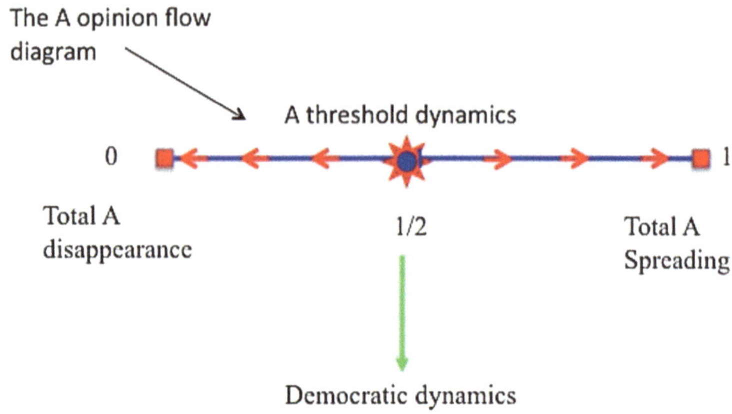

Fig. 3.2 The tipping dynamics produced by repeated local updates. *Arrows* show the direction of the flow

To analyse the model we restrict the local groups of discussion to the same size 3. The first update is thus obtained by distributing randomly the whole population in groups of size 3, and subsequently having the 3 agents in each group people to adopt the initial local majority, i.e., 3 A and (2 A − 1 B) end 3 A and vice versa for B. After one update, at time $(t + 1)$ the new proportion of people supporting opinion A is,

$$p_A(t+1) = p_A(t)^3 + 3p_A(t)^2\{1 - p_A(t)\}. \tag{3.1}$$

Repeating the same process again and again yields,

$$p_A(t+k) = p_A(t+k-1)^3 + 3p_A(t+k-1)^2(1 - p_A(t+k-1)), \tag{3.2}$$

where $p_A(t+k-1)$ is the proportion of supporters A at a distance of $(k-1)$ updates from time t.

The update function exhibits two attractors $p_{A,0} = 0$ and $p_{A,1} = 1$. First one corresponds to a total opinion polarization along the refutal of the claim rr=RR asserting that indeed rr $\neq$ RR. At the other extremity the second attractor $p_{A,1}$ represents the case of a total support for the claim that rr=RR with no one supporting the view that rr $\neq$ RR. Both attractors are sitting on each side of a separator $p_{c,3} = \frac{1}{2}$. The separator makes the dynamics a tipping point dynamics with $p_A(t) < \frac{1}{2}$ leading towards $p_{A,0}$ and $p_A(t) > \frac{1}{2}$ towards $p_{A,1}$ as shown in Fig. 3.2.

According to our ideal model of opinion dynamics, opinion dynamics obeys the main criterion of a democratic practice sharing its main expected features. Any initial supporter proportion $p_A(t) < 50\,\%$ is weakened by on going mutual exchanges of arguments among agents to eventually disappear. However this polarization effect requires a sufficient number of updates to be completed. For instance we have the series $p_A(t) = 0.48$, $p_A(t+1) = 0.47$, $p_A(t+2) = 0.45$, $p_A(t+3) = 0.43$, $p_A(t+4) = 0.40$, $p_A(t+5) = 0.35$, $p_A(t+4) = 0.28$ down to $p_A(t+9) = 0.03$ and $p_A(t+10) = 0.00$. Within 10 successive updates, 48 % of support to some issue has turned against it.

Though the numbers of required updates are of the order of less than 10, apart for the special case $p_A(t) \approx 0.50$, it is worth the stress that to make an application to a real public debate, one update should be mapped to some numbers of days, whose value has to be evaluated using a series of polls. Notice that the dynamics is perfectly symmetric with respect to both opinions with a threshold of initial support at (50 %).

It is worth to notice that increasing the size of the discussing group to odd values does not change the main properties of the dynamics with yet a separator at 50 % and the two attractors $p_{A,0}$ and $p_{A,1}$. The only difference is the shrinking of the number of required updates to reach an attractor. The case of even size is more subtle. While choosing not change opinion at a tie preserves the odd-size properties, it is possible to consider a bias at a tie, which in turn can produce some interesting minority spreading situations.

3.3 The Disturbing Effect of Liars

While the ideal model of opinion dynamics reproduces the expected features of a democratic paradigm, the assumption of ideal agents who follow nicely the majority of arguments given during an informal discussion is a bit at most very naive but at least a little excessive. While many agents are certainly open minded and are sincerely looking for the best choice, some other agents adopt a totally different attitude. For all kind of reasons, they do have a solid interest in making one choice to be dominant. Accordingly, in case of difficulty in a local debate, they will lie claiming their opinion is not an opinion but the truth, despite an eventual majority of opposite arguments.

To grasp the drastic effect of liars, we consider that a proportion a of agents assert that the equality rr=RR has been proved scientifically although only limited evidence has been established. For instance, in the debate about global warming some agents do believe that it has been scientifically proven that the anthropic production of carbon dioxide is the main cause of the increase of global temperatures. At contrast, other agents claim there is no scientific proof of human responsibility but could convinced of the opposite if given enough evidence.

The update Eq. (3.1) has to be rewritten as,

$$p_A(t+1) = p_A(t)^3 + 3p_A(t)^2(1 - p_A(t)) + a(1 - p_A(t)^2), \qquad (3.3)$$

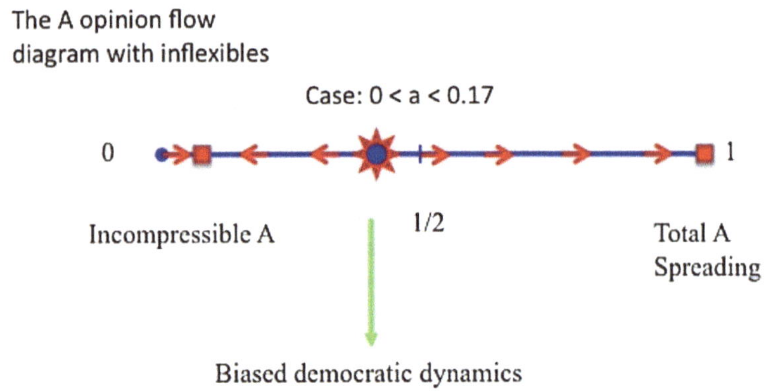

Fig. 3.3 The tipping dynamics produced by repeated local updates with a proportion of inflexibles in the range $0 \leq a < a_c \equiv 3 - 2\sqrt{2} \approx 0.172$. *Arrows* show the direction of the flow

where last term accounts for the configurations where one inflexible A is against two B agents with both B agents keep their B opinion since it is majority while the A inflexible preserves its A choice.

Given a support $p_A(t)$ for opinion A means a subgroup of a inflexibles and $(p_A(t) - a)$ open minded agents with the constraint $p_A(t) \geq a$. Solving the associated fixed point equation preserves the total A spreading attractor with $p^a_{A,1} = p_{A,1} = 1$ and yields the shifts,

$$p_{A,0} \rightarrow p^a_{A,0} = \frac{1}{4}\left(1 + a - \sqrt{1 - 6a + a^2}\right), \tag{3.4}$$

and

$$p_{c,3} \rightarrow p^a_{c,3} = \frac{1}{4}\left(1 + a + \sqrt{1 - 6a + a^2}\right). \tag{3.5}$$

From above formulas it is seen that both $p^a_{A,0}$ and $p^a_{c,3}$ are real only when $0 \leq a < a_c \equiv 3 - 2\sqrt{2} \approx 0.172$. Associated with opinion flow diagram is shown in Fig. 3.3, which becomes identical to Fig. 3.2 at $a = 0$. Having $a \neq 0$ gives some substantial advantage to the A claim that rr=RR since to invade the full population the initial support does not need to be higher that 50 %. However opinion B can still wins the debate provided it starts from an initial support larger than $p^a_{c,3}$.

The variations of $p^a_{A,0}$ and $p^a_{c,3}$ as a function of a are exhibited in Fig. 3.4. From the figure it is seen that both values get closer one to another as a increases to eventually coalesce at a critical value of $a_c \equiv 3 - 2\sqrt{2} \approx 0.172$. At $a = a_c$ we have $p^{a_c}_{A,0} = p^{a_c}_{c,3} \approx 0.293$. The area in the lower part represents the incompressible A inflexible zone.

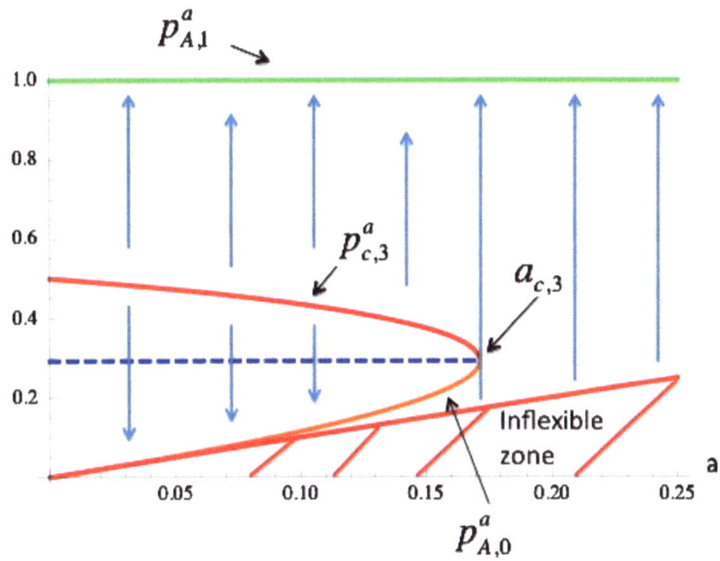

Fig. 3.4 The tipping dynamics produced by repeated local updates with a proportion of inflexibles in the range $0 \leq a < a_c \equiv 3 - 2\sqrt{2} \approx 0.172$. *Arrows* show the direction of the flow

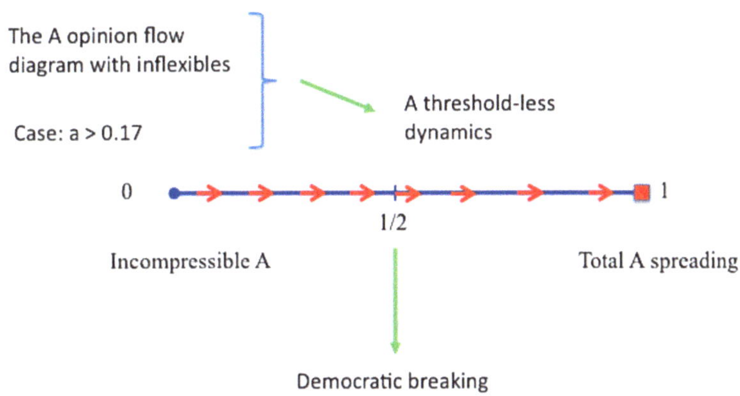

Fig. 3.5 In the range $a \leq a_c$ the dynamics is totalitarian with no tipping point. A opinion always wins the public debate. *Arrows* show the direction of the flow

In the range $a \leq a_c$ the dynamics is put upside down with a dramatic breaking of the democratic balance. There, even a huge support for B opinion of the order of 80 % will systematically be shifted to adopt the opinion A. Such a phenomena is triggers by the fact that the dynamics is monitored by a unique attractor located at $p^a_{A,1} = 1$ as seen from both Figs. 3.4 and 3.5.

Chapter 4
Anticipating Stock Market Movements with *Google* and *Wikipedia*

Helen Susannah Moat, Chester Curme, H. Eugene Stanley, and Tobias Preis

Abstract Many of the trading decisions that have led to financial crises are captured by vast, detailed stock market datasets. Here, we summarize two of our recent studies which investigate whether Internet usage data contain traces of attempts to gather information before such trading decisions were taken. By analyzing changes in how often Internet users searched for financially related information on *Google* (Preis et al., Sci Rep 3:1684, 2013) and *Wikipedia* (Moat et al., Sci Rep 3:1801, 2013), patterns are found that may be interpreted as "early warning signs" of stock market moves. Our results suggest that online data may allow us to gain new insight into early information gathering stages of economic decision making.

4.1 Introduction

Stock market data provide extremely detailed records of decisions that traders have made, in an area in which disasters have a widespread impact. As a result, these stock market records have generated considerable scientific attention [7, 8, 11–13, 18–24, 26–28].

H.S. Moat (✉) • T. Preis
Behavioural Science, Warwick Business School, University of Warwick, Scarman Road, Coventry, CV4 7AL, UK
e-mail: Suzy.Moat@wbs.ac.uk; Tobias.Preis@wbs.ac.uk

C. Curme
Center for Polymer Studies, Department of Physics, Boston University, 590 Commonwealth Avenue, Boston, MA 02215, USA
e-mail: ccurme@bu.edu

H.E. Stanley
Boston University, Department of Physics, Boston, MA, USA
e-mail: hes@bu.edu

D. Matrasulov and H.E. Stanley (eds.), *Nonlinear Phenomena in Complex Systems: From Nano to Macro Scale*, NATO Science for Peace and Security Series C: Environmental Security, DOI 10.1007/978-94-017-8704-8_4,

Decisions, such as trading decisions, do not however consist solely of the final execution of a chosen action, such as a trade recorded at the stock exchange. Instead, humans often begin by gathering information to help identify what the consequences of possible actions might be [33].

Nowadays, the Internet has greatly extended human capabilities to distribute and gather information [1, 6, 14, 15, 31]. As a result, online resources have become the first port of call in many quests for new information. Providers of such online resources often collect extensive data on their usage, adding to a range of new large-scale measurements of collective human behavior [5, 17]. These new Internet derived datasets open up new avenues for scientists to investigate the early information gathering stages of decision making processes.

Previous studies have demonstrated that analysis of search data can provide insight into current or even subsequent behavior in the real world. For example, changes in the frequency with which users look for certain terms on search engines such as *Google* and *Yahoo!* have been correlated with changes in the numbers of reports of flu infections across the USA [9], the popularity of films, games and music on their release [10], unemployment rates [2, 4], tourist numbers [4], and trading volumes in the US stock markets [3, 25]. A recent study showed that Internet users from countries with a higher per capita gross domestic product (GDP) search for proportionally more information about the future than information about the past, in comparison with Internet users from countries with a lower per capita GDP [29].

In the two studies summarized here and described in [16] and [30] in full length, we ask whether online searches for information might contain information relevant not only to the current state of the stock market, but also to subsequent trends. Specifically, can we find any evidence that changes in the volume of searches for financial information on *Google* and *Wikipedia* may provide insight into the information gathering process of investors before they make decisions to buy or sell?

4.2 *Google* Searches and Subsequent Stock Market Moves

To investigate whether changes in information gathering behavior as captured by *Google Trends* data were related to later changes in stock price in the period between 2004 and 2011, in [30] we implemented a hypothetical investment strategy for a portfolio using search volume data, called '*Google Trends* strategy' in the following. In this strategy, as described in both [30] and [16], we quantify changes in information gathering behavior by using the relative change in search volume: $\Delta n(t, \Delta t) = n(t) - N(t-1, \Delta t)$ with $N(t-1, \Delta t) = (n(t-1) + n(t-2) + \ldots + n(t-\Delta t))/\Delta t$, where t is measured in units of weeks. We sell the DJIA at the closing price $p(t+1)$ on the first trading day of week $t+1$ if search volume has increased in week t such that $\Delta n(t, \Delta t) > 0$. We then close the position by buying

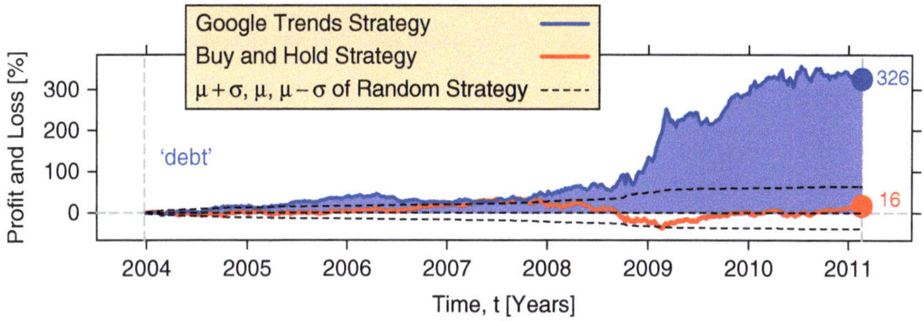

Fig. 4.1 Cumulative performance of an investment strategy based on *Google Trends* data (Reproduced from [30]). Profit and loss for an investment strategy based on the volume of the search term *debt*, the best performing keyword in our analysis, with $\Delta t = 3$ weeks, plotted as a function of time *(blue line)*. This is compared to the "buy and hold" strategy *(red line)* and the standard deviation of 10,000 simulations using a purely random investment strategy *(dashed lines)*. The Google Trends strategy using the search volume of the term *debt* would have yielded a profit of 326 %

the DJIA at price $p(t+2)$ at the end of the first trading day of the following week $t+2$. If instead search volume has decreased or remained the same in week t such that $\Delta n(t, \Delta t) \leq 0$, then we buy the DJIA at the closing price $p(t+1)$ on the first trading day of week $t+1$, and sell the DJIA at price $p(t+2)$ at the end of the first trading day of the coming week $t+2$ to close the position.

In [30], we analyzed the performance of a set of 98 *Google* search terms. We included terms related to the concept of stock markets, with some terms suggested by the *Google Sets* service, a tool which identifies semantically related keywords.

In Fig. 4.1, taken from [30], we depict the performance of our strategy between 2004 and 2011 using the search term *debt*, a keyword with an obvious semantic connection to the most recent financial crisis, and overall the term which performed best in our analyses. The performance of the *Google Trends* strategy based on the search term *debt* is depicted by a blue line, whereas dashed lines indicate the standard deviation of the cumulative return from a strategy in which we buy and sell the market index in an uncorrelated, random manner ('random investment strategy'). The standard deviation is derived from simulations of 10,000 independent realizations of the random investment strategy. Figure 4.1 shows that the use of the *Google Trends* strategy, based on the search term *debt* and $\Delta t = 3$ weeks, would have increased the value of a portfolio by 326 %. The performance of *Google Trends* strategies based on all other search terms that we analyze is depicted in a similar manner in [30].

We rank the full list of the 98 investigated search terms by their trading performance when using search data for U.S. users only (Fig. 4.2a) and when using globally generated search volume (Fig. 4.2b). In order to ensure the robustness of our results, the overall performance of a strategy based on a given search term

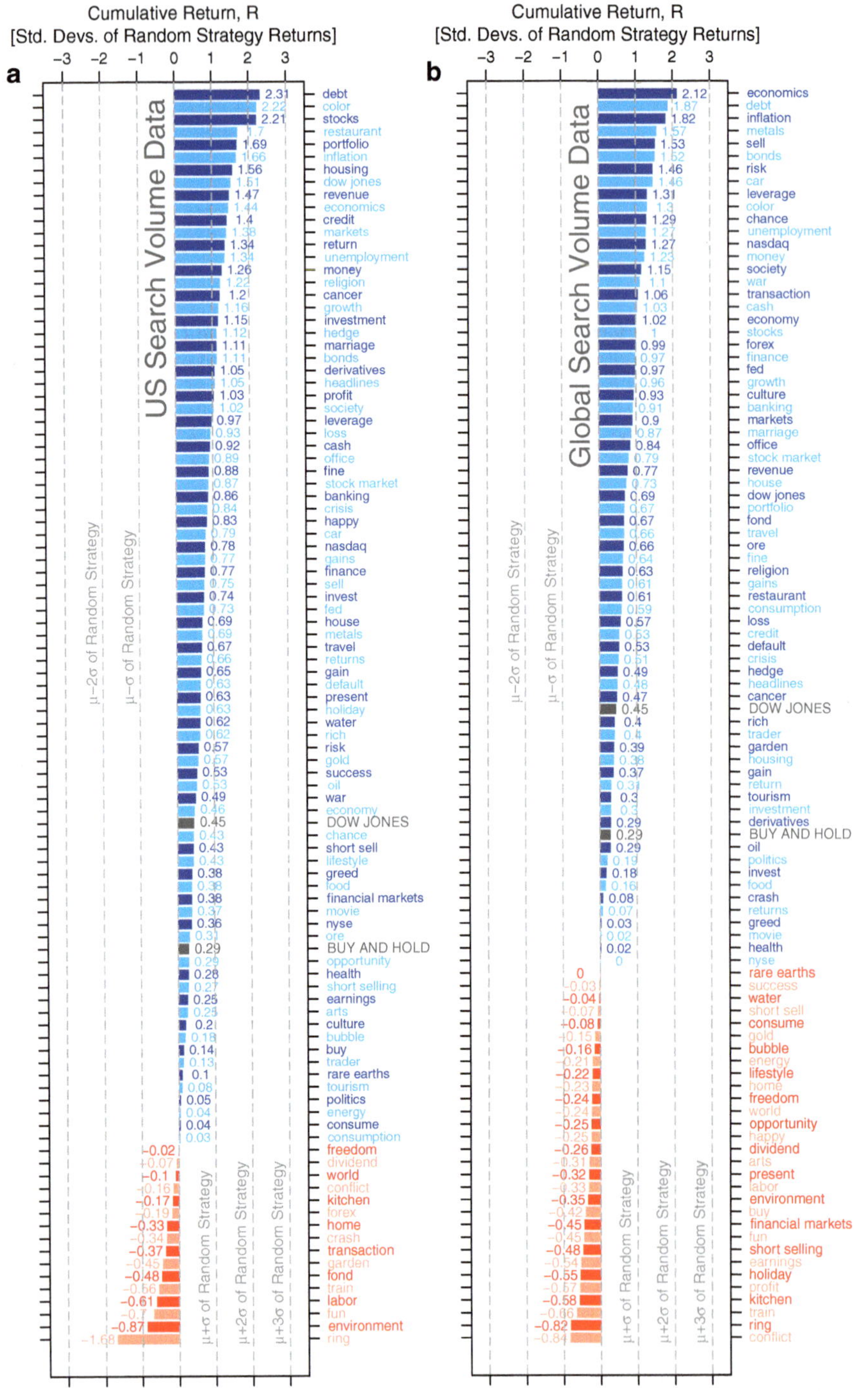
Cumulative Return, R
[Std. Devs. of Random Strategy Returns]
a
US Search Volume Data
μ−2σ of Random Strategy
μ−σ of Random Strategy
μ+σ of Random Strategy
μ+2σ of Random Strategy
μ+3σ of Random Strategy
debt
color
stocks
restaurant
portfolio
inflation
housing
dow jones
revenue
economics
credit
markets
return
unemployment
money
religion
cancer
growth
investment
hedge
marriage
bonds
derivatives
headlines
profit
society
leverage
loss
cash
office
fine
stock market
banking
crisis
happy
car
nasdaq
gains
finance
sell
invest
fed
house
metals
travel
returns
gain
default
present
holiday
water
rich
risk
gold
success
oil
war
economy
DOW JONES
chance
short sell
lifestyle
greed
food
financial markets
movie
nyse
ore
BUY AND HOLD
opportunity
health
short selling
earnings
arts
culture
bubble
buy
trader
rare earths
tourism
politics
energy
consume
consumption
freedom
dividend
world
conflict
kitchen
forex
home
crash
transaction
garden
fond
train
labor
fun
environment
ring
b
Global Search Volume Data
economics
debt
inflation
metals
sell
bonds
risk
car
leverage
color
chance
unemployment
nasdaq
money
society
war
transaction
cash
economy
stocks
forex
finance
fed
growth
culture
banking
markets
marriage
office
stock market
revenue
house
dow jones
portfolio
fond
travel
ore
fine
religion
gains
restaurant
consumption
loss
credit
default
crisis
hedge
headlines
cancer
DOW JONES
rich
trader
garden
housing
gain
return
tourism
investment
derivatives
BUY AND HOLD
oil
politics
invest
food
crash
returns
greed
movie
health
nyse
rare earths
success
water
short sell
consume
gold
bubble
energy
lifestyle
home
freedom
world
opportunity
happy
dividend
arts
present
labor
environment
buy
financial markets
fun
short selling
earnings
holiday
profit
kitchen
train
ring
conflict

is determined as the mean value over the six returns obtained for $\Delta t = 1 \ldots 6$ weeks. Returns of the strategies are calculated as the logarithm of percentage profit, following the usual definition of returns. Here we report R, the cumulative returns of a strategy, in standard deviations of the cumulative returns of these uncorrelated random investment strategies. In [30], we find that returns from the *Google Trends* strategies we tested are significantly higher overall than returns from the random strategies ($\langle R \rangle_{US} = 0.60$; $t = 8.65$, $df = 97$, $p < 0.001$, one sample t-test).

We compare the performance of these search terms with two benchmark strategies. The 'buy and hold' strategy is implemented by buying the index in the beginning and selling it at the end of the hold period. This strategy yields 16 % profit, equal to the overall increase in value of the DJIA in the time period from January 2004 until February 2011. We further implement a 'Dow Jones strategy' by using changes in $p(t)$ in place of changes in search volume data as the basis of buy and sell decisions. In [30] we find that this strategy also yields only 33 % profit with $\Delta t = 3$ weeks, or when determined as the mean value over the six returns obtained for $\Delta t = 1 \ldots 6$ weeks, 0.45 standard deviations of cumulative returns of uncorrelated random investment strategies (Fig. 4.2a, b).

It is widely recognized that investors prefer to trade on their domestic market, suggesting that search data for U.S. users only, as used in analyses so far, should better capture the information gathering behavior of U.S. stock market participants than data for *Google* users worldwide. Indeed, in [30] we find that strategies based on global search volume data are less successful than strategies based on U.S. search volume data in anticipating movements of the U.S. market ($\langle R \rangle_{US} = 0.60$, $\langle R \rangle_{Global} = 0.43$; $t = 2.69$, $df = 97$, $p < 0.01$, two-sided paired t-test).

Fig. 4.2 Performances of investment strategies based on search volume data (Reproduced from [30]). (**a**) Cumulative returns of 98 investment strategies based on search volumes restricted to search requests of users located in the United States for different search terms, displayed for the entire time period of our study from 5 January 2004 until 22 February 2011 – the time period for which *Google Trends* provides data. We use *two shades of blue* for positive returns and *two shades of red* for negative returns to improve the readability of the search terms. The cumulative performance for the "buy and hold strategy" is also shown, as is a "Dow Jones strategy", which uses weekly closing prices of the Dow Jones Industrial Average (DJIA) rather than *Google Trends* data (see *gray bars*). Figures provided next to the bars indicate the returns of a strategy, R, in standard deviations from the mean return of uncorrelated random investment strategies, $\langle R \rangle_{RandomStrategy} = 0$. *Dashed lines* correspond to $-3, -2, -1, 0, +1$, +2, and +3 standard deviations of random strategies. We find that returns from the *Google Trends* strategies tested are significantly higher overall than returns from the random strategies ($\langle R \rangle_{US} = 0.60$; $t = 8.65$, $df = 97$, $p < 0.001$, one sample t-test). (**b**) A parallel analysis shows that extending the range of the search volume analysis to global users reduces the overall return achieved by *Google Trends* trading strategies on the U.S. market ($\langle R \rangle_{US} = 0.60$, $\langle R \rangle_{Global} = 0.43$; $t = 2.69$, $df = 97$, $p < 0.01$, two-sided paired t-test). However, returns are still significantly higher than the mean return of random investment strategies ($\langle R \rangle_{Global} = 0.43$; $t = 6.40$, $df = 97$, $p < 0.001$, one sample t-test)

4.3 *Wikipedia* Views and Edits and Subsequent Stock Market Moves

In [16], we investigate whether data from the popular online encyclopedia *Wikipedia* may hold similar insights. We consider data on both how often pages on the English language *Wikipedia* have been viewed, and how often pages on the English language *Wikipedia* have been edited. We calculate our weekly measure of information gathering behavior, $n(t)$, as previously described, but using either view or edit data for *Wikipedia* in place of search volume data from *Google*. Data on *Wikipedia* page views were downloaded from the online service http://stats.grok.se, and data on *Wikipedia* page edits were obtained by parsing the *Wikipedia* "Revision history" page associated to the article. In [16], we then implement the same trading strategy described above using data generated between 10th December 2007, the earliest date for which *Wikipedia* views data are available from http://stats.grok.se, and 30th April 2012.

Figure 4.3, taken from [16], shows the distributions of returns from two portfolios of 30 hypothetical strategies, trading weekly on the DJIA. These trading strategies are based on changes in how often the 30 Wikipedia pages describing the companies in the DJIA were viewed (*blue*)) and edited (*red*) during the period December 2007–April 2012, with $\Delta t = 3$ weeks. The distribution of returns from 10,000 independent realizations of a random strategy is also shown (*gray*).

We find that there are significant differences between these three distributions ($\chi^2 = 10.21$, $df = 2$, $p = 0.006$, Kruskal-Wallis rank sum test). Our analysis shows that the returns of *Wikipedia* page view based strategies for this period are significantly higher than the returns of the random strategies ($\langle R \rangle_{Views} = 0.50$; $W = 199{,}690$, $p = 0.005$, two-tailed two-sample Wilcoxon rank-sum test, Bonferroni correction applied). There is however no statistically significant difference between the returns from the Wikipedia edit based strategies and the random strategies ($\langle R \rangle_{Edits} = -0.09$; $W = 140{,}781$, $p > 0.9$, two-tailed two-sample Wilcoxon rank-sum test, Bonferroni correction applied).

We investigate whether these results extend to *Wikipedia* articles on more general financial topics. To address this question, we make use of the fact that *Wikipedia* contains lists of pages relating to specific topics. In [16], we examine view and edit data for 285 pages relating to general economic concepts, as listed in the subsection "General Economic Concepts" on the English language *Wikipedia* page "Outline of Economics".

Figure 4.4 shows the results of an analysis of the distribution of returns from two portfolios of 285 hypothetical strategies, trading weekly on the DJIA. These strategies are based on changes in how often these 285 financially related *Wikipedia* pages were viewed (*blue*) and edited (*red*) during the same period, again with $\Delta t = 3$ weeks. As before, we find that there is a significant difference between the returns generated by the random strategies, the *Wikipedia* view based strategies and the *Wikipedia* edit based strategies ($\chi^2 = 307.88$, $df = 2$,

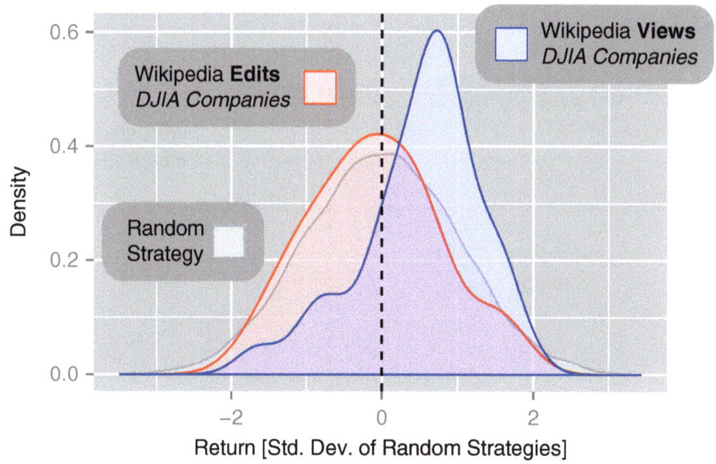

Fig. 4.3 Returns from trading strategies based on Wikipedia view and edit logs for articles relating to the companies forming the Dow Jones Industrial Average (DJIA) (Reproduced from [16]). The distributions of returns from two portfolios of 30 hypothetical strategies, trading weekly on the DJIA, based on changes in how often the 30 *Wikipedia* articles describing the companies listed in the DJIA were viewed (*blue*) and edited (*red*) during the period December 2007–April 2012, with $\Delta t = 3$ weeks. The distribution of returns from 10,000 independent realizations of a random strategy is also shown (*gray*). Data is displayed using a kernel density estimate and the *ggplot2* library [35], with a Gaussian kernel and bandwidth calculated using Silverman's rule of thumb [32]. Whereas we show in the text that random strategies lead to no significant profit or loss, we find that the returns of *Wikipedia* article view based strategies for this period are significantly higher than the returns of the random strategies ($\langle R \rangle_{Views} = 0.50$; $W = 199{,}690$, $p = 0.005$, two-tailed two-sample Wilcoxon rank-sum test, Bonferroni correction applied). There is however no statistically significant difference between the returns from the Wikipedia edit based strategies and the random strategies ($\langle R \rangle_{Edits} = -0.09$; $W = 140{,}781$, $p > 0.9$, two-tailed two-sample Wilcoxon rank-sum test, Bonferroni correction applied)

$p < 0.001$, Kruskal-Wallis rank sum test). Again, the returns of *Wikipedia* page view based strategies are significantly higher than the returns of random strategies for this period ($\langle R \rangle_{Views} = 1.10$; $W = 2{,}286{,}608$, $p < 0.001$, two-tailed two-sample Wilcoxon rank-sum test, Bonferroni correction applied). In contrast, we find no evidence of a statistically significant difference between the returns from the *Wikipedia* edit based strategies, and the random strategies ($\langle R \rangle_{Edits} = 0.12$; $W = 1{,}516{,}626$, $p = 0.19$, two-tailed two-sample Wilcoxon rank-sum test, Bonferroni correction applied).

We note in [16] that the lack of relationship found for the data on *Wikipedia* edits may simply reflect the substantial difference in the volume of data available for views and for edits, despite the much larger number of pages considered in this second analysis, where further relevant statistics on views and edits of *Wikipedia* pages are provided in [16]. For the purposes of these investigations, we therefore do not consider edit data further.

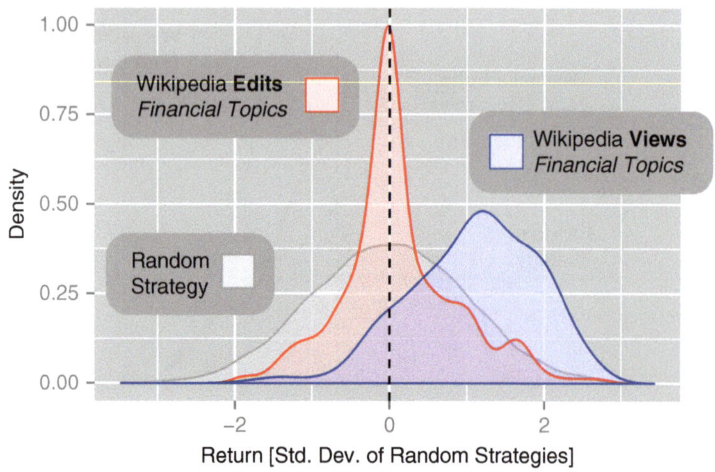

Fig. 4.4 Returns from trading strategies based on *Wikipedia* access and edit logs for pages relating to finance (Reproduced from [16]). Parallel analysis of the distribution of returns from two much larger portfolios of 285 hypothetical strategies, based on changes in how often a set of 285 financially related *Wikipedia* pages were viewed (*blue*) and edited (*red*) during the same period as Fig. 4.3, again with $\Delta t = 3$ weeks. Our analysis shows that the returns of *Wikipedia* page view based strategies are significantly higher than the returns of random strategies for this period ($\langle R \rangle_{Views} = 1.10$; $W = 2{,}286{,}608$, $p < 0.001$, two-tailed two-sample Wilcoxon rank-sum test, Bonferroni correction applied). Once again however, we find no evidence of a statistically significant difference between the returns from the *Wikipedia* edit based strategies, and the random strategies ($\langle R \rangle_{Edits} = 0.12$; $W = 1{,}516{,}626$, $\alpha = 0.05$, two-tailed two-sample Wilcoxon rank-sum test, Bonferroni correction applied)

4.4 Financial Relevance of Information Searched for Before Stock Market Falls

Our assumption so far was that only *Google* and *Wikipedia* usage data relating to financial topics would provide any insight into information gathering processes before trading decisions, and therefore future changes in the DJIA. To verify this assumption, in [16] we carry out a further analysis of view data relating to 233 *Wikipedia* pages describing actors and filmmakers, where further details of these pages are provided in [16]. We suggest that such pages have less obvious financial connotations.

We analyze the distribution of returns for a portfolio of 233 hypothetical trading strategies based on changes in how often these pages were viewed, trading weekly on the DJIA with $\Delta t = 3$ weeks during the period December 2007–April 2012, as in the previous *Wikipedia* analyses. We ensured that this set of pages, of similar size to the set of pages relating to financial topics, had at least equivalent traffic during the period of investigation, to ensure that any failure to find a relationship was not due to power issues caused through lack of data on *Wikipedia* views.

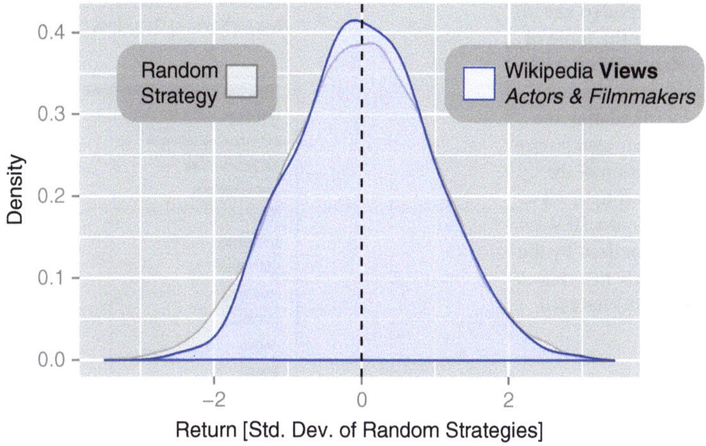

Fig. 4.5 Returns from trading strategies based on *Wikipedia* access logs for pages relating to actors and filmmakers (Reproduced from [16]). Parallel analysis of the distribution of returns for another portfolio of 233 hypothetical strategies based on changes in how often a set of 233 Wikipedia pages relating to actors and filmmakers were viewed (*blue*). Here, we find that there is no significant difference between the returns generated by the random strategies and the *Wikipedia* view based strategies ($\langle R \rangle_{Views}$ = 0.04; W = 1,189,114, p = 0.59, two-tailed two-sample Wilcoxon rank-sum test)

In Fig. 4.5, we show the returns from these 233 strategies based on changes in the number of views of *Wikipedia* articles on actors and filmmakers (*blue*), alongside returns from the random strategies (*gray*). We find that there is no significant difference between the returns generated by the random strategies and the *Wikipedia* view based strategies ($\langle R \rangle_{Views} = 0.04$; $W = 1{,}189{,}114$, $p = 0.59$, two-tailed two-sample Wilcoxon rank-sum test).

Similarly, in [30], we investigate whether differences in performance of the 98 *Google Trends* strategies we tested can be partially explained using an indicator of the extent to which different terms are of financial relevance. We quantify financial relevance by calculating the frequency of each search term in the online edition of the *Financial Times* from August 2004 to June 2011, normalized by the number of *Google* hits for each search term (Fig. 4.6). We find that the return associated with a given search term is correlated with this indicator of financial relevance (Kendall's tau $= 0.275$, $z = 4.01$, $N = 98$, $p < 0.001$) using Kendall's tau rank correlation coefficient.

4.5 Discussion

In the investigations described in [16] and [30], summarized here, we find evidence of increases in searches for financially related information before stock market falls. These results are consistent with the hypothesis that historic usage data from *Google*

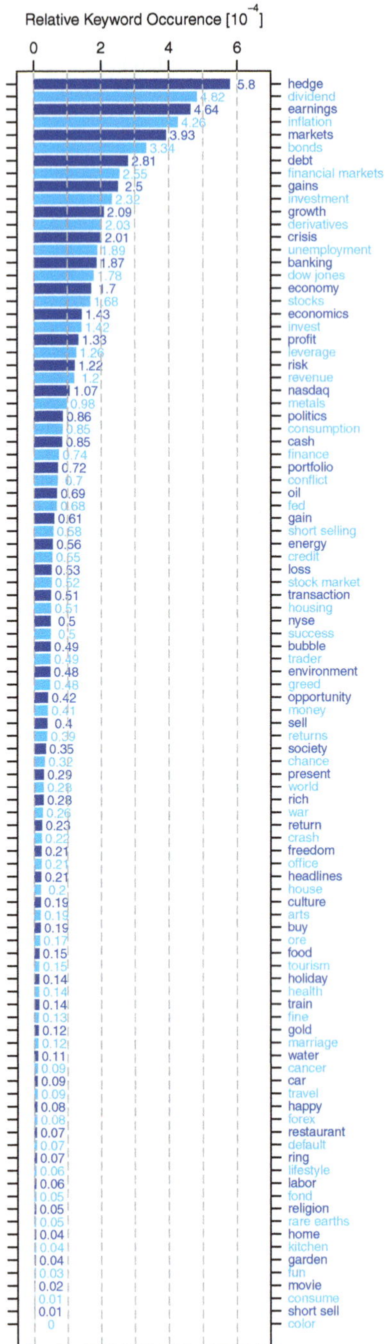

Fig. 4.6 How related are search terms to the topic of finance? (Reproduced from [30]). We quantify financial relevance by calculating the frequency of each search term in the online edition of the *Financial Times* (http://www.ft.com) from August 2004 to June 2011, normalized by the number of *Google* hits (http://www.google.com) for each search term

and the online encyclopedia *Wikipedia* may have provided some insight into future trends in the behavior of financial market actors.

In [16], we have proposed one potential explanation in line with these results. We first suggest that *Google* and *Wikipedia* records may provide a proxy measurement of the information gathering process of a subset of investors for the investigated period. We further note that previous studies in behavioral economics have demonstrated that humans are loss averse [34]: that is, they are more concerned about losing \$5 than they are about missing an opportunity to gain \$5. By this logic, it could be argued that the trading decision of greatest consequence for a trader would be to sell a stock at a lower price than they had previously believed it was worth. If we assume that investors may be willing to invest more efforts in information gathering before making a decision which they view to be of greater consequence, then it would follow that increases in information gathering would precede falls in stock market prices, in line with our results.

Our results suggest that Internet usage data may offer a window into the information gathering processes which precede real world actions captured in large behavioral data sets. By combining these new data sets, we may be able to gain new insight into different stages of collective economic decision making.

Acknowledgements The results summarized here were first reported in Preis, T., Moat, H.S., Stanley. H.E.: Quantifying trading behavior in financial markets using Google Trends. Sci. Rep. 3, 1684 (2013), http://www.nature.com/srep/2013/130425/srep01684/full/srep01684.html, and Moat, H.S. et al.: Quantifying Wikipedia usage patterns before stock market moves. Sci. Rep. 3, 1801 (2013), http://www.nature.com/srep/2013/130508/srep01801/full/srep01801.html. We thank Adam Avakian and Dror Y. Kenett for comments. H.S.M. and T.P. acknowledge the support of the Research Councils UK Grant EP/K039830/1. This work was also supported by the Intelligence Advanced Research Projects Activity (IARPA) via Department of Interior National Business Center (DoI/NBC) contract number D12PC00285. The U.S. Government is authorized to reproduce and distribute reprints for Governmental purposes notwithstanding any copyright annotation thereon. Disclaimer: The views and conclusions contained herein are those of the authors and should not be interpreted as necessarily representing the official policies or endorsements, either expressed or implied, of IARPA, DoI/NBC, or the U.S. Government.

References

1. Alanyali M, Moat HS, Preis T (2013) Quantifying the relationship between financial news and the stock market. Sci Rep 3:3578
2. Askitas N, Zimmermann KF (2009) Google econometrics and unemployment forecasting. Appl Econ Q 55:107–120
3. Bordino I et al (2012) Web search queries can predict stock market volumes. PLOS One 7:e40014
4. Choi H, Varian H (2012) Predicting the present with Google Trends. Econ Rec 88:2–9
5. Conte R et al (2012) Manifesto of computational social science. Eur Phys J Spec Top 214:325–346

6. Corley M, Brocklehurst PH, Moat HS (2010) Error biases in inner and overt speech: evidence from tongue twisters. J Exp Psychol: Learn Mem Cognit 37:162–175
7. Feng L, Li B, Podobnik B, Preis T, Stanley HE (2012) Linking agent-based models and stochastic models of financial markets. Proc Natl Acad Sci USA 109:8388–8393
8. Gabaix X, Gopikrishnan P, Plerou V, Stanley HE (2003) A theory of power-law distributions in financial market fluctuations. Nature 423:267–270
9. Ginsberg J et al (2009) Detecting influenza epidemics using search engine query data. Nature 457:1012–1014
10. Goel S, Hofman JM, Lahaie S, Pennock DM, Watts DJ (2010) Predicting consumer behavior with web search. Proc Natl Acad Sci USA 107:17486–17490
11. Hommes CH (2002) Modeling the stylized facts in finance through simple nonlinear adaptive systems. Proc Natl Acad Sci USA 99:7221–7228
12. Lux T, Marchesi M (1999) Scaling and criticality in a stochastic multi-agent model of a financial market. Nature 397:498–500
13. Mantegna RN, Stanley HE (2002) Scaling behaviour in the dynamics of an economic index. Nature 376:46–49
14. Moat HS (2010) Modelling subphonemic information flow: an investigation and extension of Dell's (1986) model of word production. Doctoral dissertation, University of Edinburgh
15. Moat HS, Corley M, Hartsuiker RJ (2008) Connecting phonological encoding to articulation: is cascading required? A computational investigation. In: Love BC, McRae K, Sloutsky VM (eds) Proceedings of the 30th annual conference of the Cognitive Science Society, Washington, DC, pp 1320–1325. Cognitive Science Society, Austin
16. Moat HS, Curme C, Avakian A, Kenett DY, Stanley HE, Preis T (2013) Quantifying Wikipedia usage patterns before stock market moves. Sci Rep 3:1801
17. Moat HS, Preis T, Olivola CY, Liu C, Chater N (in press) Using big data to predict collective behavior in the real world. Behav Brain Sci
18. Preis T (2010) Quantifying and modeling financial fluctuations. Doctoral dissertation, University of Mainz
19. Preis T (2011) Econophysics—complex correlations and trend switchings in financial time series. Eur Phys J Spec Top 194:5–86
20. Preis T (2011) GPU-computing in econophysics and statistical physics. Eur Phys J Spec Top 194:87–119
21. Preis T, Stanley HE (2010) Switching phenomena in a system with no switches. J Stat Phys 138:431–446
22. Preis T, Golke S, Paul W, Schneider JJ (2006) Multi-agent-based order book model of financial markets. EPL 75:510
23. Preis T, Golke S, Paul W, Schneider JJ (2007) Statistical analysis of financial returns for a multiagent order book model of asset trading. Phys Rev E 76:016108
24. Preis T, Paul W, Schneider JJ (2008) Fluctuation patterns in high-frequency financial asset returns. EPL 82:68005
25. Preis T, Reith D, Stanley HE (2010) Complex dynamics of our economic life on different scales: insights from search engine query data. Philos Trans R Soc A 368:5707–5719
26. Preis T, Schneider JJ, Stanley HE (2011) Switching processes in financial markets. Proc Natl Acad Sci USA 108:7674–7678
27. Preis T, Virnau P, Paul W, Schneider JJ (2011) Accelerated fluctuation analysis by graphic cards and complex pattern formation in financial markets. New J Phys 11:093024
28. Preis T, Kenett DY, Stanley HE, Helbing D, Ben-Jacob E (2012) Quantifying the behavior of stock correlations under market stress. Sci Rep 2:752
29. Preis T, Moat HS, Stanley HE, Bishop SR (2012) Quantifying the advantage of looking forward. Sci Rep 2:350
30. Preis T, Moat HS, Stanley HE (2013) Quantifying trading behavior in financial markets using Google Trends. Sci Rep 3:1684
31. Preis T, Moat HS, Bishop SR, Treleaven P, Stanley HE (2013) Quantifying the digital traces of Hurricane Sandy on Flickr. Sci Rep 3:3141

32. Silverman BW (1986) Density estimation. Chapman and Hall, London
33. Simon HA (1955) A behavioral model of rational choice. Q J Econ 69:99–118
34. Tversky A, Kahneman D (1991) Loss aversion in riskless choice: a reference-dependent model. Q J Econ 106:1039–1061
35. Wickham H (2009) ggplot2: elegant graphics for data analysis. Springer, New York

Chapter 5
Nonequilibrium Quantum Dynamics of Biomolecular Excitons

Cesar A. Mujica-Martinez, Peter Nalbach, and Michael Thorwart

Abstract We investigate the effect of equilibrium and nonequilibrium localized vibrations on the excitation energy transfer efficiency of the Fenna-Matthews-Olson complex. By means of numerically exact real-time path-integral simulations of the transfer dynamics we find that equilibrium vibrations do not enhance coherence times. On the other hand, nonequilibrium vibrations induce prolonged coherence times and increased transfer efficiency. By quantifying the transfer dynamics in terms of a non-Markovianity measure based on the time evolution of the trace distance of two quantum states we find, in all cases, a monotonic decrease of the trace distance with increasing time which implies that the exciton transfer follows a Markovian dynamics.

5.1 Introduction

The directed transport of excitation energy is at the heart of photosynthesis which is one of the most important biochemical processes on earth. It is a typical nonequilibrium transfer process and funnels the energy provided by the solar photons down to usable chemical energy in photoactive living systems. Photosynthesis allows the living systems to harvest an enormous amount of energy, for instance, the energy captured by photosynthesis per year on our planet is of the order of 100 TW which should be compared to roughly 15 TW of human energy consumption per year [1]. Along with the harvest of energy goes the production of about 105 billion tons of biomass per year or 12 million tons per hour on the planet. Roughly half of the

C.A. Mujica-Martinez • P. Nalbach • M. Thorwart (✉)
I. Institut für Theoretische Physik, Universität Hamburg, Jungiusstr. 9, 20355 Hamburg, Germany
e-mail: cesar.mujica@physik.uni-hamburg.de; peter.nalbach@physik.uni-hamburg.de; michael.thorwart@physik.uni-hamburg.de

D. Matrasulov and H.E. Stanley (eds.), *Nonlinear Phenomena in Complex Systems: From Nano to Macro Scale*, NATO Science for Peace and Security Series C: Environmental Security, DOI 10.1007/978-94-017-8704-8_5,

photosynthesis happens in organisms in water, i.e., mainly in the ocean, while the other half is performed by systems on the shore.

Since long, the molecular mechanism of the light harvesting has been of interest for many different fields in science. For instance, green plants contain chloroplasts in their leaves which contain a stack of membranes whose molecular structure is highly nontrivial. On a length scale of nanometer, many different biomolecular clusters act in concert to harvest the energy of a solar photon and to convert it into chemical energy which starts a complicated sequence of chemical reactions [2]. As physicists, we are predominantly interested in the early steps of the photosynthetic chain of reaction, in particular, in the ultrafast processes when the excitation energy is intermittently stored in an exciton which forms a quasiparticle of a strongly bound electron-hole pair. These initial steps happen in particular ring-shaped parts of the biomolecular complexes which are denoted as antenna structures. After the exciton is formed at a certain molecular site, it transfers its energy by a radiationless Coulomb dipole coupling to neighboring sites such that the energy eventually finds its way through a network of few molecular sites. Finally, the energy ends up in the reaction center which is an energy sink and which is the place where an electron transfer is initiated [3].

A central role during this cascade of excitation energy transfers is played by the fluctuations provided by the environment in which the biomolecule is hosted. By nature, the transfer processes are quantum mechanical transfer processes. However, the presence of strong fluctuations has led to the orthodox picture of the Förster theory of excitation energy transfer which is based on the assumptions that the excitons are well localized quasiparticles and the transfer is a classical hopping-like dynamics along the molecular sites. The exciton population of each site is described by a classical probability and quantum mechanics in this description enters only when the transfer rates are determined by Fermi's Golden Rule, being perturbative in the dipole coupling strength.

Recent experiments [4–8] though have challenged this picture and have described signatures of nontrivial quantum effects in biological matter under ambient conditions. Particular active compounds have been studied such as the Fenna-Matthews-Olson (FMO) complex in the green sulfur bacterium *Chlorobium tepidum* by ultrafast two-dimensional optical spectroscopy [9]. It was found that the time-resolved spectral signal associated to the excitonic coherence displayed beating oscillations which lasted over longer than 1 ps at 77 K [5]. Even at room temperature, rather long-lasting oscillations for a few hundreds of fs were described [6]. Similar features were found in marine cryptophyte algae [7] and in the light-harvesting complex 2 of purple bacteria [8].

These reports posed various challenges for an accurate theoretical description due to the complexity of the systems under study. One of the best characterized biomolecular systems is the FMO complex [10, 11]. It is sufficiently simple for an effective but still accurate modelling and consists of seven relevant active molecular sites (see below). Most importantly, also the frequency spectrum of the fluctuations which act on the molecular levels is rather well known [12]. One problem is that it exhibits a rather non-typical and non-trivial form and it is a priori not clear whether simple Redfield-like quantum master equations are an appropriate tool to the describe the transfer dynamics. This is because such an approach is by construction based on a weak-coupling assumption which goes along with a Markovian approximation [13]. Such an assumption is valid only for a pure and structureless Ohmic spectral distribution of the bath fluctuations. Below, we will show by numerically exact path-integral simulations and by calculating the appropriate quantum information theoretical measure of the non-Markovianity that despite these issues, the full exciton dynamics follows a Markovian dynamics.

In addition, we show that the observed long quantum coherence times of the electronic states cannot be explained by a non-trivial role of fluctuations coming from a continuous frequency distribution of the fluctuations generated by the pigment-protein-solvent environment. We show that, instead, discrete vibrational modes of the molecular backbone may yield increased coherence times. Their effect, however, depends on which molecular site the vibration couples to. In the same way, we show that the presence of discrete vibrational states can also enhance the efficiency of the exciton transfer through the FMO complex.

5.2 The Model of the FMO Complex

The FMO complex consists of three identical subunits, each of which contains eight bacteriochlorophyll *a* (BChl *a*) molecular sites [14]. The recently discovered eighth site [11] is irrelevant for our discussion below, since it is only weakly coupled to the other sites. The seven sites can be reduced to its two lowest electronic levels with their excited states electronically coupled along the complex. The excitation dynamics can be described within the single exciton subspace due to the different times scales of ∼ps exciton transfer as compared to the ∼ns recombination. Eventually, the Hamiltonian [15]

$$H_{\mathrm{FMO}} = \begin{pmatrix} 240 & -87.7 & 5.5 & -5.9 & 6.7 & -13.7 & -9.9 \\ & 315 & 30.8 & 8.2 & 0.7 & 11.8 & 4.3 \\ & & 0 & -53.5 & -2.2 & -9.6 & 6.0 \\ & & & 130 & -70.7 & -17.0 & -63.3 \\ & & & & 285 & 81.1 & -1.3 \\ & & & & & 435 & 39.7 \\ & & & & & & 245 \end{pmatrix} \tag{5.1}$$

in units of cm^{-1} in site representation for *Chlorobium tepidum* is the basis for our analysis below. Notice that the BChl 3 has been defined as the site with the lowest energy [14].

The exciton energy eigenstates of the FMO complex are obtained by diagonalizing Eq. (5.1), with the squares of the eigenvector elements of the seven exciton states schematically shown in Fig. 5.1. Experimental results [14] indicate that the BChls 1

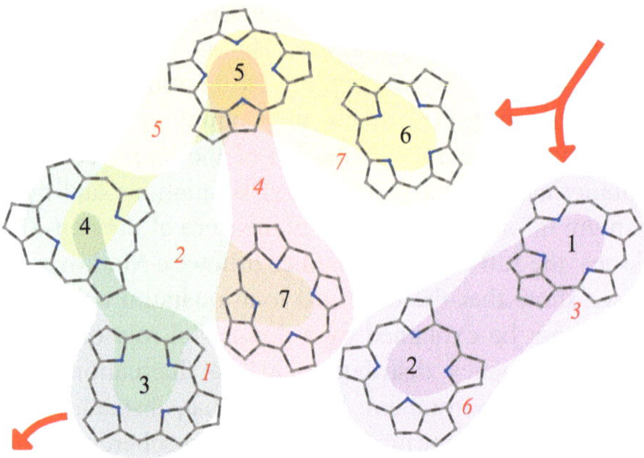

Fig. 5.1 Structural arrangement of the seven BChl molecules (*black numbers*) in FMO (*Chl. tepidum*) [11] superposed with a schematic representation of the delocalization patterns of the different excitons (*coloured shading*, *italic red numbers*). The exciton numeration is in ascending energy order. Entrance and exit sites are indicated by *red thick arrows*

and 6 are oriented towards the baseplate protein. Therefore, it is believed that these are the initially excited sites. As indicated in Fig. 5.1, exciton *1* is almost completely localized at BChl 3 which is the energy sink towards the reaction center. The exciton pairs *3* and *6* are mainly localized on the BChls 1 and 2. These two BChls are the ones which are most strongly coupled in the FMO Hamiltonian (Eq. (5.1)). Meanwhile, the pair of excitons *5* and *7* is mainly localized on the BChls 5 and 6, which is the second most strongly coupled BChl pair.

5.3 The Spectral Distribution of the Environmental Fluctuations

The vibrational pigment-protein-solvent environment induces fluctuations on the exciton dynamics. These fluctuations are most efficiently described in terms of an open quantum system approach [16], where the Gaussian fluctuations are described by harmonic modes [17]. This gives rise to the total Hamiltonian

$$H = H_{FMO} + \sum_{j=1}^{7} |j\rangle\langle j| \sum_{k} \upsilon_k^{(j)} q_{j,k} + \sum_{j=1}^{7} \frac{1}{2} \sum_{k} \left(p_{j,k}^2 + \omega_{j,k}^2 q_{j,k}^2 \right), \qquad (5.2)$$

with momenta $p_{j,k}$, displacement $q_{j,k}$, frequency $\omega_{j,k}$ and coupling $\upsilon_k^{(j)}$ of the environmental vibrations at site j. The fluctuations acting on the FMO sites are described by a spectral frequency distribution $G(\omega) = \sum_{j,k} \left(|\upsilon_k^{(j)}|^2 / 2\omega_{j,k} \right)$

$\delta(\omega - \omega_{j,k})$ [16], which contains two relevant features. On the one hand, a continuous background refers to the broad frequency distribution of solvent modes providing fluctuating electric fields. On the other hand, well localized vibrational modes of the host protein are present as peaks in the frequency distribution $G(\omega)$. Again two possibilities to capture the localized vibrations arise: (i) they can either be taken as part of the environment or (ii) their quantum dynamics can explicitly be followed as part of the "system". We will follow both routes below. The difference between both is that in (i), the fluctuations are always assumed to be thermal, i.e., resulting from an equilibrated environment, while in (ii), the fluctuations are effectively non-thermal.

Adolphs and Renger [15] have derived a spectral density function by parametrization of the experimental results in Ref. [12]. It contains both a broad low frequency contribution $S_0 g_0(\omega)$ by the protein and solvent vibrations with Huang-Rhys factor S_0 and a single effective high-energy vibrational mode of the pigments with Huang-Rhys factor S_H. It is determined as

$$G(\omega) = S_0 g_0(\omega) + S_H \delta(\omega - \omega_H). \tag{5.3}$$

The Huang-Rhys factor $S_0 = 0.5$ was estimated from temperature-dependent absorption spectra of FMO complexes and the low-frequency function $g_0(\omega)$ has the same form as the spectral density that was originally extracted from calculations of fluorescence line narrowing spectra of B777-complexes [18]. It is given by

$$g_0(\omega) = 6.105 \cdot 10^{-5} \times \frac{\omega^5}{\omega_1^4} e^{-\sqrt{\frac{\omega}{\omega_1}}} + 3.8156 \cdot 10^{-5} \times \frac{\omega^5}{\omega_2^4} e^{-\sqrt{\frac{\omega}{\omega_2}}}, \tag{5.4}$$

with $\omega_1 = 0.575\,\mathrm{cm}^{-1}$ and $\omega_2 = 2\,\mathrm{cm}^{-1}$. In addition, a vibrational mode of the individual pigments, with the Huang-Rhys factor $S_H = 0.22$, is included at frequency $\omega_H = 180\,\mathrm{cm}^{-1}$ [15]. For our analysis below, the artificial δ-peak will be broadened with a Lorentzian line shape since the protein is embedded in water, which, as a polar solvent, gives rise to an additional weak Ohmic damping of the protein vibrations [19]. The form of a Lorentzian peak ensures that the Huang-Rhys factor S_H is kept constant when varying its width γ. In Fig. 5.2, it is shown the resulting spectral density function in Eq. (5.3) for several widths γ.

5.4 The Tool of the Quasiadiabatic Path Integral

In order to determine the time evolution of the excitonic dynamics in the FMO complex under the action of environmental fluctuations, we calculate the time-dependence of the reduced density matrix $\rho(t)$. It characterizes the system part and is obtained after tracing out the bath degrees of freedom. It is calculated using the QUAPI scheme [20,21], which is a numerically exact iteration scheme that has been

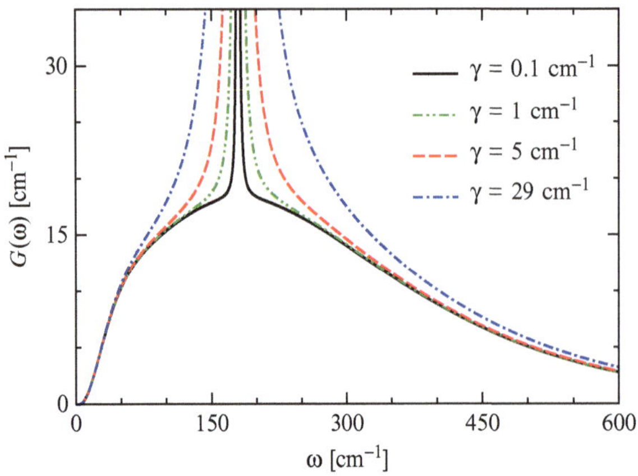

Fig. 5.2 Spectral density $G(\omega)$, Eq. (5.3), as determined by Adolphs and Renger [15] for different widths γ of the Lorentzian peak associated to the vibrational mode. The latter is centered at $\omega_H = 180\,\mathrm{cm}^{-1}$

successfully applied to many problems of open quantum systems, and it allows to treat nearly arbitrary spectral functions at finite temperatures [19, 22–26].

In this numerical scheme, it is possible either (i) to include the localized vibrational mode as part of the bath in the spectral density, or, (ii) associate it to the system resulting in an increased dimension of the Hilbert space related to $\rho(t)$. In the latter case, an additional projection of the increased reduced density operator on the excitonic states is necessary to obtain the exciton dynamics only.

5.5 The FMO Exciton Dynamics with Equilibrated Vibrations

In the following, we present results for the time evolution of the population of the FMO sites under different environmental conditions, in particular for different ways of including the vibrational mode present in Eq. (5.3). We fix the width of the mode to $\gamma = 29\,\mathrm{cm}^{-1}$ and notice that the results do only weakly depend on this parameter. We find (not shown) similar results for smaller values of γ, the reason being that the large continuous background already provides the dominating broadening of the corresponding excitonic level [27].

We start in this section by assuming that the vibrational mode equilibrates very rapidly to its thermal state and only provides fluctuations around the latter which act on the FMO excitonic states. In Fig. 5.3, we show the results for $T = 77\,\mathrm{K}$ (right column) and $T = 300\,\mathrm{K}$ (left column) for the initial preparation $\rho_{11}(0) = 1$ (upper row) and $\rho_{66}(0) = 1$ (lower row). We observe no enhanced coherence times in almost all cases apart from the case at low temperature $T = 77\,\mathrm{K}$ and for

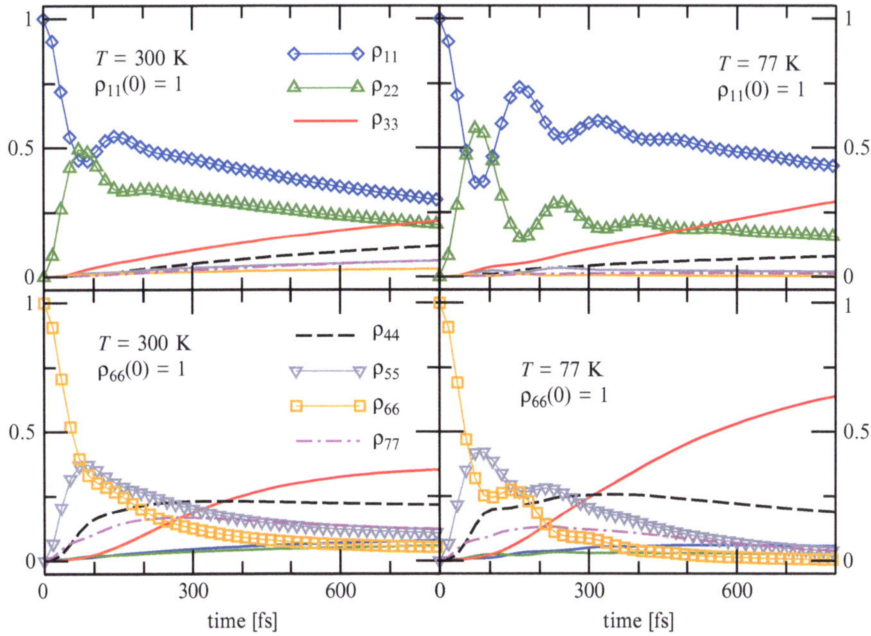

Fig. 5.3 Time-dependent occupation probabilities of all seven FMO sites for $T = 300\,\text{K}$ and $T = 77\,\text{K}$ with $\rho_{11}(0) = 1$ and $\rho_{66}(0) = 1$ for the FMO spectrum in Eq. (5.3) [15]. The equilibrated vibrational mode provides additional fluctuations around its thermal state

$\rho_{11}(0) = 1$ (upper right), where damped coherent oscillations arise up to $\approx 500\,\text{fs}$. This illustrates that the fluctuations of an equilibrated vibrational mode are not "coherent" enough to be responsible for the observed long-lasting coherent signals [5, 6] and other mechanisms have to be invoked to explain such a long-lasting non-trivial quantum coherence.

5.6 The Effect of Nonequilibrium Vibrations on Coherence

In the previous section, we have assumed that the vibrational mode equilibrates so rapidly that we can treat it as part of the fluctuational spectrum for a thermal bath. In this section, we turn to the other possibility and consider the case when the vibrational mode is not instantaneously in its equilibrium state but evolves as part of the "system" in a quantum coherent manner. It approaches its equilibrium state by a coupling with the environment characterized by Eq. (5.3). To determine the population of the exciton states, we average the reduced density operator for the FMO complex plus vibrational mode over the vibrational degrees of freedom in all cases treated in this section.

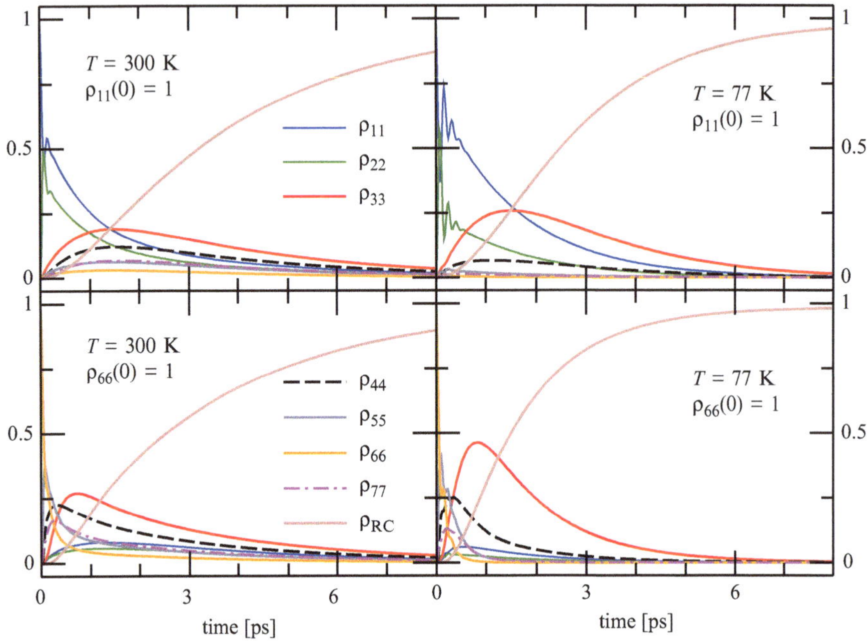

Fig. 5.6 Time evolution of the FMO site populations and the trapping or sink state in absence of any vibrational mode for the same parameters as above

the complex once it has arrived at site 3. As usual and since we are not interested in the explicit details of the dumping process, we include a coupling term in the Hamiltonian which is imaginary and which corresponds to a large and constant decay rate of the exciton population at site 3. This also implies that no back transfer is possible. In the following, we study the impact of the vibrational mode on the total transfer efficiency.

5.7.1 *Transfer Efficiency of the Static FMO Complex*

To begin, we first set the stage by avoiding any vibrational mode and study the pure excitonic dynamics. Figure 5.6 shows the results for the individual site populations as above and in addition the population ρ_{RC} of the energy sink. Notice that the time window on the horizontal axis has been increased. Clearly, the populations of the FMO sites alone no longer add up to 1 since energy (and thus site populations) is (are) constantly removed from the complex. Apart from few oscillations of selected populations at very short times, no coherence in the populations is found. The population of the energy sink increases in a monotonous manner.

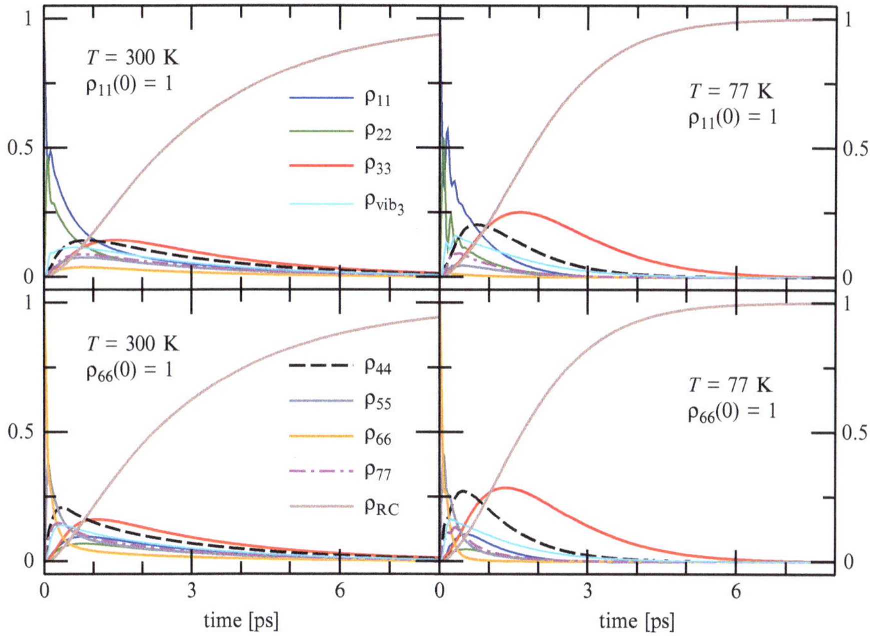

Fig. 5.7 Time evolution of the FMO site populations, the vibrational mode and the trapping or sink state in presence of the vibrational mode at the exit site 3 for the same parameters as above

5.7.2 *Vibrational Exit Site 3*

As a first case in this section, we couple the site 3, which is the exit site, to the vibrational mode and study the site populations, the population of the vibrational mode and of the sink. The results are depicted in Fig. 5.7. In comparison to the case without the sink, see Sect. 5.6.1 (Fig. 5.4), the coherence times are not increased which is clear since the sink only removes energy and thus could only deteriorate the coherence properties. However, the population of the sink grows faster as compared to the case where no vibrational mode is present (Fig. 5.6) and, consequently, the transfer efficiency is increased. The physical picture for this effect is that the presence of the vibration at site 3 simply offers more states which can become populated during the exciton transfer in the complex and, consequently, more states also can dump their energy in the sink. This then increases the sink population faster and yields to an overall increased transfer efficiency.

5.7.3 *Vibrational Entrance Site 1*

Next, we add the explicit vibrational mode at the entrance site 1. The results are shown in Fig. 5.8. As in the case without the sink (see Sect. 5.6.2, Fig. 5.5), the

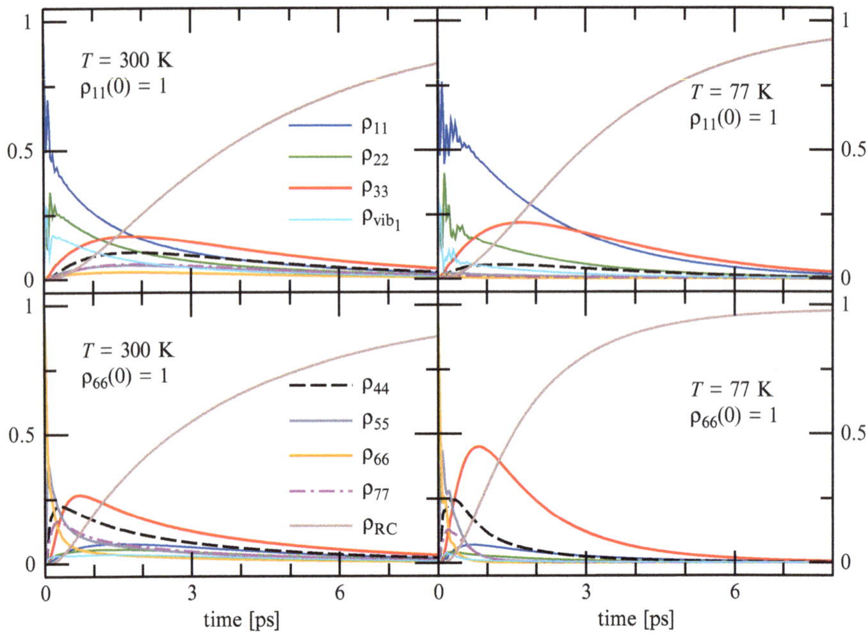

Fig. 5.8 Same as in Fig. 5.7, but with the vibrational mode coupled to the entrance site 1

coherent oscillations of the populations of the sites 1 last longer due to the presence of the vibrational mode. Naturally, also the population of the vibrational mode oscillates at short times, see light blue line in Fig. 5.8. Despite these features, the transfer efficiency is not substantially altered in comparison to the case without the vibrational mode, see Fig. 5.6.

5.7.4 Vibrational Modes at All Sites

Having elucidated the role played by the vibrational states at the relevant individual sites 1 and 3, we can now turn to the case when all FMO sites are coupled to different but equal vibrational modes. The limitations of the QUAPI method consists in an exponential growth of the array sizes and the computational times for growing system Hilbert space dimensions. We have thus to restrict the Hilbert space dimensions to vibrational states with energies up to 450 cm^{-1} above the energy of site 3. However, we have good reasons to believe that this technical restriction has no severe implications: the relevant FMO system Hamiltonian shows that the FMO BChl pairs which are strongly coupled typically have smaller energy gaps. Then, the included vibrational states typically have comparable or larger energies so that the

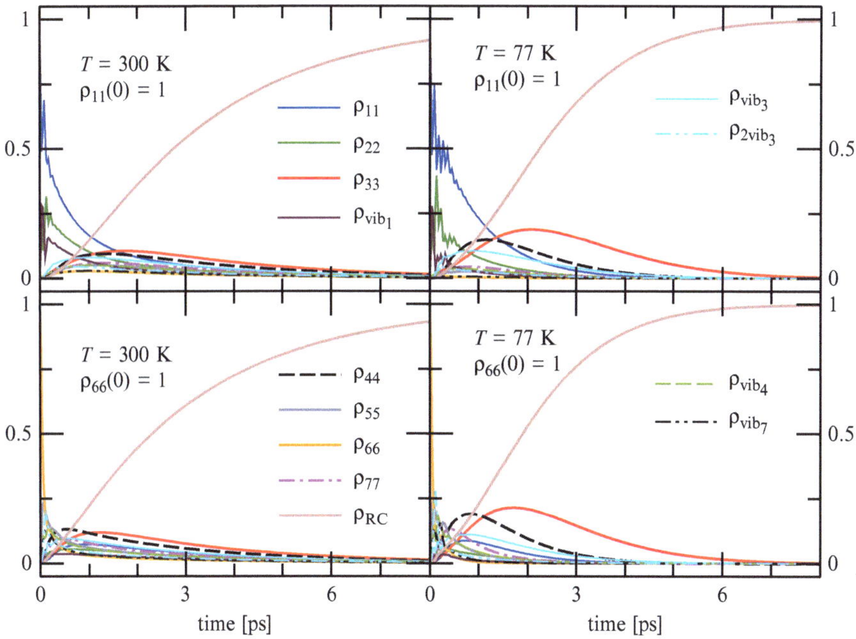

Fig. 5.9 Same as in Fig. 5.7, but when the vibrational mode is coupled to all sites. Notice that due to computational restrictions, only vibrational energies up to $450\,\mathrm{cm}^{-1}$ above the energy of site 3 are included

Table 5.1 Comparison of energy transfer times with and without nonequilibrium vibrational states at 300 K

Initial excitation at	No vibrational mode (ps)	With vibrational modes (ps)	Faster by (%)
Site 1	3.7	3.1	16.2
Site 6	3.3	2.9	12.1

relevant FMO excitonic energy ranges are covered. The results are shown in Fig. 5.9. In this constellation, prolonged coherence times go along with an increased transfer efficiency.

5.7.5 *Transfer Times and Increase of Transfer Efficiency*

To summarize the discussion on the impact of nonequilibrium vibrational states on the efficiency of the energy transfer in the FMO complex, we quantify the energy transfer times for the cases with and without vibrational states, see Table 5.1.

The transfer time is thereby taken as the rise time of the exponential growth of the population of the sink. At $T = 300\,\mathrm{K}$ for the initial excitation starting at site 1, we find a transfer time of 3.7 ps in absence of vibrational states. In contrast, this value decreases to a transfer time of 3.1 ps when the vibrational states are explicitly included. This decrease corresponds to a growth of the efficiency by 16.2 %. Analogously, when the FMO dynamics is prepared to start at site 6, a transfer time of 3.3 ps without the vibrational states result. Likewise, this value decreases to a transfer time of 2.9 ps when the vibrations are present, which gives rise to an increase of the transfer efficiency by 12.1 %. To judge on the relevance of the gain, we have to keep in mind that such values are very significant for systems which have been optimized on evolutionary time scales of billions of years.

At this point, we can conclude that the presence of a nonequilibrium underdamped vibrational mode at $180\,\mathrm{cm}^{-1}$ indeed improves quantum transfer efficiency. Along with this, it can be responsible for the observed prolonged quantum coherence times [5, 6]. However, as we have seen, an enhanced transfer efficiency also may result in absence of an enhanced quantum coherence and could be simply due to the fact that vibrational states simply offer more routes to the energy transfer and thus increase the number of transport channels. This could happen with or without prolonged quantum coherence times.

5.8 Non-Markovianity of the Exciton Dynamics in the FMO Complex

Dissipative quantum dynamics only is purely Markovian for a strictly Ohmic spectral density. The environmental fluctuation spectral function $G(\omega)$ of Adolphs and Renger as given in Eq. (5.3) is highly non-Ohmic: it behaves super-Ohmically at low-frequency, includes a turn-over at intermediate frequencies, and decays algebraically at large frequencies. In addition, the vibrational mode is present so that this non-standard and structured environment, in conjunction with the prolongated coherent dynamics observed in experiments [5, 6] and simulations [28–30] could generate non-Markovian effects. To quantify them, we use the non-Markovianity measure developed in Ref. [31]. It is based on the physical features of the system-bath interaction in terms of information backflow from the environment to the system, which has been experimentally measured [32, 33].

In an open quantum system, any two initial states $\rho_{1,2}(0)$ evolve over time according to a family of trace preserving and completely positive quantum dynamical maps $\Phi(t,0)$ such that $\rho_{1,2}(t) = \Phi(t,0)\rho_{1,2}(0)$. Because of the different dynamical evolution of these two quantum states, they can be distinguished in terms of the trace distance, which is a metric in the space of physical states [34]. The dynamical change of the trace distance can be interpreted as a changing distinguishability of the states which is accompanied by an information exchange between the system and its environment [34]. The very small – and in fact zero – correlation time between the system and environmental dynamics in a Markovian process leads to a monotonic flow of information from the system to the environment. Conversely, temporal correlations in a non-Markovian process may generate a backflow of information from the environment to the system and can induce memory effects. This feature can be quantified in terms of the time evolution of the trace distance between the pair of quantum states $\rho_{1,2}$ of the open system which is defined as [35]

$$D(\rho_1, \rho_2) = \frac{1}{2} \mathrm{tr} |\rho_1 - \rho_2|, \tag{5.5}$$

where $|O| = \sqrt{O^\dagger O}$. It satisfies $0 \leq D \leq 1$ [31]. For open quantum systems, the trace distance of the states $\rho_{1,2}(t)$ (evolving under the dynamical map $\Phi(t)$) is a monotonically decreasing function of time $D(\Phi\rho_1, \Phi\rho_2) \leq D(\rho_1, \rho_2)$, which means that the distinguishability of two states always decreases. Therefore, it is useful to define the rate of change of the trace distance as

$$\sigma(t, \rho_{1,2}(0)) = \frac{d}{dt} D(\rho_1(t), \rho_2(t)), \tag{5.6}$$

which depends on the specific initial states $\rho_{1,2}(0)$. During a Markovian evolution any two initial states become less and less distinguishable for growing times, then $\sigma \leq 0$ is always satisfied. Consequently, a non-Markovian process fulfills $\sigma > 0$. The non-Markovianity measure (of the quantum process $\Phi(t)$) quantifies the total increase of the distinguishability over the whole time evolution, i.e., the *total* amount of information which flows from the environment back to the system, according to

$$\mathcal{N}(\Phi) = \max_{\rho_{1,2}(0)} \int_{\sigma>0} \sigma(t, \rho_{1,2}(0))\, dt. \tag{5.7}$$

Here, the time integration extends over all time intervals (a_i, b_i) in which $\sigma > 0$, and the maximum is taken over all pairs of initial states [31]. For this reason, $\mathcal{N}(\Phi)$ represents a functional of the family of dynamical maps $\Phi(t)$ describing the physical process [34].

In the following, we determine the trace distance $D(\rho_1, \rho_2)$ of two quantum states over their time evolution and search for the time intervals during which this quantity increases. The initial states are $\rho_1(0) = \rho_{11}$ and $\rho_2(0) = \rho_{66}$, corresponding to the BChl 1 and BChl 6 sites (see Fig. 5.1), which are the entrance sites [14, 15, 19]. The results for the trace distance for the two different temperatures $T = 77$ K and $T = 300$ K are shown in Fig. 5.10. The black dashed lines correspond to the case of FMO exciton dynamics in presence of equilibrated vibrations, as presented in Sect. 5.5 (Fig. 5.3). The red and dash-dotted blue lines correspond, respectively, to the case of FMO transfer efficiency in presence of nonequilibrium vibrations (as presented in Sect. 5.7) in the static FMO complex (Fig. 5.6) and when vibrational modes at all sites are included (Fig. 5.9). As can be seen, the trace distance monotonically decays from its starting value 1 for all relevant times. This

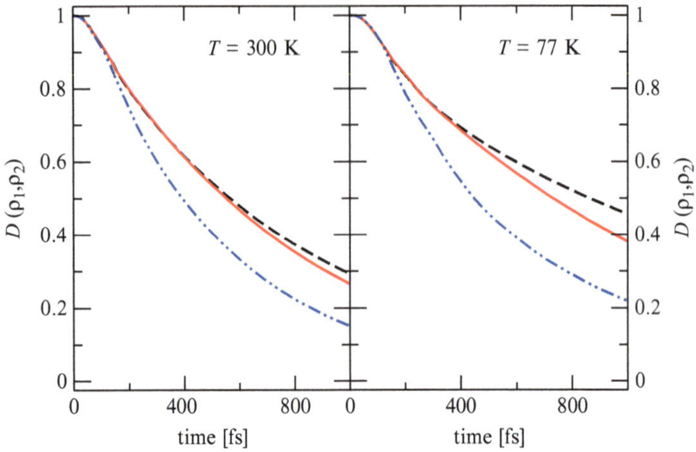

Fig. 5.10 Time evolution of the trace distance Eq. (5.5) for the temperatures $T = 300$ K (*left*) and $T = 77$ K (*right*). See text for details

happens so for both low and high temperature, with the decay been slower at lower temperature. The faster decay of the trace distance when nonequilibrium vibrations are included is a consequence of the higher amount of decoherence channels within the system, which causes a faster distinguishability lost. Accordingly, no time intervals exist for which a finite contribution to the non-Markovianity measure arises and we can conclude that the exciton dynamics in the FMO complex is fully Markovian. Similar results are obtained for different widths of the Loretzian peak, and even when it is in resonance with excitonic transitions (not shown) [27].

5.9 Conclusion

We have shown that nonequilibrium localized vibrations improve quantum energy efficiency and at the same time result in the observed prolongued quantum coherent beatings. Both are a result of the modes being underdamped, i.e., they cannot thermalize on faster time scales than compared to the electronic energy transfer dynamics. This is precisely what is assumed when treating these modes as part of the environmental fluctuation bath since a system-bath approach assumes the bath in thermal equilibrium at all times. By evaluating the time evolution of the trace distance, we found that these localized vibrational modes does not induce any non-Markovian effects in the exciton dynamics of the FMO complex.

Our results prove that the transfer dynamics of the FMO complex entirely follows a Markovian dynamics. However, the resulting Markovian dynamics is still not describable by weak system-bath coupling approaches [13]. Our results indicate the plausibility of using Markovian Redfield type equations to treat the exciton transfer dynamics only if the rates are introduced either as effective parameters or are determined by higher order treatments. This will considerably simplify the numerical effort in future investigations and, thus, larger light-harvesting complexes will be treatable.

Acknowledgements We acknowledge financial support by the DFG Sonderforschungsbereich 925 "Light-induced dynamics and control of correlated quantum systems" and by the German Academic Exchange Service (DAAD).

References

1. Blankenship RE et al (2011) Comparing photosynthetic and photovoltaic efficiencies and recognizing the potential for improvement. Science 332:805–809
2. Ke B (2003) Photosynthesis: photobiochemistry and photobiophysics Kluwer Academic Publishers, New York
3. van Amerongen H, Valkunas L, van Grondelle R (2000) Photosynthetic excitons. World Scientific, Singapore
4. Brixner T et al (2005) Two-dimensional spectroscopy of electronic couplings in photosynthesis. Nature (London) 434:625–628
5. Engel GS et al (2007) Evidence for wavelike energy transfer through quantum coherence in photosynthetic systems. Nature (London) 446:782–786
6. Panitchayangkoon G et al (2010) Long-lived quantum coherence in photosynthetic complexes at physiological temperature. Proc Natl Acad Sci USA 107:12766–12770
7. Collini E et al (2010) Coherently wired light-harvesting in photosynthetic marine algae at ambient temperature. Nature (London) 463:644–647
8. Harel E, Engel GS (2012) Quantum coherence spectroscopy reveals complex dynamics in bacterial light-harvesting complex 2 (LH2). Proc Natl Acad Sci USA 109:706–711
9. Cundiff ST, Mukamel S (2013) Optical multidimensional coherent spectroscopy. Phys. Today 66:44–49
10. Matthews B, Fenna R, Bolognesi M, Schmid M, Olson J (1979) Structure of a bacteriochlorophyll *a*-protein from the green photosynthetic bacterium *Prosthecochloris aestuarii*. J Mol Biol 131:259–285
11. Tronrud DE, Wen J, Gay L, Blankenship RE (2009) The structural basis for the difference in absorbance spectra for the FMO antenna protein from various green sulfur bacteria. Photosynth Res 100:79–87; Protein Data Bank file 3ENI
12. Wendling M et al (2000) Electron–vibrational coupling in the Fenna-Matthews-Olson complex of *Prosthecochloris aestuarii* determined by temperature-dependent absorption and fluorescence line-narrowing measurements. J Phys Chem B 104:5825
13. Nalbach P, Thorwart M (2010) Multiphonon transitions in the biomolecular energy transfer dynamics. J Chem Phys 132:194111
14. Milder MTW, Brüggemann B, van Grondelle R, Herek JL (2010) Revisiting the optical properties of the FMO protein. Photosynth Res 104:257–274
15. Adolphs J, Renger T (2006) How proteins trigger excitation energy transfer in the FMO complex of green sulfur bacteria. Biophys J 91:2778–2797
16. Weiss U (2008) Quantum dissipative systems, 3rd edn. World Scientific, Singapore
17. May V, Kühn O (2011) Charge and energy transfer dynamics in molecular systems, 3rd edn. Wiley VCH, Weinheim
18. Renger T, Marcus RA (2002) On the relation of protein dynamics and exciton relaxation in pigment–protein complexes: an estimation of the spectral density and a theory for the calculation of optical spectra. J Chem Phys 116:9997–10019

19. Nalbach P, Braun D, Thorwart M (2011) Exciton transfer dynamics and quantumness of energy transfer in the Fenna-Matthews-Olson complex. Phys Rev E 84:041926
20. Makri N, Makarov DE (1995) Tensor propagator for iterative quantum time evolution of reduced density matrices. I. Theory. J Chem Phys 102:4600–4610
21. Makri N, Makarov DE (1995) Tensor propagator for iterative quantum time evolution of reduced density matrices. II. Numerical methodology. J Chem Phys 102:4611–4618
22. Thorwart M, Reimann P, Jung P, Fox R (1998) Quantum hysteresis and resonant tunneling in bistable systems. Chem Phys 235:61–80
23. Thorwart M, Eckel J, Mucciolo ER (2005) Non-Markovian dynamics of double quantum dot charge qubits due to acoustic phonons. Phys Rev B 72:235320
24. Eckel J, Reina JH, Thorwart M (2009) Coherent control of an effective two-level system in a non-Markovian biomolecular environment. New J Phys 11:085001
25. Nalbach P, Thorwart M (2009) Landau-Zener transitions in a dissipative environment: numerically exact results. Phys Rev Lett 103:220401
26. Mujica-Martinez CA, Nalbach P, Thorwart M (2013) Organic π-conjugated copolymers as molecular charge qubits. Phys Rev Lett 111:016802
27. Mujica-Martinez CA, Nalbach P, Thorwart M (2013) Quantification of non-Markovian effects in the Fenna-Matthews-Olson complex. Phys Rev E 88:062719
28. Christensson N, Kauffmann HF, Pullerits T, Mančal T (2012) Origin of long-lived coherences in light-harvesting complexes. J Phys Chem B 116:7449–7454
29. Kreisbeck C, Kramer T (2012) Long-lived electronic coherence in dissipative exciton dynamics of light-harvesting complexes. J Phys Chem Lett 3:2828–2833
30. Chin AW et al (2013) The role of non-equilibrium vibrational structures in electronic coherence and recoherence in pigment–protein complexes. Nat Phys 9:113–118
31. Breuer H-P, Laine E-M, Piilo J (2009) Measure for the degree of non-Markovian behavior of quantum processes in open systems. Phys Rev Lett 103:210401
32. Liu BH et al (2011) Experimental control of the transition from Markovian to non-Markovian dynamics of open quantum systems. Nat Phys 7:931–934
33. Tang JS et al (2012) Measuring non-Markovianity of processes with controllable system-environment interaction. Europhys Lett 97:10002
34. Breuer H-P (2012) Foundations and measures of quantum non-Markovianity. J Phys B At Mol Opt Phys 45:154001
35. Nielsen MA, Chuang IL (2010) Quantum computation and quantum information, 10th anniversary edn. Cambridge University Press, Cambridge

Chapter 6
Fractal Dimensions and Entropies of Meragi Songs

Adnan Aydemir and Güngör Gündüz

Abstract Melodies can be treated as time series systems with the pitches (or frequencies of the notes) representing the values in subsequent intervals. The pattern of a melody can be revealed in a scattering diagram where pitches represent vertices, and the directed pathways which connect the former pitches to the next ones signify the relations established during the performance. The pathways form a pattern which is called animal diagram (or lattice animal) in the vocabulary of graph theory. The slopes of pathways can be used to characterize an animal diagram and thus to characterize a melody; and the scattering diagram can be used to find out the fractal dimension. In addition, the entropy, the maximum entropy, and the negentropy (or the order) of melodies can be determined. The analysis of Meragi songs in terms of fractal dimension and entropy was carried out in this work. It was found out that there is not a correlation between the fractal dimension and the entropy; therefore, the fractal dimension and the entropy each characterizes different aspects of Meragi songs.

6.1 Introduction

Melodies display a pattern, because they are constructed from entities (i.e. notes) with finite frequencies which follow each other with a kind of order. The ratio of notes can be expressed with rational fractions first expressed by the great philosopher/mathematician Pythagoras, notably, $9/8, 5/4, 4/3, 3/2, 5/3, 15/8$, and $2/1$.

A. Aydemir
Kimya Mühendisliği Bölümü, Yeditepe Üniversitesi, İstanbul, Turkey,

G. Gündüz (✉)
Kimya Mühendisliği Bölümü, Orta Doğu Teknik Üniversitesi, Ankara, Turkey
e-mail: ggunduz@metu.edu

D. Matrasulov and H.E. Stanley (eds.), *Nonlinear Phenomena in Complex Systems: From Nano to Macro Scale*, NATO Science for Peace and Security Series C: Environmental Security, DOI 10.1007/978-94-017-8704-8_6,

The value of frequencies (i.e. pitches) can be mathematically expressed more or less by $2^{n/12}$ with $n = 1 - 12$. For instance, the note 'do' (or 'C') with a frequency of 261.6 Hz is followed by 'do-sharp' (i.e. do# or C#) with a frequency of $261.6 \times 2^{1/12} = 277.16$ Hz. The next note is 're' which has a frequency of $261.6 \times 2^{2/12} = 293.63$ Hz, but the ratio-tuned frequency is, $261.6 \times (9/8) = 294.3$ Hz. Since only certain frequencies are used while composing a melody the randomness allowed is pretty much restricted by discarding other possible frequencies. Actually musical instruments are a kind of harmonic oscillators which create sound waves with frequencies limited to integer multiples or harmonics of lowest pitch. Another important thing is that the frequencies usually follow an order and there are some kinds of symmetries in the time series structure of melodies. Some notes or groups of notes are repeatedly played at varying sequences, and sometimes the groups are altered by changing one or two of the notes.

The analyses show that melodies have strong autocorrelation indicating that they have memory. This is actually expected as the harmony and the repetitions of some notes deliver memory to a system. In the overall melodies are not perfectly ordered systems and order-disorder competition prevail in the entire melody. Music gives enthusiasm which can be achieved only if the steady state behavior is broken and some unsteadiness is introduced. In other words, each melody owns entropy as well as order [1–3].

The repetition of some notes or groups of notes introduces similarity, and thus nonlinearity. In other words not all notes are equally probable and the preferential use of some of the notes introduces nonlinearity. Therefore each melody also has a characteristic fractal dimension.

The pattern of a melody is determined by the extent of order and randomness that it has, and its fractal dimension and entropy gives us a way of characterizing its structure. Melodies have different properties and therefore different types of fractal dimensions can be used in the characterization [4–10]. It is known that the frequency (f) spectrum of music exhibits $1/f$ behavior, and it belongs to what is called 'pink noise' [11, 12]. The 'white noise' is independent of frequency (i.e. $1/f^0$) and the 'brown or red noise' shows $1/f^2$ dependence. The data point of red noise is pretty stuck to the diagonal of the scattering diagram without much fluctuation. The Gaussian noise shows a radial distribution around the central point of the scattering diagram and the number density decreases as the radius increases. In a scattering diagram, the white noise covers the entire area and the Gaussian spectrum is highly gathered around the diagonal. However the pink spectrum spreads around the diagonal. Therefore the scattering diagrams can be used to find out the fractal dimension of melodies by box counting method [13].

Music as a complex system has rich number of variables and parameters and there are very different approaches to characterize it either in terms of graph theoretical methods [14], viscoelastic properties [15], or thermodynamics [16].

In this work the relation between fractal dimension and entropy were searched for melodies. In order to reduce different psychological effects the songs composed by one single composer, namely, Meragi (1365–1435) was studied. He is one of the earliest known composers and he influenced the music made in the Middle East and in the Central Asia [17]. His music can be considered to have more basic and fundamental structure than those produced then after. He is regarded as one of the cornerstone in Turkish 'art music' also. The traditional Turkish music is broadly categorized as 'art music' and 'folk music'. The former evolved in İstanbul as the 'Ottoman Palace Music', and it actually is an excellent synthesis of Central Asian Turkish music with Byzantine, Balkan, Arabic, and Persian music. The Empire's music evolved with the cultural contribution of her all other members. So to characterize the songs of Meragi can constitute a framework to analyze the historical evolution of 'Turkish art music' till twenty-first century with further studies in future. The current study covers only the characterization of his melodies with fractal dimension and entropy.

6.2 Fractal Dimension

Although the fractal dimension of melodies can be found from the scattering diagram by box counting method, it is crucial to determine the size of box, or the stick size. The box size can be decided by considering the minimum and maximum pitch values used in the songs. After a series of calculations it was found out that the fractal dimensions can be determined using two scattering diagrams with 5×5 and 8×8 grids. Other sizes used in the calculations did not give persistently dependable and constant values. An example of scattering diagrams used in the calculations is given in Fig. 6.1.

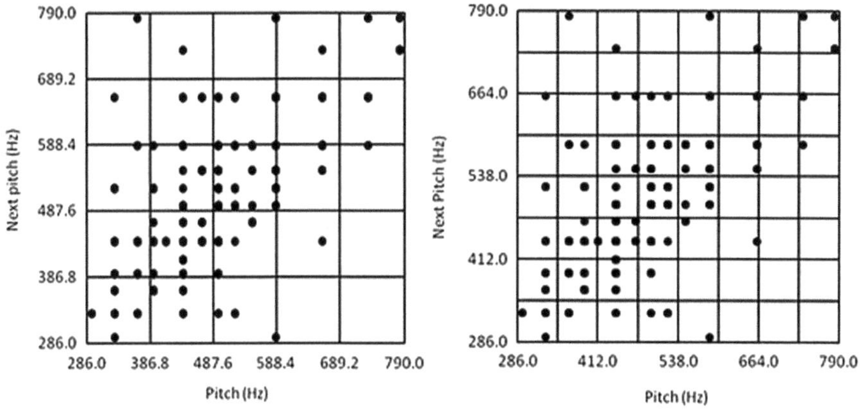

Fig. 6.1 Scattering diagrams (name of the song: *Buti darem ki...*)

The fractal dimension is calculated from the following values as done in an earlier work [3].

	1st figure (5 × 5)	2nd figure (8 × 8)
# of boxes that enclose the points, n	22	40
# of total boxes, N	25	64

The fractal dimension 'd' is then simply found out as,

$$d = \frac{\ln(n_2/n_1)}{\ln(N_2/N_1)} = \frac{\ln(40/22)}{\ln(64/25)} = 0.636$$

6.3 Entropic Terms

The entropy 'S' of songs is simply calculated from the Boltzmann entropy equation given by,

$$S(p) = -\sum p_i \log_2 p_i$$

where entropy is defined on '$\log_2$' basis so that it can be interpreted also in terms of 'information' or 'order'. The 'p_i' designates the probability of the occurrence of certain pitch used in the song. The calculation of entropy is also given in the former work [3]. The maximum entropy is given by,

$$S_{max} = \log_2 n$$

where 'n' denotes the number of notes. The total number of 'n' is equal to the total number of notes used in the song, therefore S_{max} changes during the performance of a song. In other words, both S and S_{max} change in time in the course of playing the song.

The order or the negative entropy 'S_{neg}' is found from,

$$S_{neg} = S_{max} - S$$

The change of entropic terms (i.e. S, S_{max}, and S_{neg}) is given in Fig. 6.2 for the song, of which scattering diagram is given in Fig. 6.1.

6.4 General Features of Meragi Songs

In this study 21 known songs of Meragi were studied, and the fractal and entropic values were investigated. The change of fractal dimensions with the number of notes is given in Fig. 6.3. It is seen that there is not a correlation between the fractal

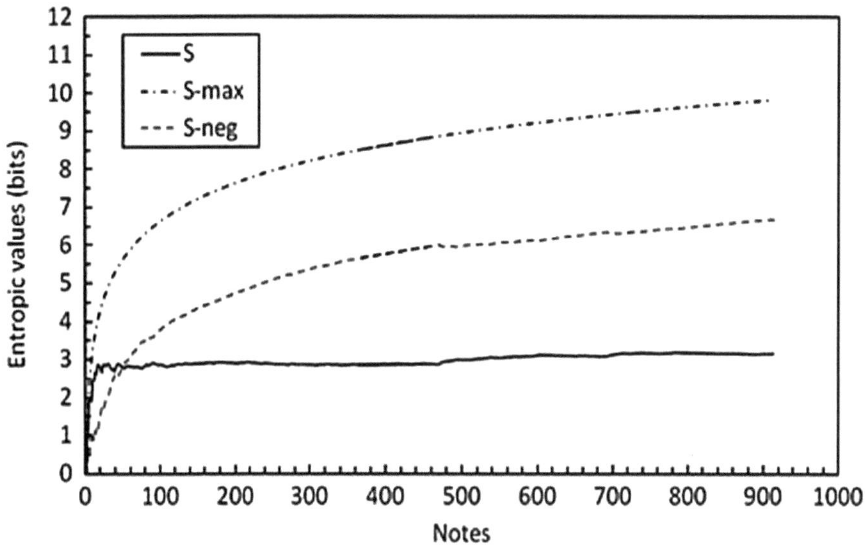

Fig. 6.2 The entropic values of the song

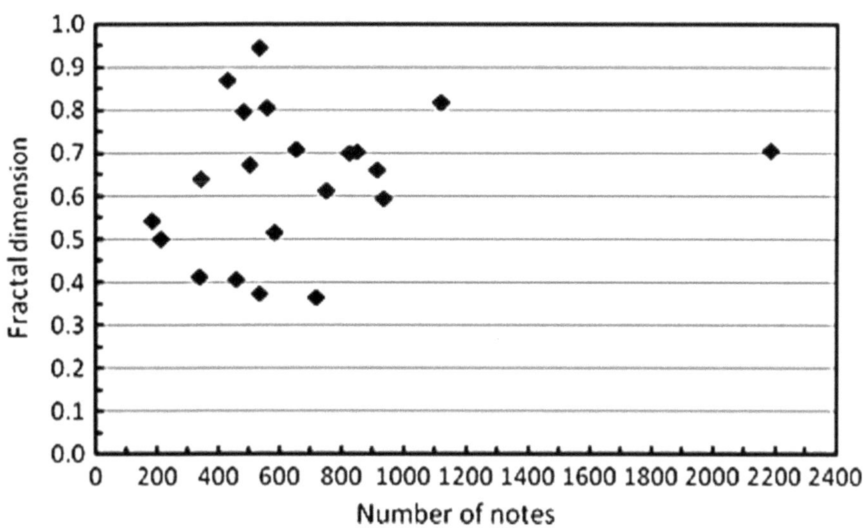

Fig. 6.3 Fractal dimensions of songs with respect to the total number of notes

dimension and the number of notes included in songs. The fractal dimension is a measure of the extent of nonlinearity existing in a system, but it is not dependent on the number of notes used. In other words the nonlinearity of the sequence of notes does not depend on the number of notes. A melody has its own pattern which does not change much with the number of notes it has. Some songs have fractal dimensions around 0.35 and some around 0.9. The large fractal dimension is due to

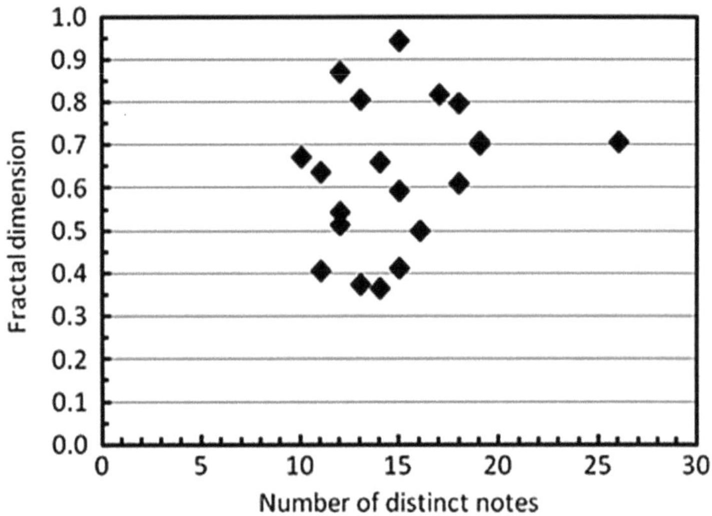

Fig. 6.4 The change of fractal dimension with the number of distinct pitches

more number of filled boxes in the second picture of Fig. 6.1. The more filling means the more distinct pathways occurring in the song. That means what comes after a certain note is more uncertain if the fractal dimension comes out to be a large value.

The increase in the number of distinct notes (i.e. pitches) changes the way the points are scattered in Fig. 6.1. Therefore it may be worth to investigate the change of fractal dimension with the number of distinct notes. This is shown in Fig. 6.4.

It is again seen that the fractal dimension does not quite correlate with the number of distinct notes, i.e. with the number of constituent elements of the songs. As the number of distinct notes is small (i.e. between 10 and 12) the fractal dimension also is relatively small, but very small and very large fractal dimensions occur when it is between 13 and 15. The fractal dimensions are equal when the number of distinct notes is 19 and 26, and they are not too big either.

The entropy of a song does not significantly change after some number of notes played as seen from Fig. 6.2; it reaches a saturation value. In fact no matter how long a song is its total entropy does not change as seen from Fig. 6.5.

The very long song with around 2,200 notes seen in Fig. 6.3 is excluded in Fig. 6.5 for more precise view. The maximum entropy keeps always increasing with the number of notes included in a song, and in the same may the negative entropy also keeps increasing as seen from Fig. 6.5. Both the fractal dimension and the overall entropy of a melody do not depend on the total number of notes.

In Fig. 6.6 the change of the overall entropy values with respect to the fractal dimension is depicted. It is seen that the overall entropy does not depend on fractal dimension, and all entropy values are around 3 while fractal dimension changes between 0.35 and 0.95. However, the overall maximum entropy and the negative entropy both fluctuate significantly.

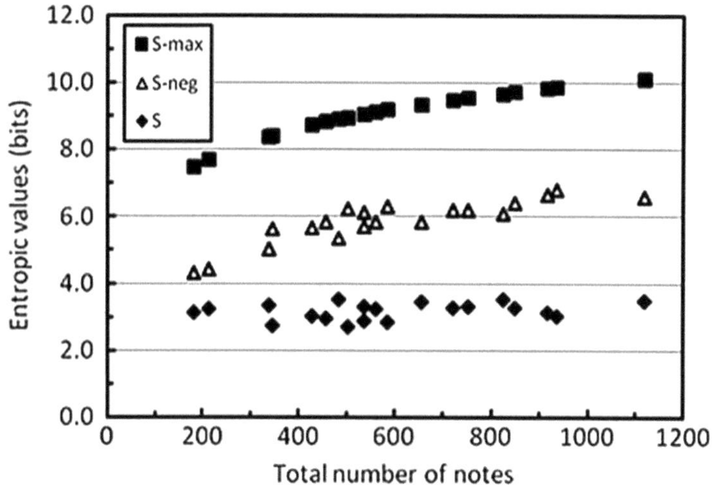

Fig. 6.5 Change of entropic values with total number of notes

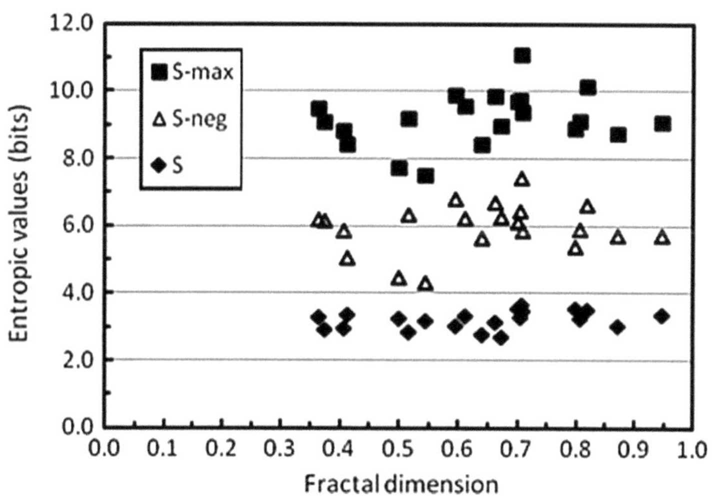

Fig. 6.6 The change of overall entropic values with fractal dimensions for the songs

An important point to mention is that after a certain number of the notes the number of points in the boxes seen in Fig. 6.1 does not change. That means, there is no unexpected change in the order of consecutive notes. After certain number of notes, the 'present note – next note' pattern follows one of the earlier patterns. In other words, the pathways in the scattering diagram are repeatedly used and no new pathway is generated after certain number of notes played. Therefore, it may be meaningful to plot Fig. 6.6 in terms of the entropic values calculated at the

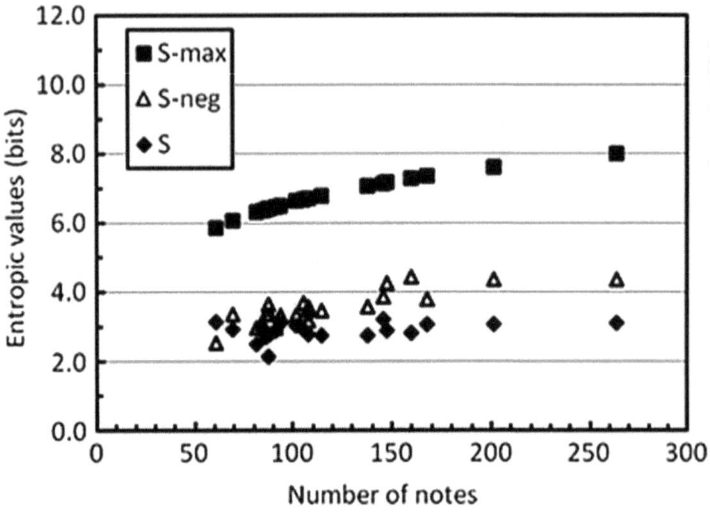

Fig. 6.7 The change of entropic values at reduced number of notes

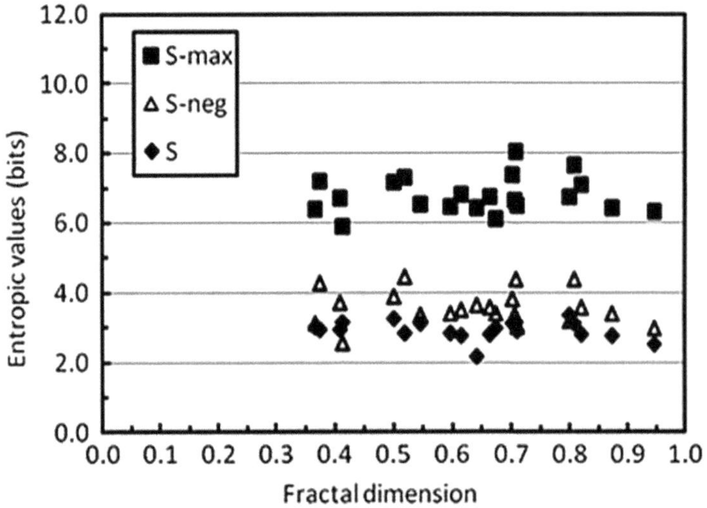

Fig. 6.8 The change of entropic values with fractal dimensions for the songs where the same number of notes were used in both calculations

notes where no new point is introduced into the scattering diagram. The entropy calculations done accordingly are shown in Fig. 6.7.

Compared to Fig. 6.5 the tendency of the change of entropic values is same, but smaller S_{max} and S_{neg} values are achieved. The fractal dimension corresponding to the entropic values given in Fig. 6.7 is given in Fig. 6.8.

It is seen from Fig. 6.8 that the scatterings in S_{max} and S_{neg} values are pretty much reduced compared to the case given in Fig. 6.6. So it is not only S but also S_{max} and S_{neg} values also do not correlate with the fractal dimension in melodies.

6.5 Conclusion

The investigation done using 21 songs of Meragi showed that his songs do not indicate the existence of a correlation between the fractal dimension and the entropy, and also with the maximum and the negative entropies. Fractal dimension and entropy characterize different aspects of melodies.

References

1. Knopoff L, Hutchinson W (1983) Entropy as a measure of style: the influence of sample length. J Music Theory 27(1):75–97
2. Manzara LC, Witten IH, Jones M (1992) On the entropy of music: an experiment with Bach chorale melodies. Leonardo Music J 2(1):81–88
3. Gündüz G, Gündüz U (2005) The mathematical analysis of the structure of some songs. Physica A 357:565–592
4. Hsü KJ, Hsü AJ (1990) Fractal geometry of music. Proc Natl Acad Sci 87:938–941
5. Miller SL, Miller WM, McWhorter PJ (1993) Extremal dynamics: a unifying physical explanation of fractals, 1/f noise, and activated processes. J Appl Phys 73(6):2617–2628
6. Bigerelle M, Iost A (2000) Fractal dimension and classification of music. Chaos Solitons Fractals 11:2179–2192
7. Su ZY, Wu T (2006) Multifractal analyses of music sequences. Physica D 221:188–194
8. Su ZY, Wu T (2007) Music walk, fractal geometry in music. Physica A 380:418–428
9. Das A, Das P (2006) Fractal analysis of different eastern and western musical instruments. Fractals 14(3):165–170
10. Das A, Das P (2010) Fractal analysis of songs: performer's preference. Nonlinear Anal Real World Appl 11:1790–1794
11. Voss RF, Clarke J (1975) '1/f' in music and speech. Nature 258:317–318
12. Voss RF, Clarke J (1978) "1/f noise" in music: music from 1/f noise. J Acoust Soc Am 63:258–263
13. Madden C (1999) Fractals in music. High Art Press, Salt Lake City
14. Hazama F (2011) Spectra of graphs attached to the space of melodies. Discret Math 311:2368–2383
15. Gündüz G (2009) Viscoelastic properties of networks. Int J Mod Phys C 20:1597–1615
16. Gündüz G (2012) Thermodynamics of relation-based systems with applications in econophysics, sociophysics, and music. Physica A 391:4637–4653
17. Elinç E (2005) M.Sc. thesis (in Turkish), Abdülkâdir Merâği'nin eserlerinin makamsal açıdan incelenmesi (Study of the melodies of Abdülkâdir Merâği from the view point of their makams) Afyon Kocatepe Üniversitesi, Sosyal Bilimler Enstitüsü (Afyon Kocatepe University, Institute of Social Sciences)

Part III
Network Dynamics in Macroscale Systems

Chapter 7
Large-Scale Connectivity vs. Spreading Efficiency: Spectral Analysis on Explosive Percolation

N.N. Chung, L.Y. Chew, and Choy Heng Lai

Abstract Some of the results of a spectral analysis of the process of explosive percolation on complex networks (Chung et al., Europhys Lett 101:66003, 2013) are presented here. More specifically, we explore how the maximum eigenvalues of the adjacency matrix of a network, which governs the spreading efficiency, evolves as the density of network connections are increased. We find that for networks with connectivity that grow in an explosive way, information spreading and mass transport are carried out inefficiently. When we looked into the conventional explosive percolation models, the sudden emergency of large connectivity comes with relatively lowered efficiency of spreading. Nevertheless, the spreading efficiency of the explosive models can be improved by introducing heterogeneous structures into the networks.

7.1 Introduction

The phenomenon of an abrupt development of large-scale connectivity in networks is the subject of intensive research in recent years, after Achlioptas et al. [1] introduced their percolation model that exhibits an unexpected sharp transition.

N.N. Chung
Temasek Laboratories, National University of Singapore, 117508, Singapore
e-mail: phycnn@nus.edu.sg

L.Y. Chew
Division of Physics and Applied Physics, Nanyang Technological University, 637371, Singapore

C.H. Lai (✉)
Department of Physics and Yale-NUS College, National University of Singapore, 119260, Singapore
e-mail: phylaich@nus.edu.sg

D. Matrasulov and H.E. Stanley (eds.), *Nonlinear Phenomena in Complex Systems: From Nano to Macro Scale*, NATO Science for Peace and Security Series C: Environmental Security, DOI 10.1007/978-94-017-8704-8_7,

The essential element in the model is that when one increases progressively the number of connections between nodes in a network following a certain suppression principle, a giant connected cluster emerges suddenly above a certain critical threshold. The resulting transition was originally thought to be discontinuous, but some models can be constructed where the transitions are continuous. Later works show that the abrupt creation of several stable giant components is also a possible outcome, and the study of explosive percolation has since been extended to complex network with different characteristics: scale-free networks, real-world networks with community structure, two- and higher-dimensional lattices, etc.

Some excitement is associated with applying the mechanism to explain the growth process of several real networks, including the human protein homology network and nanotube clustering. For real-world systems, such behavior can have vital consequences when the additional of single links may drastically change macroscopic connectivity and hence the dynamics and function of the networks. For instance, in a neuronal circuit, this would mean that the growth of one or a few additional synaptic connections might drastically alter the information processing function in the brain. Similarly, the establishment of a small number of specific social relations may significantly increase the possible spreading extent of infectious diseases or rumor.

Here we present our study of the evolution of the spectral properties of the network during the process of explosive percolation. In particular, we study how the maximum eigenvalue of the adjacency matrix of the network evolves as the density of connection increases. The reason for this focus is that the maximum eigenvalue of the network is known to influence the efficiency in the spread of information or disease on networks [2]. Specifically, the spreading threshold below which the information or disease would not spread widely has been shown to be inversely proportional to the largest eigenvalue of the network adjacency matrix. So the question is: in a process that displays explosive percolation, how does the maximum eigenvalue of the network adjacency matrix change when the system evolves towards the critical threshold?

In this talk, we use a simple model to illustrate the ideas and present the results of our investigation.

7.2 Building a Network

We can approach the building or growth of a network in different ways. We can start with a set of N nodes and establish links/edges one at a time according to some algorithm, or we can start with some finite set of nodes and links/edges, and have some specific rules to add new nodes to the set. For the purpose of this talk, we shall focus on what happens to the network when the number of connections between an initial set of nodes is increased following a *suppression principle*: the growth of all clusters are suppressed.

The smallest cluster (SC) model is one of the simplest models that show strongly discontinuous percolation transition. We start with an initial set of $N = 2^n$ isolated nodes. At each step of the network growth, the two *smallest* clusters – the suppression principle at work – are identified and merged into a larger cluster through the creation of a link between them. In this case, $N/2$ clusters of size 2 are created during the first stage, which involves the first $N/2$ steps. In the second stage, $N/4$ links are added, each connecting two clusters of size 2 into a cluster of size 4. The process continues and by the end of the stage $y = n - 1$, only two components remain, each with size $N/2$. Then the next step connects these two components, resulting in a sudden jump in the largest cluster size from $N/2$ to N.

We can formulate the process more concretely. If the nodes to be connected in each step are chosen randomly from the two smallest clusters, then at the end of stage y, we would have $N/2^y$ edges are added to the network and the total degree is increased by $N/2^{(y-1)}$. From this we deduce that the evolution of the degree distribution is described by (where k is the degree of a node)

$$\partial P(k, y)/\partial y = \frac{1}{2^{y-1}}(-P(k, y) + P(k-1, y)). \tag{7.1}$$

With this we obtain the probability of nodes with degree $k = y$ at stage y as

$$P(y, y) = 1/2^{\sum_{j=0}^{y-1} j}. \tag{7.2}$$

We can make several observations:

1. $P(k, y) = 0$ for all $k > y$: we never have a node with a degree k larger than the stage y. This is just the result of the suppression principle.
2. In the earlier stages, the maximum degree of the network, k_m, increases linearly with the stage y. Nonetheless, since $P(y, y)$ drops rapidly as y increases, the probability for k_m to increase becomes very small beyond a critical stage.

We proceed to define y_c to be the critical stage where $P(y_c, y_c) = 1/N$, the value of which can be obtained by solving the equation:

$$y_c(y_c - 1) = 2n. \tag{7.3}$$

So, from the above discussion, the conclusion is that the probability for the maximum degree k_m to increase is very small beyond the critical stage y_c. This observation already has an important implication for what we are looking for. We know that the maximum eigenvalue λ_m of a network's adjacency matrix is *inversely proportional* to the square root of the network's maximum degree! In other words, λ_m is *not* expected to grow explosively at the percolation threshold. This is exactly what we see in our numerical computations (Fig. 7.1).

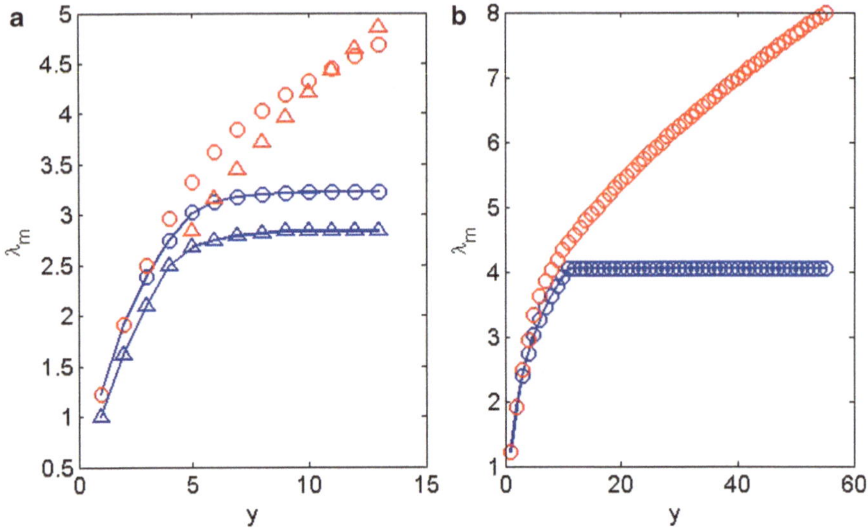

Fig. 7.1 The maximum eigenvalue at thend of each stage for the smallest cluster model (*lower curves*) and the modified smallest cluster model (*upper curve*) of size (**a**) $N = 2^{13}$ and (**b**) $N = 2^{55}$. Note that the ensemble averages obtained by simulation of the SC model with 2^{13} nodes are shown as *triangle* whereas the approximations obtained through numeral iteration of the evolution equations of the degree distributions are shown as *circles*

There is an obvious variation of the model that we can explore further: if the nodes to be connected in each step are chosen to be the largest-degree nodes in the smallest cluster, then we have the following evolution of the degree distribution:

$$\partial P(k,y)/\partial y = \begin{cases} -1/2^{y-1}, & \text{for } k = k_m, \\ 1/2^{y-1}, & \text{for } k = k_m + 1, \\ 0, & \text{for all other } k. \end{cases} \tag{7.4}$$

In this modified SC model, the maximum degree increases linearly with the stage. Thus, as shown in Fig. 7.1, larger maximum eigenvalues and more efficient spreading are obtained at the percolation threshold for giant clusters that emerge explosively, due to the heterogeneous or hub structure of the network introduced by the modified algorithm.

7.3 Summary

The explosive percolation model has altered the traditional understanding on percolation transitions. While the sudden emergence of a large-scale connectivity may significantly change the dynamics and function of a network, this alone is not a sufficient condition for information and mass transport in real-world networks. The fact that most of the real-world networks are small-world networks, in which the average path length scales as $\log(N)$, suggests that navigation efficiency in a network is a critical consideration. However, as shown by our results, the sudden emergence of a giant component in an explosive model comes at a cost of lowered spreading efficiency, which is undesirable in communication networks. There are however strategies that promote the formation and growth of network hubs in the explosive models that can allow us to have both the efficiency in spreading and large-scale connectivity in the network concurrently. Similarly, when spreading efficiency and large-scale connectivity are not needed at the same time, one may explore that possible trade-off between the size of the giant connected components and the increase of the maximum eigenvalue (in other words, spreading efficiency) at the percolation threshold.

References

1. Achlioptas D, D'Souza RM, Spencer J (2009) Explosive percolation in random networks. Science 323:1453
2. Chung NN, Chew LY, Lai CH (2013) Spectral analysis on explosive percolation. Europhys Lett 101:66003

Chapter 8
Power Grids, Smart Grids and Complex Networks

Antonio Scala, Guido Caldarelli, Alessandro Chessa, Alfonso Damiano, Mario Mureddu, Sakshi Pahwa, Caterina Scoglio, and Walter Quattrociocchi

Abstract We present some possible Complex Networks approaches to study and understand Power Grids and to improve them into Smart Grids. We first sketch the general properties of the Electric System with an attention to the effects of Distributed Generation. We then analyse the effects of renewable power sources on Voltage Controllability. Afterwords, we study the impact of electric line overloads on the nature of Blackouts. Finally, we discuss the possibility of implementing Self Healing capabilities into Power Grids through the use of Routing Protocols.

A. Scala (✉)
ISC-CNR Physics Department, Univ. La Sapienza, Piazzale Moro 5, 00185 Roma, Italy
e-mail: antonio.scala@cnr.it

G. Caldarelli • A. Chessa
IMT Alti Studi Lucca, piazza S. Ponziano 6, 55100 Lucca, Italy
e-mail: guido.caldarelli@imt.it; alessandro.chessa@imt.it

A. Damiano
Dipartimento di Ingegneria Elettrica ed Elettronica, Univ. di Cagliari, Cagliari, Italy
e-mail: alfio@diee.unica.it

M. Mureddu
Linkalab, Complex Systems Computational Laboratory, 09129 Cagliari, Italy
e-mail: mario.mureddu@linkalab.it

S. Pahwa • C. Scoglio
Department of Electrical and Computer Engineering, College of Engineering,
Kansas State University, Manhattan, KS, USA
e-mail: sakship@ksu.edu; caterina@ksu.edu

W. Quattrociocchi
London Institute of Mathematical Sciences, 22 South Audley St, Mayfair, W1K 2NY
London, UK
e-mail: walter@londoninstitute.org

D. Matrasulov and H.E. Stanley (eds.), *Nonlinear Phenomena in Complex Systems: From Nano to Macro Scale*, NATO Science for Peace and Security Series C: Environmental Security, DOI 10.1007/978-94-017-8704-8_8,

8.1 Introduction

Nowadays, one of the most pressing and interesting scientific challenges deals with the analysis and the understanding of processes occurring on complex networks [1–9]; one of the most important target for applying the results of such a field are real infrastructural networks. Our society critically depends on the continuity of functioning of Network Infrastructures like power, gas or water distribution; securing such critical infrastructures against accidental or intentional malfunctioning is a key issue both in Europe and in the US [10, 11]. Among those infrastructures, the electrical power grid is perhaps the most crucial one as many other facilities like telecommunications, banking systems, oil and gas pumping, and even water depend on the electric power system [12].

In Sect. 8.2 we give an overview of the electric power systems and of the effects of the recent introduction of distributed renewable sources.

In Sect. 8.3 we study the effects of the allocation of distributed renewable generation on the resilience of power grids.

In Sect. 8.4 we investigate the phenomenon of abrupt breakdown of an electric power-system under two scenarios: load growth (mimicking the ever-increasing customer demand) and power fluctuations (mimicking the effects of renewable sources).

In Sect. 8.5 we introduce the concept of resilience by exploitation of redundant links to recover the connectivity of the system. The introduced self-healing capabilities through the application of distributed communication protocols grants the "smartness" of the system.

Finally, we summarise our results in Sect. 8.6.

8.2 Power Networks and Distributed Generation

Historically, electric power systems have developed bottom-up by the integration of local networks into regional and national ones. Such a trend is nowadays still continuing via an integration on an international scale.

The three primary functions of the electric utility are Generation, Transmission, and Distribution (Fig. 8.1). The Distribution system is the most readily perceived part of the electric power system since it contributes most directly to providing electric power to the customers. It can be distinguished in Primary distribution operating at Medium voltage ($\sim 10^3$ V) and in secondary distribution operating at Low voltage ($\sim 10^2$ V). Industries are generally served by Medium voltage; residential customers by Low voltage. Distribution networks are local networks and have mostly a radial (tree-like) structure in order to optimize the economic costs and to calculate the power consumption of each user on an simple basis. The Transmission system dispatches large flows of electric power at long distances; in order to minimise dissipation, it operates at High voltages ($\sim 10^4$–10^5 V). As a network, it has a

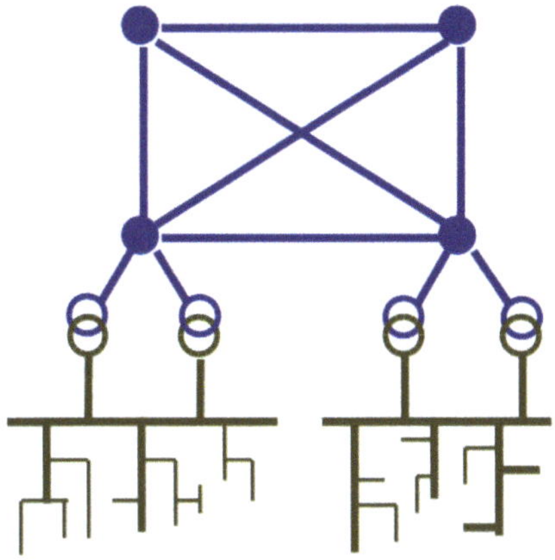

Fig. 8.1 The electric power system is separated in Transmission (High voltage lines, $\sim 10^4$–10^5 V) and Distribution (Medium voltage $\sim 10^3$ V and Low voltage $\sim 10^2$ V, also referred to as Primary and Secondary distribution). Transmission dispatches along long distances high quantities of power generated by large generators; for the sake of robustness, it is a meshed network. Distribution networks are local networks optimised for power dispatch and have mostly a radial (tree-like) structure in order to optimize the economic costs and to calculate the power consumption of each user on an simple basis. Industries are generally served by Medium voltage; residential customers by Low voltage

meshed configuration that allows to increase its robustness through redundancy. In the first years the electric power system was a collection of local, disconnected networks and Generation was distributed. Subsequently, due to economy of scale, Generation has concentrated in large facilities, requiring the creation of high-voltage lines for long-distance dispatch (Transmission). Nevertheless, Distributed Generation has never disappeared since it has an important ancillary role both in ensuring backup power in case of malfunctioning and in sustaining the variations in loads due to the customer demand dynamics. In fact, large facilities have long reaction times (it can take a day or even more to start up or shut down a large fossil-fuel power station) while smaller facilities can be as fast as fraction of hours.

Electric power systems are mostly designed to operate at a sinusoidal voltage of a given frequency (typically 50 or 60 Hz) and magnitude. Any significant deviation in the waveform magnitude ($\pm 5\,\%$) or frequency ($\pm 1\,\%$) is a potential problem for the quality of the dispatched power. To operate properly, a continuous and accurate balance between power demand and generation has to be maintained in the system; up to know, daily forecasts and the high-frequency electric power market have allowed to keep the system mostly in a stable state.

Nowadays we are experiencing an increase of Distributed generation due to the introduction of low cost green generators (mostly solar and wind). Such "green" generators, while very advisable for the sake of emission decrease and

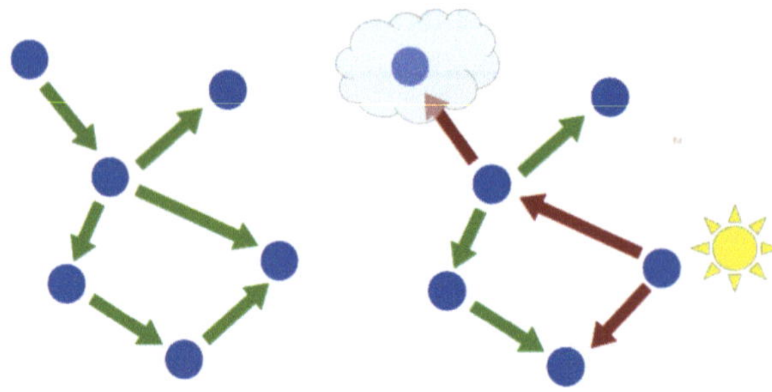

Fig. 8.2 The erratic nature of renewable sources introduces fluctuations in the power production that can push the power system away not only from its stable state, but even to the region of parameters where the system has been designed to work safely. As an example, we sketch the effects of the introduction of renewable generators on a isolated grid like the ones that can be found on isolated islands like Guadalupe. On the *left panel* we show an hypothetical power flow configuration along which the actual grid has been engineered: not only the possible amounts of power flowing along the lines have been considered, but also their directions. On the *right panel* we show the effect of the introduction of "*green*" generation: a weather instability can switch the direction of the power flows, eventually causing automatic protections to trip the lines

of environmental sustainability, introduces several problems in the electric power system. In fact, at difference with "classical" distributed generators that can be controlled to enhance system's stability, the erratic nature of renewable sources introduces fluctuations in the power production that can push the power system away not only from its stable state, but even to the region of parameters where the system has been designed to work safely. As an example, consider the effects of the introduction of renewable generators at the distribution level. Most distribution grids have been engineered considering not only the possible amount of power flowing along the lines, but also its direction. "Green" power can eventually switch the direction of the power flows, eventually causing automatic protections to trip the lines (Fig. 8.2).

The difficult task of integrating the stochastic and often volatile renewable sources into a the grid designed with a power-on-demand paradigm could perhaps solved leveraging on distributed storage [13]; nevertheless, massive and economic power storage is not yet readily available.

As a consequence of the introduction of renewable sources, many studies have concentrated on the dynamic behaviour of power grids to understand how to ensure stability and avoid loss of synchronization during typical events like the interconnection of distributed generation. The large number of elements present into real grids calls for simplifications like the mapping among the classic swing equations [14] and Kuramoto models [15–17] that allows to study numerically or analytically the synchronization and the transient stability of large power networks. Even simple models [18] akin to the DC power flow model [19] show that the network topology can dynamically induce a complex size probability distributions of blackouts (power-law distributed), both when the system is operated near its limits [20] or when the system is subject to erratic disturbances [21]. New realistic metrics to assess the robustness of the electric power grid with respect to the cascading failures [22] are therefore needed.

8.3 Voltage Control

In this section we will concentrate on the stability of the system respect to voltage fluctuations. Deviations of the voltage amplitude on a power grid can cause system operation problems; although voltage constraints are not as restrictive as frequency constraints, it can be regulated or controlled by generation or other connected equipment as long as it is in the range of the $\pm 5\,\%$ allowed fluctuations around its nominal value.

8.3.1 AC Power Flow

To model power grids, we use the more computational intensive AC power flow algorithms since, although DC flows are on average wrong by a few percent [23,24], error outliers could distort our analysis.

The AC power flow is described by a system of non-linear equations that allow to obtain complete voltage angle and magnitude information for each bus in a power system for specified loads [25]. A bus of the system is either classified as Load Bus if there are no generators connected or as a Generator Bus if one or more generators are connected. It is assumed that the real power P_i and the reactive power Q_i at each Load Bus i are given, while for Generator Buses the real generated power P_i and the voltage magnitude $|V_i|$ are given. A particular Generator Bus, called the Slack Bus, is assumed as a reference and its voltage magnitude $|V|$ and voltage phase Θ are fixed. The branches of the electrical system are described by the bus admittance matrix Y with complex elements Y_{ij}s. Figure 8.3 sketches the complex network associated with the AC power flow description of a grid.

8.3.2 Model for the Erratic Renewables

To model distributed renewable sources, we will introduced a skewed probability distribution of load demands representing a crude model of reality that ignores the effects like the correlations (due for examples to weather conditions) between different consumers or distributed producers. Thus, the effects of "green generators" on a power grid are considered to be stochastic variations in the power requested

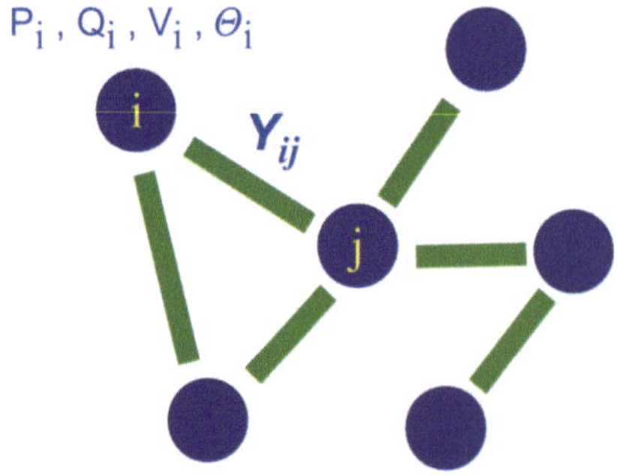

Fig. 8.3 Even in the simplified AC power flow description where the system is assumed to be stationary (the signal is a stationary sinusoidal), the network corresponding to a power grid is a weighted networks where both nodes (buses) and edges (lines) have multiple characteristics. In particular, P_i is the net real power and Q_i is the net reactive power injected at the ith bus, Y_{ij} is the complex impedance of the $i - j$ line, V_i is the voltage amplitude and θ_i is the voltage angle at the ith bus. Moreover, additional parameters like the maximum capacity of a line (i.e. the maximum amount of power that can flow) must be considered when analysing system failures

by load buses. Load buses with a green generator will henceforth called green buses. We will consider the location of green buses to be random; the fraction p of green buses will characterize the penetration of the distributed generation in a grid.

If the power dispatched by distributed generation is high enough, loads can eventually become negative: this effect can be related to the efficiency of green generators. We model such an effect by considering the load on green buses described by the skew-normal distribution [26], a pseudo-normal distribution with a non-zero skewness:

$$f(x, \alpha) = 2\phi(x)\,\Phi(\alpha x)$$

where α is a real parameter and

$$\phi(x) = \exp\left(-x^2/2\right)/\sqrt{2\pi} \qquad \Phi(\alpha x) = \int_{-\infty}^{\alpha x} \phi(t)\,dt$$

Thus, the parameter α will characterize the level of the distributed generation: to positive α correspond loads positive on average, while for negative α green nodes will tend to dispatch power.

8.3.3 Analysis

Our model grids will therefore consist of three kind of buses: N_G generators (fixed voltage), N_l pure loads (fixed power consumption) and N_g green buses (stochastic

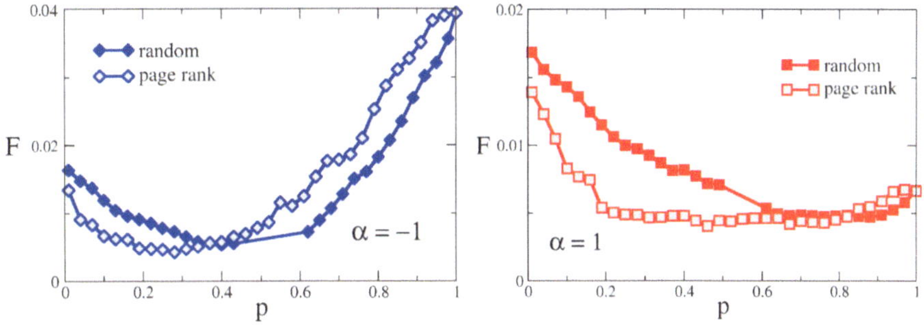

Fig. 8.4 Comparison between random placement (*filled symbols*) and page-rank placement (*empty symbols*) of green generators in the Polish grid, both for surplus production of renewable energy (*left panel*, $\alpha = -1$) and for levels of renewable energy production below the normal load request (*right panel*, $\alpha = 1$). The page-rank placement of renewable sources allows to attain lower values of the fraction F of buses operating near their nominal tension (and hence a higher resiliency) at lower values of the penetration p. The best case is realized for levels of renewable energy production below the normal load request, where a plateau to low values of F is quickly attained

power consumption) with $N_G + N_l + N_g = N$ the total number of buses and $N_g + N_l = N_L$ the number of load nodes. The fraction $p = N_g/N_L$ measures the penetration of renewable sources in the grid.

Hence, a steady state analysis has been carried out and the transient phenomena connected to the power flow control have been neglected. Under this hypothesis the frequency variation connected to power flow control has been considered stabilized and the system has been considered characterized by a constant steady state supply voltage frequency. Therefore, if all the nodes are near their nominal voltage, it is much easier to control the system and to avoid reaching infeasible levels of power flow. Consequently, to measure the effects of power quality of a power grid under distributed generation we measure the fraction F of load buses whose tension goes beyond $\pm 5\,\%$ of its nominal voltage. Notice that real networks are often operated with some of the buses beyond such parameters so that (especially for large networks) it is expected to be $F \neq 0$ under operating conditions. The maximum of the resilience for a power grid (intended as the capability of restoring full feasible flows) is expected to be for $F = 0$.

In Fig. 8.4 we show the effects on the voltage stability of the penetration of the renewables on the Polish Grid. We model the penetration both in the case where the locations of the renewables are chosen at random and in the case where the locations are chosen according to a policy. In particular, we analyse the case in which new renewables are introduced according to their Page-Rank metrics [27]. As already shown in [24], we find that the Page-Rank policy makes the system less unstable respect to voltage fluctuation both in the case of traditional distributed generation ($\alpha > 0$) and green distributed generation ($\alpha < 0$). Notice that Page Rank is strictly related to several invariants occurring in the study of random walks and electrical networks [28].

8.4 Cascading Failures

High standards for the reliability of interconnected Electric Power Systems (EPSs) are being developed both in Europe and the US by councils and associations of EPS operators [29, 30]. Nevertheless, not only do power outages occur, but also large outages are more likely than what would be naively expected; in fact, the analysis of historical data reveals that their occurrence is power-law distributed [31], implying a significant risk of system-wide failures. Given the disruption and economic damages caused by major outages, understanding the nature of such occurrences is a major problem to be addressed.

An important general question is whether EPSs are subject to emergent behaviour or not. In fact, EPSs are aggregations of large number of simple units; it therefore makes sense to ask if EPSs, as a whole, exhibit additional complexity beyond what is dictated by the simple sum of its parts. To this aim, we investigate if an abrupt breakdown transition could emerge in a simple yet realistic model of power grids.

In the context of power systems, a cascading outage is a sequence of failures and automatic disconnections consequent to an initiating event; a system-wide outage is also called "black-out". The rapid succession of automatic reactions in an EPS happens in a time-scale that is typically too short to stop the process by human intervention. Reactions following an initiating event or events include sequential tripping (disconnection) of transmission lines and generators. Initiating events can be due to natural causes (like a line sagging into vegetation, or high wind or lightning shorting a line) but also to human actions (or inaction) or due to imbalances between load and generation.

While no two cascading outages are the same [32], we will study a class of possible outages and analyse their characteristics. In particular, we will consider the fragility of EPSs with respect to outages due to cascades of line overloads causing lines to trip. To this end, we will put under stress a realistic EPSs to understand the nature of systemic outages. The nature of the stress will be twofold: first, we will consider the case in which an increasing demand on a fixed infrastructure leads to line overloads and subsequent outages. This would correspond to the (hopefully) unrealistic case of EPS that are operated to the limit of their capacities in order to maximize profits. Second, we will consider the important case of fluctuations in demands and generation; this is a particularly relevant case as the steady penetrations of renewable sources is introducing in the grids new erratic sources whose effects and consequences on existing power grids have not yet been fully understood.

We consider the model introduced by Pahwa et al. [33]. In such a model, the initial distribution of loads and sources represents the stress imposed to the power grid. The initial power flows on lines are calculated using the DC power flow model (see Materials). If the load on a line goes beyond its capacity, the line trips (disconnects) and power flows are recalculated on the new topology (i.e. the grid *minus* the tripped lines). Such procedure is repeated until convergence; we will refer to such a model as the Overload Cascade Model (*OCM*).

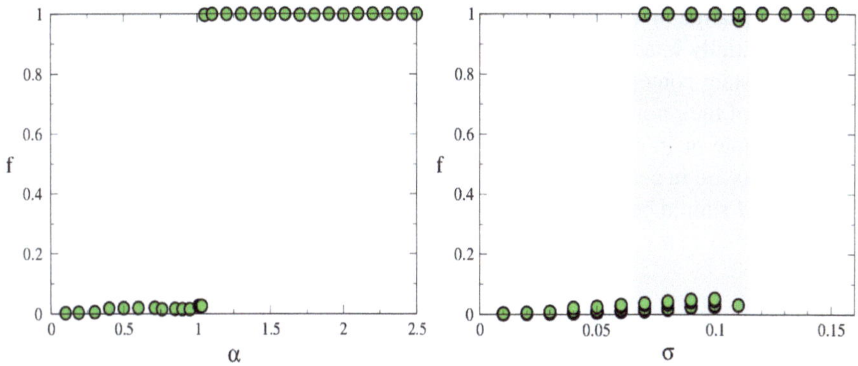

Fig. 8.5 *Left panel*: Effects of a constant increase in loads. Notice the abruptness of the transition. *Right panel*: Effects of fluctuations in loads. The points corresponds to 10 different realizations of the noise. The transition is abrupt (the fraction of tripped lines is either ∼0 or ∼1), but whether a cascading failure happens depends on the actual realization of the noise. Therefore, the transition does not happen at a particular critical value of σ but can occur in a region where fluctuations range between 6 and 12 % of the nominal loads (*shaded area*)

Our model does not account for the whole complexity of an EPS; in particular, it disregards both transients and the dynamics of phase angles. Nevertheless, it allows to sort out the role of a class of events always present in any blackout, i.e. line overloads. Another important property of the *OCM* is that, due to the long-range nature of the interaction, it is amenable to analytic approximations that lead to predict the universal behaviour of the system [34].

For our scopes, we consider a network of 2,746 nodes, consisting of a snapshot of the national high-voltage Polish power grid obtained from the data collected and used by Polish transmission system operators.

We first stress the Polish network by considering a growth in the power demand while keeping the network fixed. Such mechanism is not so far from reality, as in recent years, the economic competition and deregulation has led the power systems to be operating fairly close to their limits. We model such growth of the demand as a simultaneous increase of all the loads by a factor α and record the fraction f of branches that fail at the end of a cascade.

We then stress the Polish network via are flow fluctuations mimicking both the stochastic components in customers' behaviour and the effects of erratic renewable energy sources. We parametrize the size of fluctuation by allowing the initial loads to fluctuate uniformly by a fraction σ.

The left panel of Fig. 8.5 shows that by increasing all the loads the breakdown is abrupt as in a first order transition. The situation is more complicated in the case of random loads. The right panel of Fig. 8.5 shows the results for different realization of the noise (fluctuations). We find that, for a given realization of the noise, the system is either in a safe state (the fraction of tripped links is ∼0) or in a systemic failure state (the fraction of tripped links is ∼1). At difference with the case of uniform load increase, the transition does not happens at a given σ, but can happen in a whole range of values. In particular, for the Polish grid we find that the grid stays essentially intact when the loads are allowed to fluctuate less than ~6 % and that the system comes in a blackout state when loads are allowed to fluctuate more than 12 % of their nominal values. For intermediate values, the system can either be in a safe state or in a black-out state depending on the realization of the disorder. Such results are in accordance with mean field arguments indicating that black-outs in the *OCM* should be a first order phenomenon.

8.5 Distributed Algorithms for Smart Grids

The continuous growth and development of technological networks has introduced complex connectivity (as well as dependence) patterns within their constitutive components that can often trigger systematic side effects like system-wide cascades of failures. Hence, our current and future networks need healing mechanisms that are able to cope with systemic effects and multiple failures in an automatic and possibly distributed way. Such mechanisms are at the core of the process of building up Smart Grids, i.e. networks that are able to self-sustain and optimise in a distributed way to customers' needs.

At the Transmission level, electric power networks are already "smart", but in a centralised way: Supervisory Control and Data Acquisition (SCADA) systems allow operators to monitor, control, and dispatch generation. A failure either of the SCADA or of the supporting telecommunications can cause control operators to make incorrect system adjustments.

We consider instead the power system at the Distribution level, where the structure is mostly tree-like. We introduce a healing strategy based on the activation of fixed redundant resources (backup links) and study the resilience of the networks to multiple failures. Since the presence of backup links is customary in technological networks and our strategy can be implemented via routing protocols, our self-healing procedure is within the reach of current technology. For sake of simplicity, we will consider a single node to be the source of the quantity to be distributed on the network. Moreover, at each instant of time, the topology of the links in the network distributing the power is assumed to be a tree (the *active tree*). As a further simplification we will not take into account the magnitudes of flows – i.e., all links and sources are assumed to have infinite capacity – but we will focus on maximizing the connectedness of the system in order to serve as many nodes as possible.

In order to implement our strategy and its self-healing capabilities, we consider the presence of *dormant* backup links – i.e., a set of links that can be switched on. Nodes are assumed to be able to communicate with their neighbours by means of a suitable distributed interaction protocol with a limited amount of knowledge: the set of neighbouring nodes connected either via active or via dormant links. Then, when either a node or a link failure occurs, all the nodes below the failure will disconnect from the active tree and become unserved. Such unserved nodes can now try to

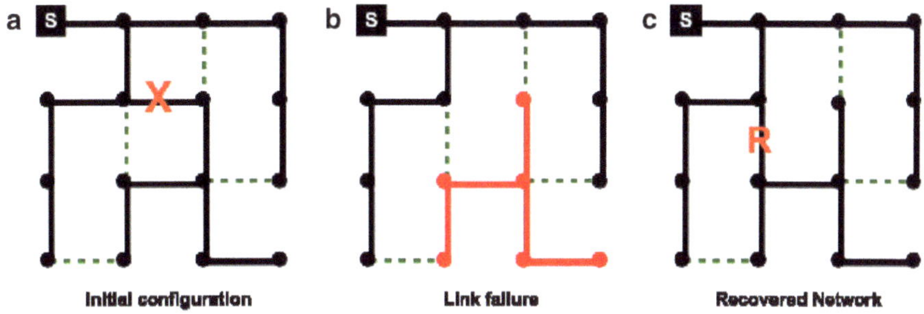

Fig. 8.6 Example of healing after single link failure. Notice that failure of a single node can be modelled as the failure of all its links; hence, multiple links failure are the more general event to be considered. (*Left panel*) In the initial state, the source node (*filled square*, *upper left corner*) is able to serve all 16 nodes through the links of the active tree. The *four dashed lines* (*green online*) represent dormant backup links that can be activated upon failure. The redundancy of the system is $p = 4/9$ as only 4 of the 9 possible backup links are present. The link marked with an X is the one that is going to fail. (*Central panel*) A single link failure disconnects all the nodes of a sub-tree; in the example, a sub-tree of 6 nodes (*red online*) is left isolated from the source – i.e., the system has a damage $\Delta = 6$. (*Right panel*) By activating a single dormant backup link, the self-healing protocol has been able to recover connectivity for the whole system, in this case bringing back the number of served nodes at its maximum value 16. The link that has recovered the connectivity is marked with an R

reconnect the active tree by waking up through the protocol some dormant backup links. Such a process will reconstruct a new active-tree that can restore totally or partially the flow, i.e. heal the system (Fig. 8.6).

By measuring the fraction *FoS* of served nodes after multiple random failures of k links, we study the effects of the redundancy r (the fraction of backup links added to the initial spanning tree). We investigate both the case which best resembles the actual situation – i.e. nodes disposed over a grid – and the role of the networks' connectivity patterns by generating different underlying topologies. In particular, we generate small-world and scale free networks [35].

We use square grids as an example of planar grids topologies. Small-world networks [36] are generated starting from planar square grids and rewiring with a probability p a link with a randomly selected node. Scale-free networks are generated through the Barabasi-Albert model [37]. In the case of technological networks, small world networks are important as they can show the effects of introducing long-range links in a planar topology.

To produce suitable initial configurations of our model distribution networks, we generate random spanning trees [38] associated to each kind of network structures. The links not belonging to the spanning trees form the set of the possible backup links of our system; among such links, we choose a random fraction r of *dormant* links that can be used to heal the system. We then simulate the occurrence of uncorrelated multiple failures by deleting at random k links of the initial active tree. Subsequently, we activate dormant links according to our self-healing procedure and calculate the *FoS*.

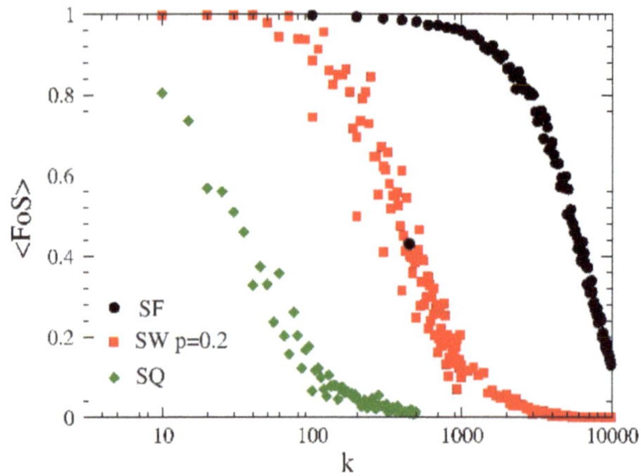

Fig. 8.7 (Color online) Comparison among different network structures. Here we show the performances of our self-healing algorithm with respect to the quality of service for increasing number of removed links with the redundancy r fixed to 0.3; for SW networks, the rewiring probability is $p = 0.2$. The average fraction of nodes$\langle FoS \rangle$ of served nodes is plotted against the number of failures k

Figure 8.7 shows that distribution grids based on the scale free topologies are the more robust. Nevertheless, they should be disregarded when considering the case of technological networks since economic and geometric constraints make scale free networks unfeasible on planar topologies. Therefore, the most relevant results are the ones on small-world networks, showing that introducing a small fraction of long-range links can enhance robustness by orders of magnitudes respect to planar grids with the same redundancies.

8.6 Conclusions

An unconstrained allocation and growth of the distributed generation can drive a power grid beyond its design parameters. In order to overcome such a problem, we propose in Sect. 8.3 a topological algorithm derived from the field of Complex Networks to allocate distributed generation sources in an existing power grid.

Several trans-national projects aim to integrate national power-grids into "super-grids". The results of Sect. 8.4 indicate that increasing the system size causes breakdowns to become more abrupt. Thus, the possible enhancement of the systemic risk failures (blackouts) with increasing network size is an effect that should be considered in the planning of "super-grids".

While transmission systems can already be considered "smart" (although in a centralised way), much less has been done at the level of distribution, especially at the low voltage (residential customer) level. In Sect. 8.5 we have introduced a simple distributed algorithm to keep the systems connected in the case of failures. By studying various network topologies, we have found that the introduction of some long-range connections in planar grids greatly enhances the resilience of local distribution networks to multiple failures.

Acknowledgements AS, GC and WQ thank US grant HDTRA1-11-1-0048, CNR-PNR National Project "Crisis-Lab" and EU FET project MULTIPLEX nr.317532. SP and CS acknowledge the support of the US Department of Energy grant EE-0003812: "Resourceful Kansas". The contents of the paper do not necessarily reflect the position or the policy of funding parties.

References

1. Arenas A, Diaz-Guilera A, Kurths J, Moreno Y, Zhou C (2008) Synchronization in complex networks. Phys Rep 469:93–153
2. Buldyrev SV, Parshani R, Paul G, Stanley HE, Havlin S (2010) Catastrophic cascade of failures in interdependent networks. Nature 464:1025–1028
3. Liu Y-Y, Slotine J-J, Barabasi A-L (2011) Controllability of complex networks. Nature 473:167–173. http://dx.doi.org/10.1038/nature10011
4. Dorogovtsev SN, Goltsev AV, Mendes JFF (2008) Critical phenomena in complex networks. Rev Mod Phys 80:1275–1335
5. Costa LF et al (2011) Analyzing and modeling real-world phenomena with complex networks: a survey of applications. Adv Phys 60:329–412
6. Perra N, Gonçalves B, Pastor-Satorras R, Vespignani A (2012) Activity driven modeling of time varying networks. Sci Rep 2. http://dx.doi.org/10.1038/srep00469
7. Barrat A, Barthelemy M, Vespignani A (2008) Dynamical processes on complex networks Cambridge University Press, Cambridge/New York
8. Alvarez-Hamelin JI, Fleury E, Vespignani A, Ziviani A (2012) Complex dynamic networks: tools and methods. Comput Netw 56:967–969
9. Casteigts A, Flocchini P, Quattrociocchi W, Santoro N (2012) Time-varying graphs and dynamic networks. Int J Parallel Emergent Distrib Syst IJPEDS 27:387–408
10. 2003 Homeland Security Presidential Directive HSPD-7 (2003). http://georgewbush-whitehouse.archives.gov/news/releases/2003/12/20031217-5.html
11. EU COM (2006) 786 EU directive on European Programme for Critical Infrastructure Protection, Brussels, 12.12.2006. http://eur-lex.europa.eu/LexUriServ/site/en/com/2006/com2006_0786en01.pdf
12. Rinaldi SM, Peerenboom JP, Kelly TK (2001) Identifying, understanding and analyzing critical infrastructure interdependencies. IEEE Control Syst Mag 21:11–25
13. Baghaie M, Moeller S, Krishnamachari B (2010) Energy routing on the future grid: a stochastic network optimization approach. In: International conference on power system technology (POWERCON), Hangzhou, 2010, pp 1–8
14. Stagg G, El-Abiad A (1968) Computer methods in power system analysis. McGraw-Hill Education, New York
15. Filatrella G, Nielsen AH, Pedersen NF (2008) Analysis of a power grid using a Kuramoto-like model. Eur Phys J B – Condens Matter Complex Syst 61:485–491
16. Fioriti V, Ruzzante S, Castorini E, Marchei E, Rosato V (2009) Stability of a distributed generation network using the Kuramoto models. In: Setola R, Geretshuber S (eds) Critical information infrastructure security. Springer, Berlin/Heidelberg, pp 14–23
17. Dörfler F, Bullo F (2012) Synchronization and transient stability in power networks and nonuniform Kuramoto oscillators. SIAM J Control Optim 50(3):1616–1642. doi:10.1137/110851584, http://epubs.siam.org/doi/abs/10.1137/110851584
18. Dobson I, Carreras BA, Lynch VE, Newman DE (2001) System sciences, 2001. In: Proceedings of the 34th annual Hawaii international conference on an initial model fo complex dynamics in electric power system blackouts, pp 710–718. doi:10.1109/HICSS.2001.926274
19. Wood AJ, Wollenberg BF (1984) Power generation, operation and control. Wiley, New York

20. Carreras BA, Lynch VE, Dobson I, Newman DE (2002) Critical points and transitions in an electric power transmission model for cascading failure blackouts. Chaos: Interdiscip J Nonlinear Sci 12:985–994
21. Sachtjen ML, Carreras BA, Lynch VE (2000) Disturbances in a power transmission system. Phys Rev E 61:4877–4882
22. Youssef M, Scoglio C, Pahwa S (2011) Robustness measure for power grids with respect to cascading failures. In: Proceedings of the 2011 international workshop on modeling, analysis, and control of complex networks, CNET '11, San Francisco. ITCP, pp 45–49
23. Stott B, Jardim J, Alsac O (2009) Dc power flow revisited. IEEE Trans Power Syst 24:1290–1300
24. Scala A, Mureddu M, Chessa A, Caldarelli G, Damiano A (2013) Distributed generation and resilience in power grids. In: Hammerli BM, Kalstad Svendsen N, Lopez J (eds) Critical information infrastructures security. Lecture notes in computer science, vol 7722. Springer, Berlin/Heidelberg, pp 71–79. doi:10.1007/978-3-642-41485-5_7, ISBN:978-3-642-41484-8, http://dx.doi.org/10.1007/978-3-642-41485-5_7
25. Grainger J, Stevenson W (1994) Power system analysis. McGraw-Hill, New York
26. Azzalini A (2005) The skew-normal distribution and related multivariate families. Scand J Stat 32:159–188
27. Brin S, Page L (1998) The anatomy of a large-scale hypertextual web search engine. Comput Netw ISDN Syst 30:107–117
28. Chung F, Zhao W Pagerank and random walks on graphs. http://www.math.ucsd.edu/~fan/wp/lov.pdf
29. NERC North America Electric Reability Corporation (2006). http://www.nerc.com/
30. ENTSO-E European Network of Transmission System Operators for Electricity (2008). https://www.entsoe.eu/
31. Report 1010701 (2005) Mitigating cascading outages on power systems: recent research approaches and emerging methods. Technical report, EPRI
32. Blackout in the United States and Canada, April 2004 (2004). https://reports.energy.gov/
33. Pahwa S, Hodges A, Scoglio CM, Wood S (2010) Topological analysis of the power grid and mitigation strategies against cascading failures. In: Proceedings of the 4th international IEEE systems conference, San Diego, pp 272–276
34. Quattrociocchi W, Caldarelli G, Scala A (2014) Self-healing networks: redundancy and structure. PLoS ONE 9(2):e87986. doi:10.1371/journal.pone.0087986. Public Library of Science. http://dx.doi.org/10.1371%2Fjournal.pone.0087986
35. Caldarelli G (2007) Scale-free networks Oxford University Press, Oxford
36. Watts DJ, Strogatz SH (1998) Collective dynamics of /'small-world/' networks. Nature 393:440–442. http://dx.doi.org/10.1038/30918
37. Barabási A-L, Albert R (1999) Emergence of scaling in random networks. Science 286:509–512
38. Wilson DB (1996) Generating random spanning trees more quickly than the cover time. In: Proceedings of the 28th annual ACM symposium on the theory of computing, Philadephia. ACM, pp 296–303

Chapter 9
A Spectral Approach to Synchronizability of Interdependent Networks

Gregorio D'Agostino

Abstract The quality of life in full developed countries depends on the cooperative functioning of different infrastructures. One of the most striking problems is to understand in simple terms to what extent this cooperation can be assured. The complexity science provides a powerful means to analyze the interaction of such critical infrastructures at pure topological level. The application of the paradigm of complexity to the global system resulting from the interdependent infrastructures leads to the concept of "Network of Networks". The present work is devoted to understand emergent (that is collective) synchronization behaviors through the spectral analysis of the laplacian. We provide evidence that, upon increasing the number of links between the different infrastructures, the behavior of the total system experiences a drastic changes in its synchronization modes. When few links are introduced, the synchronization inside the component networks is very fast and the global synchronization takes place mainly at the boundaries; on the other side, when the number of links exceeds a threshold, the bottlenecks for the synchronization process localize mainly inside the component networks.

9.1 Introduction

The quality of life in the developed countries requires the coordinated functioning of several infrastructures such as Electric System, Aqueducts, ICT Assets, Fresh Food distribution chains, Gas-ducts, Oil Pipelines, air and overland Transports, Banks and Financial assets etc. To emphasize their vital role in the technological society, several of those infrastructures have been classified as critical.

G. D'Agostino (✉)
ENEA – CR Casaccia, via Anguillarese, 301 – 00123 Roma, Italy
e-mail: gregorio.dagostino@enea.it

D. Matrasulov and H.E. Stanley (eds.), *Nonlinear Phenomena in Complex Systems: From Nano to Macro Scale*, NATO Science for Peace and Security Series C: Environmental Security, DOI 10.1007/978-94-017-8704-8_9,

The awareness of the vulnerability of the Critical Infrastructures (CI) and thereby the need to protect them, has been growing fast during last years. The USA has been the first developed country to take an official financial commitment on the protection of CI's by emitting the celebrated American Presidential Directive PDD-63 of May 1998 under the Clinton administration, even before the terrible events of September 11. It took 10 years for the EU community to take a similar commitment through the EUDIR Council Directive 2008/114/EC of the 8th of December, 2008. Similar actions have also been taken by the other developed countries, such as Canada and Japan and even by several fast developing countries. The protection of the critical infrastructures represents one of the most significant concerns of any modern country.

The functioning of Critical Infrastructures requires control and coordination of both physical components and human actors. It is therefore important not only to employ reliable components, but also to understand human behavior both at single person and at collective level. Moreover, each infrastructure resorts to other CIs (typically, but not only, energy and ICT) to accomplish its goals: in other words, CIs are *inter-dependent*. Identifying, understanding and analyzing critical infrastructure interdependencies represents a crucial task for the policy makers and the scientific community both at the academic and at the applied level [1].

In the development of CIs, the ICT sector has played a crucial role in several respects. ICT pervades any complex activity of modern societies based on communications and represents a fundamental part for the surveillance and the governance of any complex infrastructure. The reliability and quantity of information-based services employed by our modern society has been steadily increasing during last decades (SCADA systems, web, electronic communications, e-commerce, social networking, e-banking, entertainments etc.). In order to improve their performance and to enhance their reliability, the CIs have been endowed with increasingly complex connection networks and computerized systems, thus allowing their governance optimization and reducing the humans allocated to that purpose. The wise development and use of the CI's represents the basis for the advent of the *Smart Society*, a term that is inflected in different flavors such as: *Smart Grids*, *Smart Cities*, *Smart Roads*, *Smart Roads*, *Smart Light-stops* etc. and it indicates the intelligent coordinated management of the different CIs to provide the most effective, secure and sustainable services.

All the modern infrastructure exhibit network structures and employ several networks to provide the services they are devised to. Therefore any real infrastructure has to be modeled as, at least, two interdependent networks: the ICT network to allow control and the physical network.

About a decade ago, thanks to the availability of large data and resorting to the Statistical Physics and the graph theory, a new paradigm to study large networks has been introduced: the *Network Science* [2, 3]. This discipline represents a powerful and unifying tool to understand the basic mechanisms governing different networked systems, ranging from biology, to sociology, power grids, Internet or the World Wide Web. In spite of the intrinsic differences among such networks, they do exhibit shared or universal *emergent behaviors* related to systemic effects due to the extended size. Due to such collective effects, the ordinary maintenance on each component and even on each infrastructure may not suffice to prevent malfunctioning.

The application of the paradigm of the *Network Science* to Interdependent Critical Infrastructures leads to the concept of "Networks of Networks": the *NetoNets* (*NoN*). From the Theory of Graphs perspective a *NoN* is nothing but a larger (inhomogeneous) network. However, keeping trace of the different networks provides some specific results that are useful or even necessary. Real infrastructures are governed and controlled by different operators and authorities and interaction among them are only allowed at their boundaries according to predefined rules; so that the whole system does not behave as a single entity.

To our knowledge, the first applications of the *NoN* approach to critical infrastructures was aimed at studying the propagation of failures in inter-dependent infrastructures, modeled as either trees or planar lattices [4, 5]. However, the largest impact on the scientific community came from the celebrated paper on the percolation model of cascade failures in coupled ICT/power networks [14]. Moreover, another important step toward applications of NoN to real infrastructures has been the analysis of the interdependent north-America electric grids [6] aimed at reducing the global vulnerability of the system.

A lot of efforts are nowadays devoted to develop the mathematics of *NoN*. While most of the current models are still devoted to percolation [7–10], the dynamics on *NoN* can also be approached resorting to the spectral analysis [11–13]. The basic idea is to model fundamental evolution processes on *NoN*. The largest eigenvalue of the adjacency matrix represents a powerful estimate of the epidemic threshold, while the *Algebraic Connectivity* rules several phenomena driven by an underlying diffusion mechanism. Percolation on interdependent networks [14] has become one of the most interesting and dwelling open issues in complexity science. On the other hand the application of spectral techniques to network synchronization [15] is an other long standing topic that may provide important suggestions for the management of real infrastructures and hence on the quality of our life. Along both this lines of research, this work deals with the synchronization of interdependent networks.

9.2 Synchronizability as Asymptotic Stability

The synchronization of networks has been widely studied with a special emphasis to coupled oscillators [16–19]. The following introduction is just to frame the problem within the perspective of critical infrastructures synchronization.

Let us suppose we have a homogeneous network consisting of N components and the state of each component is defined by a continuous "state variable" x_i.

Let us also suppose that, when left alone, any component experiences a dynamic evolution according to an autonomous differential equation:

$$\frac{dx_i}{dt} = f_i(x_i) + \eta_i; \tag{9.1}$$

where η_i represents some possible external forcing term due to a disturbance or noise.

For the sake of simplicity, let us also suppose that the "spaces of the phases" of the state variables are all isomorphic to an unitary "circle", that is $x_i \in [0, 1]$ with periodic boundary conditions: $x_i \sim x_i + n$ for any integer n. This intrinsic cyclicity implies the periodicity of the functions f_i's:

$$f_i(x_i + n) \equiv f_i(x_i). \tag{9.2}$$

This means that the system is devised to perform cycles, but not necessarily regularly or periodically. In other words we are assuming that the state variables are similar to the phases of a system allowed to work cyclically, which is a common condition for several real components. It is worth stressing that we are not assuming a harmonic motion and not even a periodic one. However, depending on the nature of the function f, on the initial conditions and the forcing term, the actual evolution may happen to be periodic or even harmonic.

Let us also suppose that the components of the system may influence each other. We further suppose that the evolution of the state variable s_i can be affected by the others only if they differ from it. This means that for each condition of a component there is at least one state of the rest of the system that does not induce any deviation from the autonomous evolution; that is, the system is designed to work coherently. From the mathematical point of view this means that the interaction depends on the difference of the state variables only:

$$\frac{dx_i}{dt} = f_i(x_i) + g_i(x_i - x_j) + \eta_i; \tag{9.3}$$

where the functions g_i vanish when all state variables are synchronized ($g_i(0) = 0$).

If all the state variables evolve synchronously (possibly, but not necessarily periodically) the system evolves as all components where independent. However when one or more components lapse, the system may deplete its stability and stop functioning properly. As an example, one may think at an internal combustion engine, when all parts behave properly the four separate strokes (intake, compression, power, and exhaust) must take place at the appropriate moments, during two revolutions of the engine's crankshaft. In this case the mote is periodic, but not harmonic (there are angular accelerations) and some component has a period twice longer than the others. Moreover the synchronous values of the component relative to the different cylinders do not correspond to the same physical state: $x = 0$ will represent the intake for a cylinder, the compression for an other, the power for a third and the exhaust for the last. Moreover the absolute angular position of the wheels is absolutely irrelevant for the engine. The same situation takes place for the critical infrastructures, with the variation that there are several different components. However the overall phase of the system is not specially important, while the differences in the phases can be critical.

The stability of the Eq. (9.3) around a planned trajectory can be studied by means of the Lyapunov exponents of the linearization around some stable trajectory. Again this can be a periodic attractor or not. Introducing deviations from a stable trajectory $s_i = x_i - x^0$ the equations can be linearized, thus leading to a "diffusion coupled" system of equations:

$$\frac{ds_i}{dt} = \delta\omega_i + \sum_{j=1,N} w_{ij} A_{ij} (s_i - s_j) + \eta_i;$$

that is:

$$\frac{ds_i}{dt} = \delta\omega_i - \sum_j \hat{L}_{ij} s_j + \eta_i; \tag{9.4}$$

where we have introduced the weighted adjacency matrix $\hat{A}$ (and its related laplacian $\hat{L} = \hat{D} - \hat{A}$) and frequency deviation from the average $\delta\omega_i$:

$$\begin{cases} \hat{A}_{ij} \overset{def}{=} \frac{d}{ds_j} g_i(s_i - s_j) = w_{ij} A_{ij}, \\ \hat{L}_{ij} = \sum_{k=1}^{N} \hat{A}_{ik} - \hat{A}_{ij}, \\ \delta\omega_i = f_i(x_0) - \frac{1}{N} \sum_{j=1}^{N} f_j(x_0). \end{cases} \tag{9.5}$$

The deviations from the average frequency can be further decomposed into a static part representing the intrinsic spread of component frequencies and a linear correction representing autonomous response to deviations:

$$\delta\omega_i \sim \delta\omega_i^0 + \frac{df_i(x^0)}{dx^0}(s_i) - \frac{1}{N} \sum_{j=1}^{N} \frac{df_j(x^0)}{dx^0}(s_j) = \delta\omega_i^0 + \sum_{j=1}^{N} U_{ij} s_j; \tag{9.6}$$

therefore the stability of the system results from the equilibrium between the matrix U and the matrix $\hat{L}$. The system is asymptotically stable when the eigenvalues of the linearized equations are all negative. In this case what is relevant is the stability of the relative phases since an overall phase in not defined. Both the matrix U and $\hat{L}$ exhibit a null eigenvalue that represents a common shift of phases and it is irrelevant. There will always be a null mode representing a translation of all phases by the same amount. Since the Laplacian is positive defined for vectors orthogonal to a uniform one, the system is stable when the mismatch in the frequencies does not exceed the first non trivial eigenvalue of the laplacian that is the "algebraic connectivity".

When the interaction between the different components of the system are identical all weights are the same and one may assume $\hat{L} = wL$, thus separating the topological dependence (L) from the functional one (w). In this case the asymptotic stability of Eq. (9.4) read:

$$\|\delta\omega + \eta\| \leq w\lambda_2(L)\, \|s\|\,; \tag{9.7}$$

where $\lambda_2(L)$ represents the algebraic connectivity of the laplacian and the norm is $\|v\| \overset{def}{=} \max_i |v_i|$. The former equation provides information on how strong the coupling w must be to synchronize several components given their structural spread in the frequencies and the external disturbance. Conversely, given the topological characteristics of the system, one can estimate the maximum endurable disturbance.

To turn the general equations governing a complex system into the canonical form of Eq. (9.3) may be a tremendous work, yet in several cases it can be done. The present work, does not concern the origin of Eq. (9.3) or even their solutions, the only relevant point here is that the Lyapunov criteria for stability can be written in terms of pure topological quantities and, remarkably, on the first non trivial eigenvalue (i.e. the algebraic connectivity) of the laplacian of the graph representing the network.

The original motivation of the present work has been discussed above, however, synchronization phenomena in large communities of interacting systems appear in physical, biological, chemical, and social systems. Applying the paradigm of the complexity science to the problem of synchronization consists in modeling the different entities to be synchronized as "nodes" in a graph which topology only describes the interdependence among them. The spectral analysis of the Laplacian provides a means to estimate the synchronization time and its dependence on net topology changes. The present study is framed in the context of "Networks of Networks"; some general results are provided for a set of interacting networks, while some numerical results for the case of two interacting networks are reported.

One of the most interesting outcomes of the study is the existence of a type of "phase transition" for the synchronization time upon removals (or inclusions) of interlinks between the interdependent network. There exists a critical number of interlinks at which the typical time for synchronization (i.e. the algebraic connectivity) equals that of the isolated networks. When the number of attacked links reduces down to this critical value, the synchronization of the two networks slows and more importantly it takes place along significantly different paths.

Depending on the strategy of the "attacks" (or connection inclusions), the synchronization of the composite system of systems exhibits different behaviors. Numerical simulations for interdependent systems of different sizes and component model networks are coherently interpreted by both exact calculations and perturbation theory.

9.3 Modelling Critical Infrastructures as NetoNets

From the mere topological perspective, a complex system consisting of different interdependent CI's can be modeled as a Network of Networks (NetoNets or *NoN*). The functioning of the whole system is achieved by means of the cooperation of several "components" that may belong to the same subsystem or to different ones. In the first case one says that there is "intradependence" among components of the same subsystem; while in the other case, one says that there is "interdependence" between the different subsystems. We will always apply the prefix "intra" to characteristics involving features relative to the same subsystem, and the prefix "inter" to those involving components or parts belonging to different subsystems.

The complexity science approach consists in disregarding the details of the interaction between the different components and focus on their mutual dependence and their number. From the mere topological perspective, a systems of systems, can be abstracted to a "Network of Networks" (NetoNets or *NoN*); that is the whole system can be modeled as a graph ideally partitioned into homogeneous sub-graphs, while containing also links between nodes of different sub-graphs. A homogeneous sub-graph will be referred to as a "Component Network" or simply a *cnet*. The links inside the component networks will be referred to as "intralinks", while those connecting nodes belonging to different cnets will be referred to as "interlinks". If the Network of Networks (*NoN*) consists of M *cnet*'s, from the topological point of view, it can be formally represented as:

$$PG \stackrel{def}{=} (\mathcal{N}_1, \ldots, \mathcal{N}_M; \mathcal{L}_1, \ldots, \mathcal{L}_M; \mathcal{L}_{12}, \ldots, \mathcal{L}_{M-1,M}) \tag{9.8}$$

where $G_i = (\mathcal{N}_i, \mathcal{L}_i)$ represents the i-th component network and each $\mathcal{L}_{i,j}$ contains the interlinks between the i-th and the j-th networks. The graph representing the entire network will be referred to as $G = (\mathcal{N}, \mathcal{L})$ where:

$$\begin{cases} \mathcal{N} = \bigcup_{i=1,M} \mathcal{N}_i \\ \mathcal{L} = \left(\bigcup_{i=1,M} \mathcal{L}_i\right) \bigcup \left(\bigcup_{ij} \mathcal{L}_{i,j}\right) \end{cases} \tag{9.9}$$

Figure 9.1 provides a pictorial representation of the former definitions. The cardinality of the different sets provide a simple measure of the interaction between the networks: $N_i \stackrel{def}{=} |\mathcal{N}_i|$, $L_i \stackrel{def}{=} |\mathcal{L}_i|$ and $L_{ij} \stackrel{def}{=} \left|\mathcal{L}_{i,j}\right|$. The larger the L_{ij}/N_i ratio, the higher the dependence.

9.3.1 Some Topological Notation

The *NoN* is just a graph (possibly a weighted one) were each node is labeled by two different indices identifying the component it represents and the CI it belongs to. This means that the Adjacency matrix can be rewritten as tensor depending on four

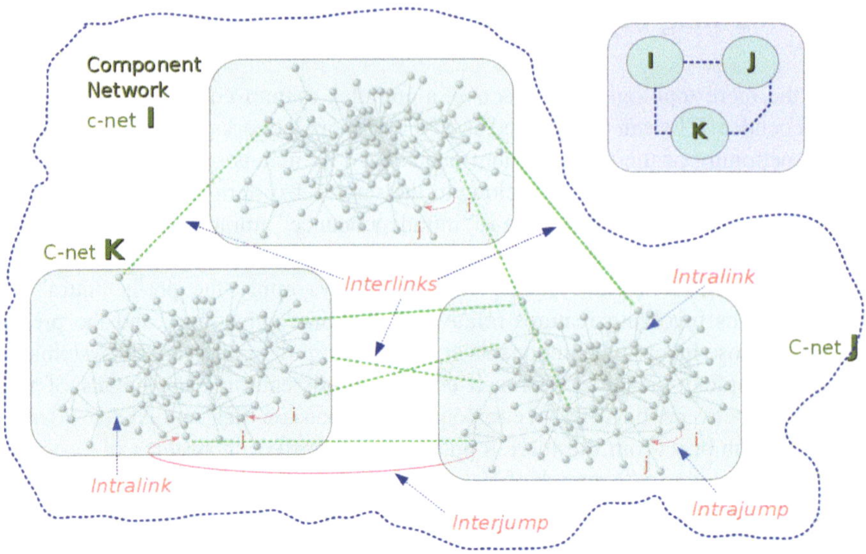

Fig. 9.1 This picture illustrates an example of three interdependent Networks in a simple complete macrotopology. The links between nodes in the same component network (cnet) are named "intralinks", while those connecting cnets belonging to different networks are named interlinks. The macrotopology results collapsing each cnet to a macro-node and linking any pair of macro-nodes if an interlink among their relative cnets exists

indices: A_{Ii}^{Jj}. The capital index will be referred to as "network index" and it labels the cnet, while the italic one will be referred to as "internal index" and it labels each component. One may provide a global index to nodes ordering the networks: $(I,i) \rightarrow (\sum_{J=1,I-1} N_J) + i$; where N_J is the size of the J-th network.

Two nodes belonging to two different networks and sharing the internal index will be named "homologous". This definition is evidently not univocal as one may perform any reordering (permutation) of the nodes; however once the choice is takes no ambiguity arises.

$$\begin{pmatrix}
A_{11}^{11} & \dots & A_{11}^{1N} & A_{11}^{21} & \dots & A_{11}^{2N} & \dots & A_{11}^{M1} & \dots & A_{11}^{MN} \\
\dots & \dots & \dots & \dots & \dots & \dots & \dots & \dots & \dots & \dots \\
A_{1N}^{11} & \dots & A_{1N}^{1N} & A_{1N}^{21} & \dots & A_{1N}^{2N} & \dots & A_{1N}^{M1} & \dots & A_{1N}^{MN} \\
A_{21}^{11} & \dots & A_{21}^{1N} & A_{21}^{21} & \dots & A_{21}^{2N} & \dots & A_{21}^{M1} & \dots & A_{21}^{MN} \\
\dots & \dots & \dots & \dots & \dots & \dots & \dots & \dots & \dots & \dots \\
A_{2N}^{11} & \dots & A_{2N}^{1N} & A_{2N}^{21} & \dots & A_{2N}^{2N} & \dots & A_{2N}^{M1} & \dots & A_{2N}^{MN} \\
\dots & \dots & \dots & \dots & \dots & \dots & \dots & \dots & \dots & \dots \\
\dots & \dots & \dots & \dots & \dots & \dots & \dots & \dots & \dots & \dots \\
\dots & \dots & \dots & \dots & \dots & \dots & \dots & \dots & \dots & \dots \\
A_{M1}^{11} & \dots & A_{M1}^{1N} & A_{M1}^{21} & \dots & A_{M1}^{2N} & \dots & A_{M1}^{M1} & \dots & A_{M1}^{MN} \\
\dots & \dots & \dots & \dots & \dots & \dots & \dots & \dots & \dots & \dots \\
A_{MN}^{11} & \dots & A_{MN}^{1N} & A_{MN}^{21} & \dots & A_{MN}^{2N} & \dots & A_{MN}^{M1} & \dots & A_{MN}^{MN};
\end{pmatrix} \tag{9.10}$$

In matrix format, the adjacency matrix for the *NoN* can be written as:

$$A = \begin{pmatrix} A_1 & B_{12} \dots & & B_{1M} \\ B_{21} & A_2 & \dots & B_{2M} \\ \dots & \dots & \dots & \dots \\ B_{M1} & \dots & B_{M-1M} & A_M \end{pmatrix}; \tag{9.11}$$

where the diagonal matrices represent the intralinks $(A_I)_i^j = A_{Ii}^{Ij}$ and the off-diagonal matrices represent the interlinks $(B_{IJ})_i^j = A_{Ii}^{Jj}$. In the following we will focus on undirected networks only and hence all adjacency matrices will be supposed to be symmetric; therefore we will assume $A_I^T = A_I$ and $B_{IJ} = B_{IJ}^T$.

Let us suppose we can ideally switch off the interaction between the cnets. In this case the adjacency matrix becomes block-diagonal; we will refer to such a matrix as the "unperturbed" adjacency matrix $A^{(0)}$. When the interaction is switched on, the total adjacency matrix will be the sum of the unperturbed one plus an interaction matrix B:

$$A = A^{(0)} + B = \begin{pmatrix} A_1 & \mathbf{0} & \dots \mathbf{0} \\ \mathbf{0} & A_2 & \dots \mathbf{0} \\ \dots & \dots & \dots \\ \mathbf{0} & \dots \mathbf{0} & A_M \end{pmatrix} + \begin{pmatrix} \mathbf{0} & B_{12} \dots & & B_{1M} \\ B_{21} & \mathbf{0} & \dots & B_{2M} \\ \dots & \dots & \dots & \dots \\ B_{M1} & \dots & B_{M-1M} & \mathbf{0} \end{pmatrix}. \tag{9.12}$$

9.4 Mean-Field Approximation

Let us suppose that we have a set of networks with similar properties, that is which topology belongs to a given model network. Let we assume that we have a rule (or a strategy) for linking nodes belonging to the different networks. We will refer to the former system (a *NoN*) as a "network of interacting homogeneous networks".

There exist two basic strategies that can be easily treated. The first linking strategy will be referred to as "Homologous Linkage Strategy" (HLS) and it corresponds to distribute links uniformly on homologous (i.e. sharing the internal index) nodes only. The second strategy will be referred to as "Random Linkage Strategy" (RLS) and it corresponds to a uniform selection of all possible node pairing (yet limited to the interdependent cnets).

The mean-field approximation of the HLS consists in replacing the ensemble of all possible L's with their average. This leads to a *NoN* where all diagonal and interacting matrices (those defining the interlinks) are the same and proportional to the unit matrix: $A^{(0)} = I \cdot \{A_1\}$ and $B = A_{macro} \cdot \alpha I$. A_{macro} is the adjacency matrix of the macrotopology, while the coefficient α represents the average number of links per interdependent pair. In this special case the matrix L_0 and the L_B commute and hence they admit a common set of eigenvectors. These can be obtained by:

$$v_k^K = \Xi_K \cdot \xi_k : \tag{9.13}$$

or in components:

$$u_{Ii}^{Kk} \stackrel{def}{=} \Xi_I^K \cdot \xi_i^k; \tag{9.14}$$

where Ξ^K represents the K-th eigenvalue of the matrix A_{macro} and ξ^K represents the k-th eigenvalue of the matrix B (if they exist). The eigenvalues relative to v_k^K will be referred to as λ_k^K. When the interaction is switched off ($\alpha = 0$), all eigenvalues are M times degenerate and coincide with the free ones: $\lambda_k^K = \lambda_i$ for all I. When the interaction is switched on, the eigenvalues change

$$\lambda_k^K = \lambda_k + \alpha \cdot \Lambda_K; \tag{9.15}$$

where Λ_K is the K-th eigenvalue of the interdependence matrix A_{macro} representing the macrotopology. It is worth stressing that the λ_i^I form a "partially ordered lattice" as for fixed i they are ranked according to the index I and vice-versa. However, the global ranking of the eigenvalues for the *NoN* depends on the value of α. For low enough α (that corresponds to weak interdependence) $\mu = \lambda_1^2 = \alpha\Lambda_1$, while for larger α's the former value may exceed $\lambda_2^1 = \lambda_2$. The λ_k^1 eigenvalues do not depend on α while λ_1^2 increases linearly. Therefore there will be a "critical value" of α at which the two eigenvalues swap:

$$\alpha_{th} = \frac{\lambda_2}{\Lambda_2}. \tag{9.16}$$

The fundamental case of two interacting networks ($M = 2$) has been extensively treated by Martin-Hernandez et al. [13]. Moreover very important results have been recently obtained by Gómez et al. [12] and Cardillo et al. [20].

Suppose we have a set of networks characterized by their laplacians $(L_1, \ldots, L_M)$ and a sequential rule to add links among them. Given the sequence of links, the algebraic connectivity (i.e. the inverse of the synchronization time) for the total interacting system is a function of the number of interlinks l. Each system will exhibit its own linkage profile depending on the type of component networks (cnets) and the linkage rule. If the linkage rules are stochastic there will be a set of possible links sequences that can be analyzed statistically. The simplest of such analysis consists in evaluating the average among all possible linkage sequences as a function of l. One expects this average to converge to a limit when the number of nodes per network N diverges while keeping the average linking $\alpha = l/N$ fixed. The average of the algebraic connectivity over the different networks (falling in a model network class) and the different linkage sequences (according to a given linkage strategy) do not coincide with the algebraic connectivity of the average laplacian, that represents a very rough approximation of it (as it disregards fluctuations). We will refer to such an approximation as the "mean-field" approach. Actually one may define different types of mean-field depending on which average is performed before the algebraic cut evaluation:

$$H \overset{def}{=} E_{L_i,B_i}[\mu] \tag{9.17}$$

$$H_l \overset{def}{=} E_{L_i}[\mu(E_{B_i}[L])] \tag{9.18}$$

$$H_{intra} \overset{def}{=} E_{B_i}[\mu(E_{L_i,B_i}[L])] \tag{9.19}$$

$$H_0 \overset{def}{=} \mu(E_{L_i,B_i}[\mu]); \tag{9.20}$$

the potential H corresponds to the real statistical average; H_l corresponds to the "fixed nets, average linkage" case; H_{intra} corresponds to the "fixed linkage, average nets" case and H_0 is the global mean-field case. If the interacting networks belong to the same class of model networks the H_0 case is equivalent to the case of identical networks interacting by a common matrix B:

$$A =\sim A_0 \cdot I + \alpha B \cdot A_{macro}; \tag{9.21}$$

where A_{macro} is the macroscopic adjacency matrix being null or unitary depending whether two component networks are interdependent or not. There are two exactly solvable cases that is worth treating: $B = \alpha \cdot I$ and $B = B^T = \alpha \cdot J$ where J is the all unitary matrix ($J_{ij} = 1 \forall ij$). The first solvable case will be referred to as "Homologous Linkage Strategy" and corresponds to distribute links uniformly on homologous nodes only. The second solvable case will be referred to as "Random Linkage Strategy" and corresponds to a uniform selection of all possible node pairing.

The phase transition occurs when the algebraic cut changes its localization from the interlinks to the intralinks. In other words when the number of interlinks is kept below a certain threshold, it is easier to disconnect the *NoN* cutting (mainly) interlinks; on the other side when the number of links exceeds the threshold the optimal cut (i.e. the Fielder cut) implies cutting mainly intralinks. From the synchronization perspective, below the threshold the cnets synchronize first internally and adiabatically they synchronize each other. Conversely beyond the threshold there are two sets of nodes belonging to different cnets that synchronize first while approaching the global synchronization on a longer scale even within the same cnet. In other words beyond the threshold the system approaches the equilibrium globally with a common time scale. In terms of the spectral properties this corresponds to compare the two lowest eigenvalues of the total laplacian:

$$\begin{aligned} \mu_{inter} &= \Lambda\alpha \\ \mu_{intra} &= \lambda_2; \end{aligned} \tag{9.22}$$

corresponding to the eigenvectors

$$\begin{aligned} v_{inter} &= v_0^1 = \Xi^2 \xi_1 \\ v_{intra} &= v_1^0 = \Xi^1 \xi_2. \end{aligned} \tag{9.23}$$

The transition point corresponds to the value where $\mu_{intra} = \mu_{inter}$ that defines the transition threshold:

$$\alpha_{th} = \frac{\lambda_2}{\Lambda_2}. \tag{9.24}$$

In the mean field approach α represents the number of interlinks per nodes $\alpha = l/N$, therefore the interpretation of the former result is that there exist a critical fraction of interlinks beyond which any further addition does not reduce the proper time of the *NoN*. Typically the number of interacting networks M is small and the corresponding Λ_2 of the macronet is larger than λ_2; however when the size of the macronet increases, $\frac{\lambda_2}{\Lambda_2}$ can be larger than one. For the HLS case, since α represents the fraction of available links and it is always smaller than one, the global system will never be able to synchronize with the same promptness of the single cnets whichever is the number of interlinks. The critical number of links in the mean-field approach for the diagonal strategy is proportional to the ratio of the algebraic connectivity of the component nets and that of the macronet:

$$l_{th}^{HLS} = N\alpha_{th} = N\frac{\lambda_2}{\Lambda_2}. \tag{9.25}$$

Same reasoning applies to the random linkage strategy:

$$\begin{aligned} \mu_{inter} &= N\Lambda\beta \\ \mu_{intra} &= \lambda_2 + K_{min}N\beta; \end{aligned} \tag{9.26}$$

where K_{min} is the minimum of macro degrees ($K_{min} \overset{def}{=} \min_{(J)}(K_J)$) and μ_{intra} is localized at the networks exhibiting such minimum macro-degree. Please note that μ_{intra} can be degenerate when more than one component network depends on K_{min} networks. The resulting critical value corresponds to the point where the two eigenvalues cross: $\beta_{th} = \frac{\lambda_2}{N(\Lambda_2 - K_{min})} = \frac{\Lambda_2}{N(\Lambda_2 - K_{min})}\alpha_{th}$. It is worth noting that when Λ_2 is smaller than K_{min}, the intra-mode is always slower than the inter-mode and hence synchronization proceeds first intranet and afterwords internet. In the mean-field approach β represents the probability for any interlink to occur in a pair of dependent networks ($\beta = l/N^2$) and hence the critical number of links is again proportional to the component net size:

$$l_{th}^{RLS} = \beta_{th}N^2 = \frac{N\lambda_2}{\Lambda_2 - K_{min}} = \frac{\Lambda_2}{\Lambda_2 - K_{min}} l_{th}^{HLS}. \tag{9.27}$$

Given the number of intralinks, the Homologous Link Strategy (HLS) always exhibits its phase transition earlier ($l_{th}^{RLS} > l_{th}^{HLS}$) than the Random Link Strategy (RLS). Moreover while in the HLS case the phase transition corresponds to a saturation of the algebraic connectivity, in the RLS case it corresponds to a change in slope of growth.

The exact solution can be explicitly written in all cases when the L_B commutes with L_A; in these cases the set of eigenvalues of the unperturbed network of Eq. (9.14) can be extended to the total interacting system.

9.5 General Results

In the general case an exact solution cannot be provided. However for small α one may resort to the perturbation theory to estimate the algebraic cut.

9.5.1 Perturbation Theory

Let us study the case of all identical interacting networks. Our merit function is the average algebraic cut:

$$H \stackrel{def}{=} E_{B_i}\left[\mu(L)\right]. \tag{9.28}$$

One can inspect some of the properties of such "thermodynamic" potential employing the perturbation theory. The problem consists in finding the minimum of the quadratic form associated with the laplacian in the unitary sphere ($u^2 = 1$), with the constraint $(\eta, u) \stackrel{def}{=} \sum_{i=1,N} u_i = 0$:

$$\mu = \inf_{u^2=1,\eta\cdot u=0} (u, Lu) = \inf_{u^2=1,\eta\cdot u=0} (u, (L_0 + \alpha L_I)u); \tag{9.29}$$

due to the min-max theorem, the problem is equivalent to solve the characteristic equations for the first (i.e. lowest) non trivial eigenvalue μ:

$$\begin{cases} (L_0 + \alpha L_I - \mu I)x & = 0, \\ u^2 = (u, u) = & = 1, \\ (\eta, u) & = 0. \end{cases} \tag{9.30}$$

When the solution is analytic in α, one may express μ and u by Taylor expansion as $\mu = \sum_{k=0}^{\infty} \mu^{(k)}\alpha^k$ and $u = \sum_{k=0}^{\infty} u^{(k)}\alpha^k$.

Substituting the expansion in the eigenvalue equation (9.30) gives the hierarchy of equations:

$$\begin{cases} L_0 u^{(k)} + L_I u^{(k-1)} & = \sum_{i=0}^{k} \mu^{(k-i)} u^{(i)} & \textit{for all } k, \\ \sum_{i=0}^{k} u^{(k-i)} u^{(i)} & = 0 & \textit{for } k \geq 1, \\ (\eta, u^{(k)}) & = 0 & \textit{for all } k. \end{cases} \tag{9.31}$$

The zero-th order of the hierarchy correspond to the kernel of the L_0 operator:

$$L_0 u^{(0)} = 0 \tag{9.32}$$

$$(u^{(0)}, u^{(0)}) = 1 \tag{9.33}$$

$$(\eta, u^{(0)}) = 0 \tag{9.34}$$

They admits a set of $M - 1$ independent solutions corresponding to eigenvectors v_0^I (for $I = 1, 2, \ldots, M - 1$); the v_0^0 eigenvector is excluded since it coincides with η. $u^{(0)}$ can be also written as a linear combination of the vectors $(\eta_I)_i = \delta_{IJ}$ that are unitary on the nodes belonging to the I-th component network and null elsewhere. $u^{(0)} = \sum_I C_I \eta_I \frac{1}{\sqrt{N}}$. The first order correction to the algebraic cut can be easily evaluated employing the first order hierarchy:

$$\mu^{(1)} = \inf_{(u^{(0)})^2 = 1, \eta \cdot u = 0} (u^{(0)}, L_I u^{(0)}). \tag{9.35}$$

The best first order approximation is hence the first non trivial eigenvector of the laplacian of a weighted adjacency matrix $\hat{V}$ defined as follows:

$$\hat{V}^{IJ} \overset{def}{=} V^{IJ} \frac{1}{N} \sum_{ij} B_{ij}^{IJ} = V^{IJ} \frac{l^{IJ}}{N} = V^{IJ} \alpha^{IJ}; \tag{9.36}$$

where α^{IJ} is the number of interlinks per node between the networks I and J. When the number of nodes tends to infinite and there is a stochastic rule for each link to exist, the arithmetic average $\frac{1}{N} \sum_{ij} B_{ij}^{IJ}$ tends with probability one to the average number of links per node:

$$\lim_{N \to \infty} \frac{1}{N} \sum_{i=1,N} \sum_{j=1,N} B_{ij}^{IJ} = E\left[\frac{l^{IJ}}{N}\right]; \tag{9.37}$$

when the rule is the same for each pair of interdependent networks all the number of links per node reach a common value $E[\frac{l^{IJ}}{N}] = \alpha$ and the matrix $\hat{V}$ factorizes: $\hat{V} = \alpha \cdot V$.

Therefore the first order correction in the perturbation theory corresponds to the mean-field approximation described earlier.

9.5.2 Intermodes

When few links exists among the cnets the synchronization is expected to take place first inside the cnets and afterwords between them. From the mathematical point of view this corresponds to a Fielder cut [21] cutting interlinks only, that is to an eigenvector constant over all the cnets. Let us first define the "inter" heuristic solution u_{inter} or the "inter" mode as the optimal (smallest μ) eigenvector cutting interlinks only:

$$(u_{inter})_i^I \stackrel{def}{=} C_I \frac{1}{\sqrt{N}}; \tag{9.38}$$

as can be seen, it depends on the cnet index only while being constant on internal components. The optimal estimate of the algebraic connectivity for an "intermode" is achieved minimizing the quadratic form given the form of the solution of Eq. (9.38):

$$\mu_{inter} \stackrel{def}{=} \min_{\{C_I : \sum_{I=1,M} C_I = 0\}} (u_{inter}, Lu_{inter}); \tag{9.39}$$

or treating the C's as a vector:

$$\mu_{inter} = \min_{\sum_{I=1,M} C_I = 0} (C, L^{inter} C) = \Lambda_2(L^{inter}); \tag{9.40}$$

where we have introduced the effective laplacian L_{inter} collapsing the internal indices:

$$(L_{inter})^{IJ} \stackrel{def}{=} \frac{1}{N} \sum_{i,j} L_{ij}^{IJ}. \tag{9.41}$$

When the number of nodes is large and the linking strategy is the same for all interdependent pairs, the effective inter-laplacian L^{inter} can be defined in terms of the macroscopic Adjacency matrix A_{macro}:

$$(A_{inter})^{IJ} \stackrel{def}{=} \sum_{i,j} A_{ij}^{IJ} = A_{macro}^{IJ} \frac{\cdot l^{IJ}}{N} \stackrel{P=1}{\rightarrow} A_{macro}^{IJ} \cdot \alpha^{IJ}. \tag{9.42}$$

In fact, the fraction of interlinks between each pairs of cnets converges with probability one to its expectation value for large N. When the linkage rules are the same for all pairs of interdependent cnets, A_{inter} is proportional to a A_{macro} and all α^{IJ} are the same:

$$(A_{inter})^{IJ} = A_{macro}^{IJ} \cdot \alpha. \tag{9.43}$$

In this case μ_{inter} depends only on the macrotopology and on the fraction of links included:

$$\mu_{inter} = \alpha \cdot \Lambda_2(L_{macro}); \tag{9.44}$$

where $L_{macro} = D_{macro} - A_{macro}$ is the laplacian of the macroscopic topology.

9.5.3 Intra Modes

When there exist a lot of links between the different cnets, one expects the total system to behave collectively. The extreme case when this takes place is when the deviation from equilibrium are the same for all cnets. That is there exist a common mode that does not depend on the index of the component network. We will refer to such collective modes, involving all cnets symmetrically, as "intramodes":

$$(u_{intra})_i^I \stackrel{def}{=} \frac{\xi_i}{M}; \tag{9.45}$$

where the cnet (capital) index is mute. The optimal "intramode" u_{intra} provides an other estimate

$$\mu_{intra} \stackrel{def}{=} \min_{(\eta,\xi)=0} (u_{intra}, Lu_{intra}) = \min_{(\eta,\xi)=0} (\xi, \frac{L}{M}\xi). \tag{9.46}$$

In this case one may distinguish between the contribute of the cnet and the contribute due to the interlinks:

$$\mu_{intra} = (\xi, L_{ave}\xi) + \phi^2 = (\xi, L_{intra}\xi); \tag{9.47}$$

where we have introduced the average laplacian of the cnets:

$$L_{ave} \stackrel{def}{=} \frac{1}{M} \sum_{I=1,M} L_I; \tag{9.48}$$

and the interactive matrix V:

$$(V_{intra}) \stackrel{def}{=} \frac{1}{M} \sum_{I=1,M;J=1,M} A_{macro}^{IJ} B_{ij}^{IJ}, \tag{9.49}$$

together with its relative laplacian counterpart: $(L_V)_{ij} \stackrel{def}{=} \delta ij(\sum_k V_{ik}) - V_{ij}$. It is worth stressing that the matrix V is positive and symmetric and hence L_V is positive semi-definite:

$$\phi^2 \stackrel{def}{=} (\xi, L_V\xi) \geq 0. \tag{9.50}$$

It is worth noting that ϕ^2 depends on the linkage strategy and it vanishes when the HLS is employed. This means that in some sense the HLS is optimal as it minimizes the algebraic connectivity of the intramode:

$$\mu_{intra} = \min_{\xi\cdot\eta=0} (\xi, (L_{ave} + L_V)\xi) \leq (\xi, L_{ave}\xi) \leq \lambda_2(L_{ave}). \tag{9.51}$$

9.6 Discussion

The phase transition from the "inter-synchronization" to the "intra-synchronization" is very general and it may generate a wide range of phenomenological behaviors depending on the different model cnets involved, on their number, size and on the linkage strategy. This will allow to design future complex systems taking into account both the vulnerability against fault propagation cascades (by epidemic threshold) and their global synchronizability. This work provides a general view on what can be inspected by a pure topological approach based on the algebraic connectivity, a similar analysis can be also performed for the maximum eigenvalue of the adjacency matrix. A lot of work should be devoted in the future to provide the decision makers with a tool to plan the developing and the protection of the interdependent critical infrastructures and, especially, their systemic behavior.

The fundamental case when two infrastructures mutually interact has been extensively studied by Wang et al. [11] and Hernández et al. [13]. Authors of [11] have provided upper and lower bounds for the epidemic threshold in a system of two interacting networks; while authors of [13] have dealt with the problem of synchronization, which we are mainly focused on. Same results from the latter paper are reported here to provide evidence of the applicability and consistency of the theory. In [13] four model cnets with different topological properties have been compared: Random Regular (**RR**) [22], Barabási-Albert (**BA**) [3]; Watts-Strogatz (**WS**) [23], and a 3D Lattice (**LA**): a deterministic three-dimensional grid. The input parameters for all models are set constraining the graphs to share the same average degree (i.e. 6), thus preventing spurious effects due to the total number of links.

From the spectral point of view the problem reduces to the study of the Laplacian associated with the Adjacency matrix of two coupled cnets: $A = A_0 + B$. Question is how the Fiedler eigenvalue does change upon topology (i.e. B) variation. Both the HLS and the RLS have been investigated. Results for the HLS simple case are reported here; however similar results hold for the general strategy.

Among several available, we have selected four main quantities (or metrics) to probe the system change upon progressive inclusion of intralinks. The "Fiedler cut" represents the number of links between the two partitions as resulting from the signs of the Fiedler eigenvector (i.e. the first non trivial eigenvector of the Laplacian). The interdependence angle (or "opening angle") represents the angle between the Fiedler eigenvector of the interdependent networks and that with the isolated ones (which eigenvalue vanishes) ($acos\left(x \cdot x^{(0)} / \|x\| \left\|x^{(0)}\right\|\right)$). The entropy is just the Shannon entropy calculated considering the squares of the Fiedler eigenvector components as occupation probabilities ($h = -\sum_i (x_i)^2 log((x_i)^2)$). The numerical results obtained averaging over 1,000 different configurations are shown in Figs. 9.2–9.5, respectively.

The drastic changes observed in all the "metrics" suggests that a "phase transition" emerges upon adding (or attacking) intralinks. Moreover the parameter profiles provide insights on the different modes of synchronization. Exact calculations in a mean field approximation and the perturbation theory provide a significant understanding of the "phase transition" origin. When a critical value of interlinks

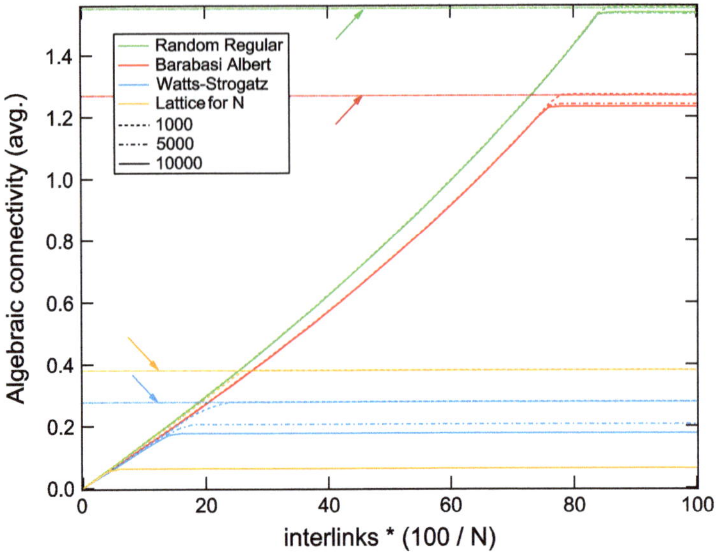

Fig. 9.2 The algebraic connectivity of different Model Interdependent Networks as a function of the number of interlinks following the homologous strategy. The flat lines (pointed by the *arrows* in the top left plot) refer to the average algebraic connectivity of the $N = 1{,}000$ respective single networks

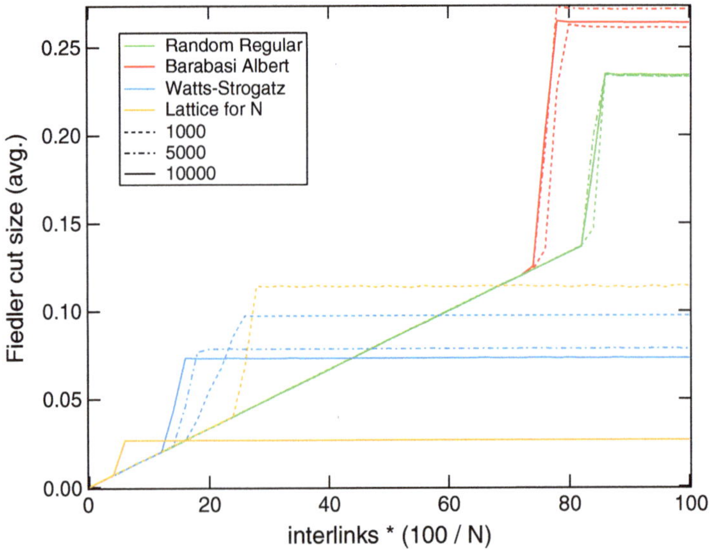

Fig. 9.3 The Fiedler cut for different Models of Interdependent Networks as a function of the fraction of homologous interlinks

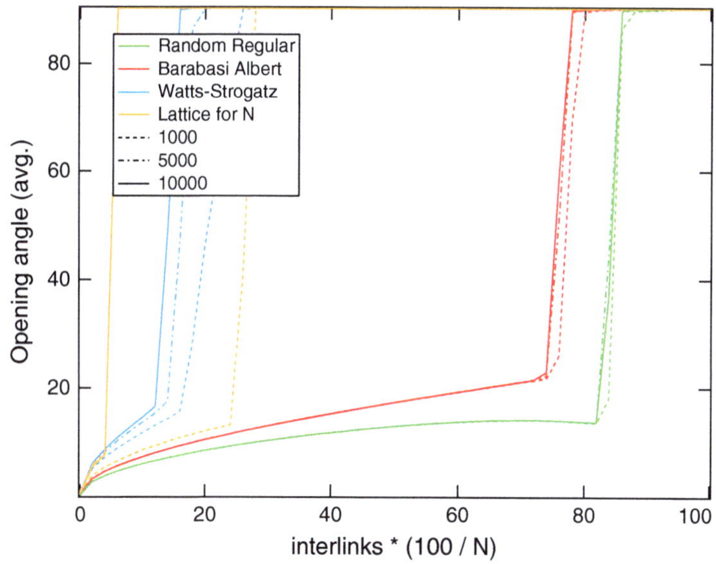

Fig. 9.4 The interdependence angle of different Model Interdependent Networks as a function of the number of interlinks following the homologous strategy. The flat lines (pointed by the *arrows* in the top left plot) refer to the average algebraic connectivity of the $N = 1{,}000$ respective single networks

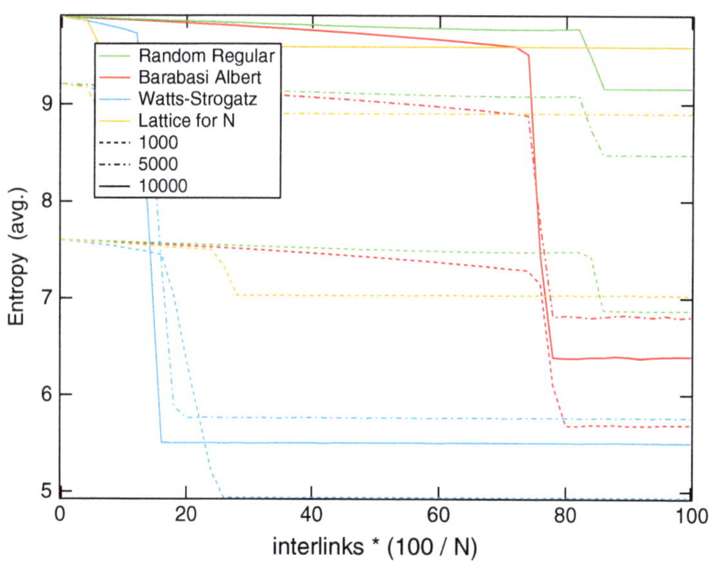

Fig. 9.5 The entropy of different Model Interdependent Networks as a function of the number of interlinks following the homologous strategy. Metrics experience a sharp transition. BA and RR graphs transitions fall around 20 % removed interlinks, whereas WS and LA graphs transition around 80 %. The flat lines (pointed by the *arrows* in the top left plot) refer to the average algebraic connectivity of the $N = 1{,}000$ respective single networks

is included, the algebraic connectivity (i.e. the typical time for synchronization) equals that of the isolated networks. This value depends on the inclusion strategy, but it is always significantly lower than the number of nodes in a single net. This means that to synchronize the networks, preserving the promptness of the isolated ones, a given value of links is required; beyond that critical value, further inclusions do not improve the performance, but just improve robustness against interlink attacks (removals). The interdependence angle provides a clear-cut indication that the observed transition corresponds to a drastic change of mode from "intra" to "inter". Similarly the drastic damp in the entropy reflects the localization of the Fiedler cut related to the "intra" mode.

The numerical experiments also confirm that the HLS provides an optimal allocation of interlinks to increase the synchronizability of the whole system. Consistently with the mean-field approach, the phase transition for the RLS takes place at about twice the number of links as the RLS; this means that the fluctuations do not compromise the average behavior.

9.7 Conclusions

This paper represents an initial step to approach the problem of infrastructure synchronization by a mere topological approach. Resorting to the spectral analysis of the laplacian, we have provided evidence that upon increasing the number of links between the different infrastructures, the synchronization mechanism of the total system experiences a drastic change. When few links are introduced, the synchronization inside the single infrastructure is very fast and the global synchronization takes place mainly at the boundaries. On the other side, there exist a threshold beyond which the bottlenecks for the synchronization process, localize mainly inside the component infrastructures.

Acknowledgements We thank Javier Martin Hernandez, Huijuan Wang and Piet Van Mieghem for their significant and useful suggestions. Interesting discussions with Antonio De Nicola, Antonio Scala and Gene Stanley are also acknowledged. This work was partly supported by the European project MOTIA (Grant JLS-2009-CIPS-AG-C1-016), Prevention, Preparedness and Consequence Management of Terrorism and other Security-related Risks Program European Commission – Directorate-General Home Affairs. This publication reflects the views only of the author, and the Commission cannot be held responsible for any use which may be made of the information contained therein.

References

1. Rinaldi SM, Peerenboom JP, Kelly TK (2001) Identifying, understanding and analyzing critical infrastructure interdependence. IEEE Control Syst Mag 21(6):11
2. Watts DJ, Strogatz SH (1998) Collective dynamics of 'small-world' networks. Nature 393(6684):440

3. Barabási AL, Albert R (1999) Emergence of scaling in random networks. Science 286(5439):509
4. Newman D, Nkei B, Carreras B, Dobson I, Lynch V, Gradney P (2005). Risk assessment in complex interacting infrastructure systems. In: Proceedings of the 38th annual Hawaii international conference on system sciences, HICSS '05, Big Island, 2005, p 63. doi:10.1109/HICSS.2005.524
5. Carreras BA, Newman DE, Gradney P, Lynch VE, Dobson I (2007). Interdependent risk in interacting infrastructure systems. In: Proceedings of the 40th annual Hawaii international conference on system sciences, HICSS '07, Waikoloa. IEEE Computer Society, Washington, DC, p 112. doi:10.1109/HICSS.2007.285
6. Brummitt CD, D'Souza RM, Leicht EA (2012) Suppressing cascades of load in interdependent networks. Proc Natl Acad Sci. doi:10.1073/pnas.1110586109
7. Lee KM, Kim J, Cho WK, Goh KI, Kim IM (2012) Correlated multiplexity and connectivity in multiplex random networks. New J Phys 14:033027
8. Leicht EA, D'Souza RM (2009) Percolation on interacting networks. ArXiv e-prints. arXiv:0907.0894
9. Parshani R, Buldyrev SV, Havlin S (2010) Interdependent networks: reducing the coupling strength leads to a change from a first to second order percolation transition. Phys Rev Lett 105:048701. doi:10.1103/PhysRevLett.105.048701
10. Dorogovtsev SN, Mendes JFF, Samukhin AN, Zyuzin AY (2008) Organization of modular networks. Phys Rev E 78:056106. doi:10.1103/PhysRevE.78.056106
11. Wang H, Li Q, D'Agostino G, Havlin S, Stanley HE, Van Mieghem P (2013) Effect of the interconnected network structure on the epidemic threshold. Phys Rev E 88:022801. doi:10.1103/PhysRevE.88.022801
12. Gómez S, Díaz-Guilera A, Gómez-Gardeñes J, Pérez-Vicente CJ, Moreno Y, Arenas A (2013) Diffusion dynamics on multiplex networks. Phys Rev Lett 110:028701. doi:10.1103/PhysRevLett.110.028701
13. Hernández JM, Wang H, Mieghem PV, D'Agostino G (2013) On synchronization of interdependent networks. Physica A. ArXiv e-prints abs/1304.4731
14. Buldyrev SV, Parshani R, Paul G, Stanley HE, Havlin S (2010) Catastrophic cascade of failures in interdependent networks. Nature 464(7291):1025. doi:10.1038/nature08932
15. Chen J, Lu J, Zhan C, Chen G (2012) Laplacian spectra and synchronization processes on complex networks. In: Thai MyT, Pardalos PM (eds) Handbook of optimization in complex networks: theory and applications. Springer optimization and its applications, chap 4. Springer, New York, pp 81–113
16. Strogatz S (2000) From Kuramoto to Crawford: exploring the onset of synchronization in populations of coupled oscillators. Physica D 143:1
17. Pecora LM, Carroll TL (1998) Master stability functions for synchronized coupled systems. Phys Rev Lett 80:2109. doi:10.1103/PhysRevLett.80.2109
18. Jadbabaie A, Motee N, Barahona M (2004). On the stability of the Kuramoto model of coupled nonlinear oscillators. In: Proceedings of the American control conference, Boston, pp 4296–4301
19. Acebrón JA, Bonilla LL, Pérez-Vicente CJ, Ritort F, Spigler R (2005) The Kuramoto model: a simple paradigm for synchronization phenomena. Rev Mod Phys 77:137
20. Cardillo A, Gómez-Gardeñes J, Zanin M, Romance M, Papo D, Pozo Fd, Boccaletti S (2013) Emergence of network features from multiplexity. Sci Rep 3. doi:10.1038/srep01344
21. Fiedler M (1975) Algebraic connectivity of graphs. Czechoslov Math J 25:619–633
22. Bollobas B (2001) Random graphs, 2nd edn. Cambridge University Press, Cambridge
23. Watts DJ, Strogatz SH (1998) Collective dynamics of small-world networks. Nature 393:440

Chapter 10
Theoretical Approaches to the Susceptible-Infected-Susceptible Dynamics on Complex Networks: Mean-Field Theories and Beyond

Claudio Castellano

Abstract Models for epidemic spreading are a paradigmatic example of the nontrivial effect of a topologically complex substrate on simple dynamics. The Susceptible-Infected-Susceptible (SIS) model has been recently the subject of considerable activity, which has uncovered that subtle effects of the quenched network structure and of dynamical correlations play a relevant role in the model behavior. Starting from the simplest mean-field theory we review the theoretical approaches which have been applied to SIS dynamics on networks, including the very recent results taking into account long-range dynamical correlations. The findings illustrate the highly nontrivial interplay between complex topology and dynamics and the need to include additional ingredients beyond the usual mean-field assumptions.

10.1 Introduction

Epidemic diseases have been a constant threat to humans throughout history. Examples such as the black death contagion during the fourteenth century, the spanish influenza pandemics in 1918, or the more recent outbreaks of AIDS, SARS and swine flu indicate that infectious diseases are one of the most important global problems that our world has to deal with. The goal of understanding how diseases spread is thus a major challenge, with huge potential implications for predicting, controlling and defeating this kind of threats.

C. Castellano (✉)
Istituto dei Sistemi Complessi (ISC-CNR), via dei Taurini 19, I-00185 Roma, Italy

Dipartimento di Fisica, "Sapienza" Università di Roma, P.le A. Moro 2, I-00185 Roma, Italy
e-mail: claudio.castellano@roma1.infn.it

D. Matrasulov and H.E. Stanley (eds.), *Nonlinear Phenomena in Complex Systems: From Nano to Macro Scale*, NATO Science for Peace and Security Series C: Environmental Security, DOI 10.1007/978-94-017-8704-8_10,

The spreading of infections is a complex collective phenomenon: the appearance of an epidemic at a global level is the macroscopic manifestation of the spreading of the disease among a large number of individuals at the microscopic level. In this respect, the study of epidemics is a natural playground for the application of concepts and methods of statistical physics, applied mathematics and probability theory, aimed at understanding emergent collective phenomena.

Mathematical epidemiology has been developed over many decades, leading to a detailed comprehension of models describing, at increasing levels of realism, the crucial features of infectious spreading [1]. However, essentially all classical results have been derived considering individuals occupying the sites of a regular lattice or assuming all to all interactions (homogeneous mixing hypothesis). The last 15 years have made clear that many social or technological networks, which constitute the contact pattern mediating most spreading phenomena, are very different from the idealized topologies considered previously [2–4]. The focus has then shifted towards the understanding of the interplay between complex network topologies and simple epidemic models, giving rise to novel, highly nontrivial phenomena.

In this paper we discuss the effects of heterogeneous topologies on the behavior of one of the simplest models for disease epidemics, with particular emphasis on the theoretical approaches used to attack the problem, their relationship and their accuracy. It turns out that subtle effects beyond the mean-field assumptions must be taken into account to have a complete understanding of the global phenomenology. This indicates that in general the validity of mean-field approaches for dynamical processes on networks should not be taken for granted and must be carefully checked.

10.2 The Susceptible-Infected-Susceptible (SIS) Model and the Classical Mean-Field Approach

The Susceptible-Infected-Susceptible (SIS) model is the simplest model describing the spreading of a disease with no acquired immunity, so that an individual can become infected many times, even immediately after recovery. In this model, individuals can be in one of two states: S = Susceptible or I = Infected. When an individual is infected, two events are possible (Fig. 10.1):

- The infected node recovers spontaneously (transition $I \rightarrow S$), with a rate μ, which is taken to be 1 with no loss of generality;
- The infected node transmits the infection to one of its contacts. This happens with a rate λ for each of the neighbors.

Depending on the single parameter of the dynamics, the spreading rate λ, the iteration of this dynamics may lead (in a system of infinite size) to two possible outcomes, characterized by the value of the order parameter, the *prevalence* ρ,

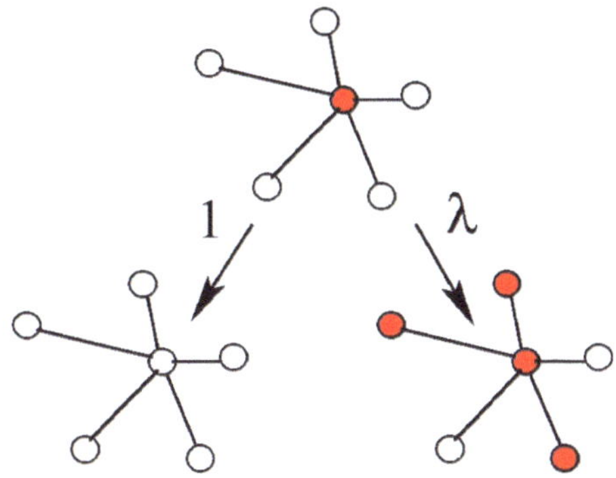

Fig. 10.1 Representation of the possible events for SIS dynamics. *White nodes* are in the S state, *red nodes* are in the I state

defined as the fraction of individuals which are infective in the stationary state. For λ smaller than a critical value λ_c (epidemic threshold) contagions are insufficient to counterbalance the recoveries and in the long-time limit no infective (active) individual remains ($\rho = 0$). Once this *absorbing state* is reached, no further evolution can take place. For $\lambda > \lambda_c$ instead, the system reaches a stationary active state with a finite fraction of the whole population in the infected state (see Fig. 10.2).

This picture is reproduced by a simple mean-field approach to the problem. Under the assumption that each individual is in contact with k other individuals chosen randomly at each time, the temporal evolution of the density of infected individuals is

$$\dot{\rho} = -\rho + \lambda k \rho (1 - \rho). \tag{10.1}$$

The stationary solution gives $\rho = 0$ for $\lambda < \lambda_c^{MF} = 1/k$, while there is a finite activity $\rho > 0$ for $\lambda > \lambda_c^{MF}$.

10.3 Heterogeneous Networks and the Heterogeneous Mean-Field Theory

One of the most relevant features which distinguish many techno-social systems is the heterogeneity of connectivity patterns, i.e. the wide variation in the number of connections individuals may have. If we denote with k the number of neighbors of a node (i.e. the *degree* of the node), in many networks the distribution of the degrees decays as a power-law $P(k) \sim k^{-\gamma}$ for large values of k. This is in stark contrast with standard (Erdös-Rényi) models for random networks where the degree distribution has an exponential tail. In many real-world examples the exponent γ is in the range between 2 and 3, hence there are few nodes (the *hubs*) which have a disproportionately large number of connections. It is natural to ask how this affects

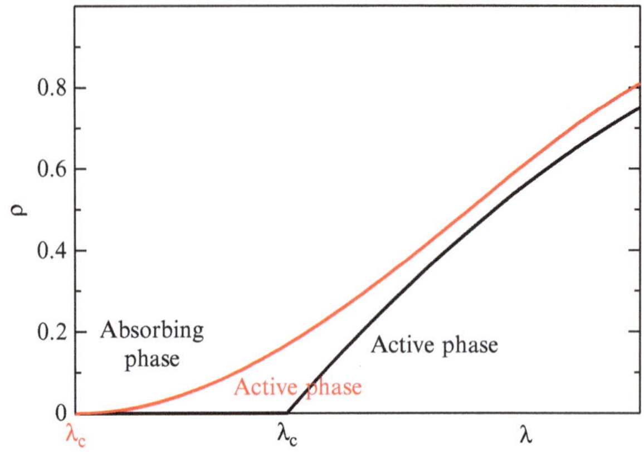

Fig. 10.2 Phase diagram of the SIS model. In *black* the value of the stationary density of infected nodes for regular lattices and the classical mean-field case, with a finite epidemic threshold separating the absorbing phase from the active phase. In *red* the same quantity for a scale-free network for $\gamma < 3$ within the HMF approach: the threshold vanishes and there is a finite prevalence for any $\lambda > 0$

the behavior of the SIS dynamics taking place on these networks. In order to take into account the effect of the heterogeneity, in a seminal paper [5] Pastor-Satorras and Vespignani proposed a modification of the mean-field approach based on the idea that the behavior of a node depends explicitly on its degree k (and on no other property of the node). According to this hypothesis the fundamental quantity describing the state of the system is ρ_k, defined as the fraction of nodes *of degree* k which are infected. The evolution of the ρ_k variables is given by a set of rate equations easily derived

$$\dot{\rho_k} = -\rho_k + \lambda k[1 - \rho_k] \sum_{k'} \frac{k' P(k')}{\langle k \rangle} \rho_{k'}, \tag{10.2}$$

where $k' P(k')/\langle k \rangle$ is the conditional probability that the neighbor of a node of degree k has degree k', assuming the absence of topological correlations in the network. By solving self-consistently this set of equations the epidemic threshold is found to be

$$\lambda_c^{HMF} = \frac{\langle k \rangle}{\langle k^2 \rangle}. \tag{10.3}$$

This expression implies very different outcomes depending on the value of the exponent γ for the degree distribution. For $\gamma < 3$ the second moment of the degree distribution diverges in the limit of infinite network size $N \to \infty$. As a consequence the threshold tends to 0 as the network size grows: for these *scale-free* networks there is asymptotically no epidemic threshold and no matter how small the spreading rate is, there is a stationary state with a finite fraction of active nodes (see Fig. 10.2). This is in clear contradiction with all results of classical epidemiology, which always predict the existence of a finite threshold. For $\gamma > 3$ (scale-rich networks) instead we recover also within HMF theory the existence of a finite threshold.

10.4 The Quenched Mean-Field Theory

HMF theory predicts that the scale-free property of a network has a critical effect on the dynamics of the SIS model on top of it. This result had a tremendous impact and spurred a wealth of investigations looking for similar effects for other types of dynamics on complex topologies.

Over the years, some activity has also been devoted to the identification of possible ways to improve the HMF approach. Interestingly, this has occurred soon within the community of computer scientists [6], while the same approach has been independently reintroduced only much later by physicists [7, 8]. In the Quenched Mean-Field (QMF) approach, the idea is to derive an equation of motion for the probability ρ_i that a specific node i is active. This means that one has to take into account not only the degree of node i but in detail the precise position of i in the network structure. The evolution of ρ_i is given then by

$$\dot{\rho_i} = -\rho_i + \lambda(1-\rho_i)\sum_{j=1}^{N} A_{ij}\rho_j, \tag{10.4}$$

where A_{ij} is the adjacency matrix, with entries A_{ij} equal to 1 if vertices i and j are directly connected and 0 otherwise. The SIS epidemic threshold is derived from Eq. (10.4) by performing a linear stability analysis of the absorbing state $\rho = 0$, yielding

$$\lambda_c^{QMF} = \frac{1}{\Lambda_N}, \tag{10.5}$$

where Λ_N is the largest eigenvalue of the adjacency matrix.

The connection between the QMF and HMF approaches is easy to understand. The Heterogeneous Mean-Field theory is equivalent to the assumption [9] that the dynamics takes place on an annealed network, i.e. a network such that all connections are rewired at a rate much faster than any dynamical process occurring on it, preserving only the number of connections k of each node. This implies that the adjacency matrix of the quenched network is effectively replaced by an annealed adjacency matrix which, for uncorrelated networks, is

$$\bar{A}(k_i, k_j) = \frac{k_i k_j}{\langle k\rangle N}. \tag{10.6}$$

The largest eigenvalue of the annealed adjacency matrix is $\langle k^2\rangle/\langle k\rangle$ and inserting this expression into Eq. (10.5) the HMF threshold Eq. (10.3) is recovered. Thus HMF theory corresponds to QMF theory with the additional approximation that the adjacency matrix of the network is replaced by its annealed counterpart. Hence the results of the QMF approach are expected to be more correct than those derived using HMF.

It is important to stress, however, that *QMF theory is not exact*. Equation (10.4) contains indeed an approximation, the disregard of dynamical correlations among the state of connected nodes. In the right hand side of Eq. (10.4), the rate of infection of node i by its neighbor j should depend on the probability $(\langle(1 - X_i)X_j\rangle)$ that i is not active and j is active, where $X_i = 0$ if the node is not active and $X_i = 1$ if it is active. Equation (10.4) is obtained by factorizing the product $\langle X_i X_j\rangle = \langle X_i\rangle\langle X_j\rangle = \rho_i\rho_j$, which is an approximation neglecting dynamical correlations.

The relevance of the QMF expression for the threshold becomes clear when one uses the explicit expression for Λ_N derived by Chung et al. [10] for power-law degree distributed networks

$$\Lambda_N = \begin{cases} c_1\sqrt{k_{max}} & \sqrt{k_{max}} > \frac{\langle k^2\rangle}{\langle k\rangle}\ln^2(N) \\ c_2\frac{\langle k^2\rangle}{\langle k\rangle} & \frac{\langle k^2\rangle}{\langle k\rangle} > \sqrt{k_{max}}\ln(N). \end{cases} \tag{10.7}$$

Here k_{max} is the largest degree in the network, i.e. the number of connections of the largest hub. Inserting the expression (10.7) into Eq. (10.5) and recalling that for a power-law degree distribution $\langle k^2\rangle \sim k_{max}^{3-\gamma}$ one finds [11]

$$\lambda_c \simeq \begin{cases} 1/\sqrt{k_{max}} & \gamma > 5/2 \\ \frac{\langle k\rangle}{\langle k^2\rangle} & 2 < \gamma < 5/2. \end{cases} \tag{10.8}$$

The formula is in striking contrast with HMF theory: the QMF approach predicts that the SIS epidemic threshold vanishes in the thermodynamic limit *for any value of the degree exponent* γ. In this framework, the fact that the threshold is asymptotically zero has nothing to do with the presence or absence of the scale-free property of the network.

The difference between the theoretical predictions of different approximations calls for a numerical investigation of their accuracy. The problem turns out to be difficult to tackle. The traditional way used to determine with precision the position of an absorbing-state phase-transition is the Finite-Size-Scaling method: the density $\rho_s(\lambda, N)$ of active nodes is computed by restricting the average only to runs which are still surviving and is plotted for fixed λ as a function of the system size N. In normal cases where the threshold goes asymptotically to a finite value, the critical point is determined as the value of λ such that ρ_S decays as a nontrivial power-law of N, separating the supercritical regime where ρ_s goes to a constant and the subcritical regime where it decays as N^{-1}. In the case of SIS, however, things are made complicated by the fact that the threshold position changes with N. It is only possible to derive indirect evidence that λ_c is much smaller than the prediction of HMF theory [11].

A more successful way to determine the position of the threshold in simulations [12] considers the peak of the susceptibility

$$\chi = \frac{\langle\rho^2\rangle - \langle\rho\rangle^2}{\langle\rho\rangle}. \tag{10.9}$$

The use of the quasi-stationary simulation method [13] allows a very precise calculation of the susceptibility and hence of its peak. It turns out that for $\gamma < 5/2$ both QMF and HMF theories (which give essentially the same prediction for the position of the threshold) are in very good agreement with numerical results and in the thermodynamic limit the two approaches tend to be exact. For larger values of γ results are less clear. In particular, for $\gamma > 3$ (where HMF and QMF give qualitatively different predictions) the susceptivity displays for large values of the system size N *two* peaks. In such a case it is not clear which of the two corresponds to the epidemic transition and should be compared with theoretical approaches.

Some insight into the meaning of the two peaks is provided by analyzing the physical origin of the different expressions for the epidemic threshold in Eq. (10.5) [14]. To understand why the threshold scales as $1/\sqrt{k_{max}}$ for $\gamma > 5/2$ it is instructive to consider the star graph centered around the largest hub in the network, composed by the hub and k_{max} leaves. The largest eigenvalue for such a subgraph is $\sqrt{k_{max}}$, implying that the threshold for the star is $\lambda_c = 1/\sqrt{k_{max}} = \lambda_c^{QMF}$. When one considers the star subgraph merged with the whole network it is clear that if the isolated star is active it will remain active when its leaves are joined with other nodes in the network. Hence the hub and its neighbors constitute a core of infected nodes able to self-sustain the activity in the network. When the hub is active also the whole network is active and this explains why the global threshold coincides with the threshold for the star graph centered around the hub. In this sense the largest hub triggers the activity in the network. For $\gamma < 5/2$ it turns out that the role of the trigger for the activity in the network is played by another subgraph, the set of most densely interconnected nodes as identified by the core with largest index in the k-core decomposition [14]. The two triggering mechanisms are present and competing for any value of γ. What changes above or below $\gamma = 5/2$ is which of the two dominates asymptotically. The presence of two distinct peaks in the numerical evaluation of the susceptibility for $\gamma > 3$ [12] is the reflection of the existence of two competing mechanisms that may lead to the active state. While QMF theory predicts that the decay $1/\sqrt{k_{max}}$ dominates asymptotically, the numerical evidence [12] is not sufficient to firmly validate this conclusion.

10.5 Localized or Endemic Activity?

The investigation about the behavior of the epidemic threshold for the SIS model was further enriched by a recent contribution by Glotsev et al. [15], considering not only the value of the largest eigenvalue in power-law degree-distributed networks, but also the associated Principal Eigenvector (PEV). It turns out that for $\gamma < 5/2$ the PEV is delocalized and this means that for $\lambda > \lambda_c^{QMF}$ the active state is endemic, i.e. a finite *fraction* ρ of the total number nodes is infected. For $\gamma < 5/2$ instead, the PEV is localized around the hub and its components decay exponentially with the topological distance from it. In such a case activity is concentrated around the hub and involves only a finite *number* of nodes: the global prevalence vanishes for large N. Finite prevalence is guaranteed only by delocalized eigenvectors, which are associated to smaller eigenvalues, so that the effective threshold for endemic activity occurs for higher values of λ, in particular for $\lambda > \lambda_c^{HMF}$. In this picture there are two transitions and three regimes, depending on the spreading rate λ. For $\lambda < \lambda_c^{QMF}$ the system reaches asymptotically the absorbing state, with no persisting activity whatsoever. For $\lambda_c^{QMF} < \lambda < \lambda_c^{HMF}$ the system is in an active state, but activity remains localized around the largest hubs in the network. For $\lambda > \lambda_c^{HMF}$ the active state is endemic, involving a finite fraction of the whole population.

This picture somehow reinstates the validity of HMF theory: the vanishing threshold predicted by QMF theory for $\gamma > 5/2$ only marks the presence of some nonendemic activity in the system. The real transition to a veritable endemic state occurs only for higher values of λ, which do not vanish as N diverges. It is however crucial to stress that this picture is derived within the QMF framework, thus completely disregarding dynamical correlations.

Pushing further the argument by Goltsev et al., in their recent work Lee and collaborators [16] pointed out the fact that activity in a star graph around an hub of degree k_i persists only over times scales $e^{a\lambda k_i}$ (with a a numerical constant), which do not grow with the system size. Hence, for $\lambda > \lambda_c$ the activity around any hub will be brought to zero by a fluctuation after a finite time. Thus, if hubs are isolated and do not interfere with each other there will be rare active clusters surviving over long but finite timescales. The resulting global activity will then slowly decay over time like in a Griffiths phase [17]. One possible way out of this Griffiths-like phase is the possibility that hubs are neighbors and reinfect each other, so that activity can globally persist. By means of an argument based on a degree-ordered-percolation process, Lee et al. show that while for $\gamma < 3$ hubs do form connected clusters, this is not the case for $\gamma > 3$, so that the Griffiths-like decay to zero of the global activity is to be expected.

10.6 A New Theoretical Approach

The QMF approach used by Goltsev et al. [15] completely neglects dynamical correlations, which are instead partially taken into account by the approach of Lee et al. [16], which considers the possibility of reinfection among nearest neighbors. The recent work of Boguñá et al. [18] goes further and considers long-range dynamical correlations, by including also the possibility of reinfection among distant hubs in the network.

The basic idea of the approach is to consider the evolution over a coarse-grained time scale. During this long temporal interval it is possible that even hubs which are far apart in the network may infect each other, via a chain of intermediate infection events of vertices along the path joining the hubs. The original evolution is thus replaced by an effective dynamics where all nodes are in contact with each other, i.e. they are on a complete graph

$$\frac{d\rho_i(t)}{dt} = -\bar{\delta}(k_i, \lambda)\rho_i(t) + \sum_{j \neq i} \bar{\lambda}(d_{ij}, \lambda)\rho_j(t)[1 - \rho_i(t)]. \tag{10.10}$$

where $\bar{\delta}(k_i, \lambda)$ and $\bar{\lambda}(d_{ij}, \lambda)$ are, respectively, the recovery and the spreading rates for this effective dynamics. The first can be straighforwardly estimated as

$$\bar{\delta}(k_i, \lambda) = e^{-a(\lambda)k_i} \tag{10.11}$$

with $a(\lambda) \sim \lambda^2$ [18]. The estimate of the spreading rate is more complicated. However it is possible to derive an expression for it

$$\bar{\lambda}(d_{ij}, \lambda) = \lambda e^{-b(\lambda)(d_{ij}-1)} \tag{10.12}$$

where d_{ij} is the topological distance between vertices i and j and $b(\lambda) = \ln(1 + 1/\lambda)$. This expression is derived under the simplifying assumption that i and j are connected only via a linear chain of nodes of degree 2, thus neglecting the possible presence of multiple paths and the fact that intermediate nodes may have degree larger than 2. For this reason Eq. (10.12) is an underestimate of the true effective spreading rate and the threshold obtained using it is an upper bound of the true epidemic threshold. For random small-world graphs the average distance between node i and j is [19]

$$d_{ij} = 1 + \frac{\ln\left(\frac{N\langle k\rangle}{k_i k_j}\right)}{\ln \kappa} \tag{10.13}$$

where $\kappa = \langle k^2\rangle/\langle k\rangle - 1$. Performing as usual a linear stability analysis one obtains the condition setting the threshold [18]

$$a(\lambda_c)k_{max} \simeq O(\ln(N)) \tag{10.14}$$

This condition implies that, for any finite γ (for which k_{max} diverges faster than $\ln(N)$) $a(\lambda_c)$ must vanish in the thermodynamic limit and hence λ_c must vanish as well. If one inserts the expression $a(\lambda) \sim \lambda^2$ the prediction for the scaling of the threshold is, apart from logarithmic corrections, the same of QMF theory. Notice that, since the effective dynamics takes place on a complete graph, this threshold corresponds to a transition to an endemic active state. It is also important to remark that if the network has no small-world property and the average distance between two nodes grows as N^{β} the same argument gives a finite or a vanishing threshold depending on the value of β [18], in agreement with the simulations of Lee et al. [16].

10.7 A New Numerical Approach

In order to validate the new theoretical results, we have devised a new way to measure the threshold in numerical simulations, overcoming the problems encountered considering finite-size-scaling or the peak of the susceptibility.

The new approach is aimed in particular at discriminating the case of an active state localized around the hubs of the network from the case of a truly endemic state which involves (although with a very small prevalence) the whole system. The idea is to perform spreading experiments, i.e. start with only one infected node (for simplicity we will consider the largest hub, but no qualitative difference occurs for other choices) and compute the lifetime of the outbreak [18]. In an infinite system, for $\lambda < \lambda_c$ all runs reach the absorbing state after a finite lifetime. Above the epidemic transition instead a fraction $P_{end}(\lambda)$ of them will have an infinite lifetime, reaching a stationary endemic state with a finite fraction of nodes infected. $P_{end}(\lambda)$ is an order parameter for the transition, while the average lifetime $\bar{T}(\lambda)$ of finite realizations plays the role of a susceptibility, being small far from the transition and having a peak at λ_c. However, in finite systems no realization has a truly infinite lifetime, since rare fluctuations lead to the absorbing state even well above the transition. Therefore, in order to identify endemic realizations in computer simulations we compute the coverage (the fraction of nodes which have been infected at least once during an outbreak) and declare the realization endemic if the coverage exceeds a predefined threshold C [18]. For computational convenience we took $C = 0.5$ but we tested that larger values of C do not significantly change the results.

We performed numerical simulations of the SIS dynamics on uncorrelated random small-world networks, built according to the uncorrelated configuration model [20]. For $\gamma = 2.25 < 5/2$ we find (Fig. 10.3) perfect agreement with

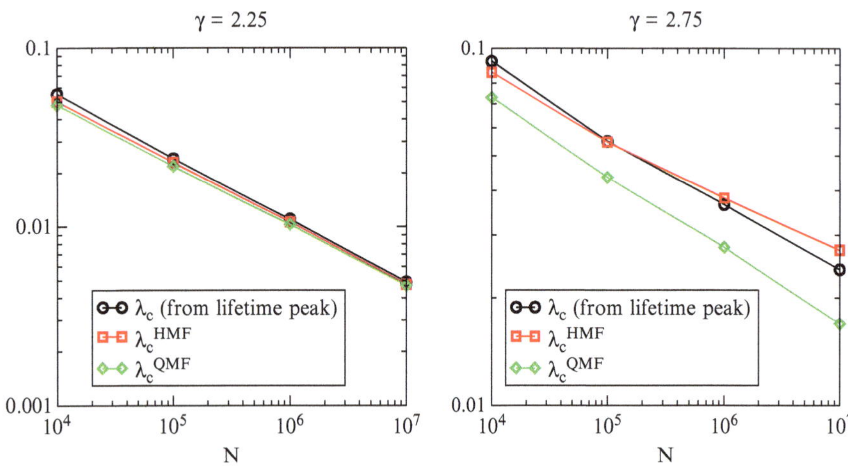

Fig. 10.3 Comparison between the numerical evaluation of the threshold using the coverage method for $\gamma = 2.25$ (*left panel*) and $\gamma = 2.75$ (*right panel*) with the results of HMF and QMF theories. The minimum value of k is $k_{min} = 3$

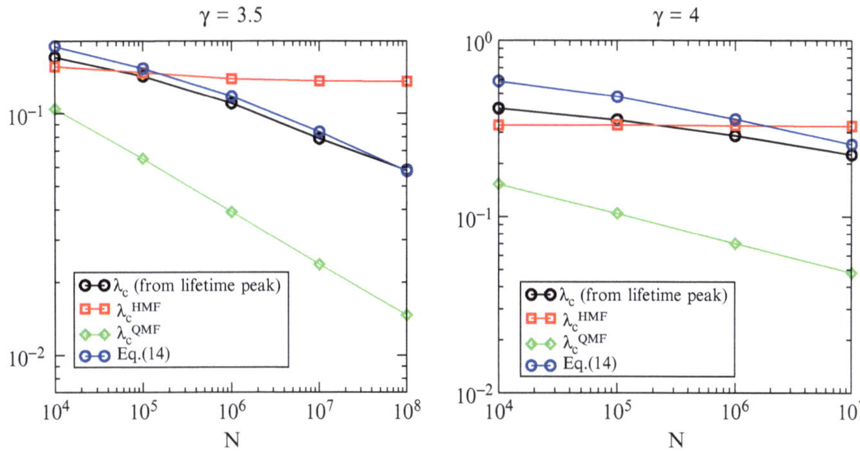

Fig. 10.4 Comparison between the numerical evaluation of the threshold using the coverage method for $\gamma = 3.5$ (*left panel*) and $\gamma = 4$ (*right panel*) with the results of HMF and QMF theories. The minimum value of k is $k_{min} = 3$ for $\gamma = 3.5$ and $k_{min} = 2$ for $\gamma = 4$

previous results obtained using the susceptibility method [12]. The same occurs for $\gamma = 2.75$, where the numerical threshold scales to zero like the QMF prediction, faster than the HMF prediction. For larger values of γ HMF and QMF theories diverge, since the former predicts a finite threshold in the limit of large N, while the latter predicts a vanishing threshold. Numerical simulations (Fig. 10.4) indicate that λ_c actually goes to zero in the thermodynamic limit. Equation (10.14) provides as expected an upper bound to the true position of the threshold and the difference is rather small. It is also worth to mention that QMF theory predicts the correct scaling for λ_c but largely underestimates the prefactor.

10.8 Conclusions

The behavior of the susceptible-infected-susceptible dynamics on networks turns out to be highly nontrivial and difficult to tackle both theoretically and numerically. Starting from the simplest mean-field theory, many approaches, of increasing level of complexity, have been applied to calculate the position of the epidemic threshold and its scaling in the thermodynamic limit. The heterogeneity of the degree distribution, the quenched structure of the network, localization of activity, short and long range dynamical correlations are the ingredients that had to be taken into account. It turns out that all of them are needed in order to reach a satisfactory understanding of the intricate SIS dynamics. The final conclusion is that the epidemic threshold vanishes in the limit of infinite network size N, for any value of γ.

Concerning more in general the problem of dynamics on complex networks, a naive expectation would be that, since networks are infinite-dimensional substrates, simple mean-field approaches should be able to correctly describe dynamical processes taking place on them. The example of SIS clearly indicates that subtle effects may occur, leading to a picture considerably more complex than the naive mean-field expectation. This example calls for a general reconsideration of the behavior of dynamical processes on complex networks and of the validity of mean-field approaches for these problems.

References

1. Anderson RM, May RM (1992) Infectious diseases in humans. Oxford University Press, Oxford
2. Albert R, Barabási AL (2002) Statistical mechanics of complex networks. Rev Mod Phys 74:47
3. Dorogovtsev SN, Mendes JFF (2003) Evolution of networks: from biological nets to the Internet and WWW. Oxford University Press, Oxford
4. Newman MEJ (2010) Networks: an introduction. Oxford University Press, Oxford
5. Pastor-Satorras R, Vespignani A (2001) Epidemic spreading in scale-free networks. Phys Rev Lett 86:3200
6. Wang Y, Chakrabarti D, Wang C, Faloutsos C (2003) Epidemic spreading in real networks: an eigenvalue viewpoint. In: Proceedings of the 22nd international symposium on reliable distributed systems, Florence, 2003. IEEE, pp 25–34
7. Van Mieghem P, Omic J, Kooij R (2009) Virus spread in networks. IEEE ACM Trans Network 17:1
8. Gómez S, Arenas A, Borge-Holthoefer J, Meloni S, Moreno Y (2010) Discrete-time Markov chain approach to contact-based disease spreading in complex networks. EPL (Europhys Lett) 89(3):38009
9. Dorogovtsev SN, Goltsev AV, Mendes JFF (2008) Critical phenomena in complex networks. Rev Mod Phys 80:1275
10. Chung F, Lu L, Vu V (2003) Spectra of random graphs with given expected degrees. Proc Natl Acad Sci USA 100:6313
11. Castellano C, Pastor-Satorras R (2010) Thresholds for epidemic spreading in networks. Phys Rev Lett 105:218701
12. Ferreira SC, Castellano C, Pastor-Satorras R (2012) Epidemic thresholds of the susceptible-infected-susceptible model on networks: a comparison of numerical and theoretical results. Phys Rev E 86:041125
13. de Oliveira MM, Dickman R (2005) How to simulate the quasistationary state. Phys Rev E 71:016129
14. Castellano C, Pastor-Satorras R (2012) Competing activation mechanisms in epidemics on networks. Nat Sci Rep 2:371
15. Goltsev AV, Dorogovtsev SN, Oliveira JG, Mendes JFF (2012) Localization and spreading of diseases in complex networks. Phys Rev Lett 109:128702
16. Lee HK, Shim PS, Noh JD (2013) Epidemic threshold of the susceptible-infected-susceptible model on complex networks. Phys Rev E 87:062812
17. Vojta T (2006) Rare region effects at classical, quantum and nonequilibrium phase transitions. J Phys A: Math Gen 39(22):R143

18. Boguñá M, Castellano C, Pastor-Satorras R (2013) Nature of the epidemic threshold for the susceptible-infected-susceptible dynamics in networks. Phys Rev Lett 111:068701
19. Hołyst JA, Sienkiewicz J, Fronczak A, Fronczak P, Suchecki K (2005) Universal scaling of distances in complex networks. Phys Rev E 72:026108
20. Catanzaro M, Boguñá M, Pastor-Satorras R (2005) Generation of uncorrelated random scale-free networks. Phys Rev E 71:027103

Part IV
Quantum Network Dynamics

Chapter 11
Physics on Graphs

Robert Schrader

Abstract This is an extended version of the talk given at the Nato Advanced Research Workshop: *New Challenges in Complex System Physics*, May 20–24, 2013 in Samarkand (Uzbekistan). We report on results on three topics in joint work with V. Kostrykin (Mainz, Germany) and J. Potthoff (Mannheim, Germany):

- Propagation of waves on graphs,
- Brownian motion on graphs,
- One particle quantum dynamics on graphs.

11.1 Basic Concepts

In this talk we will use the following notion of a *metric graph* $\mathscr{G}$. It is a finite collection of *half lines* = *external edges* and *finite intervals* = *internal edges* of given lengths with an identification of some of their endpoints (= *vertices*). The figure below shows a connected graph with 8 vertices, 7 external and 17 internal edges and 1 tadpole.

R. Schrader (✉)
Institut für Theoretische Physik, Freie Universität Berlin, Arnimallee 14, D-14195 Berlin, Germany
e-mail: robert.schrader@fu-berlin.de

D. Matrasulov and H.E. Stanley (eds.), *Nonlinear Phenomena in Complex Systems: From Nano to Macro Scale*, NATO Science for Peace and Security Series C: Environmental Security, DOI 10.1007/978-94-017-8704-8_11,

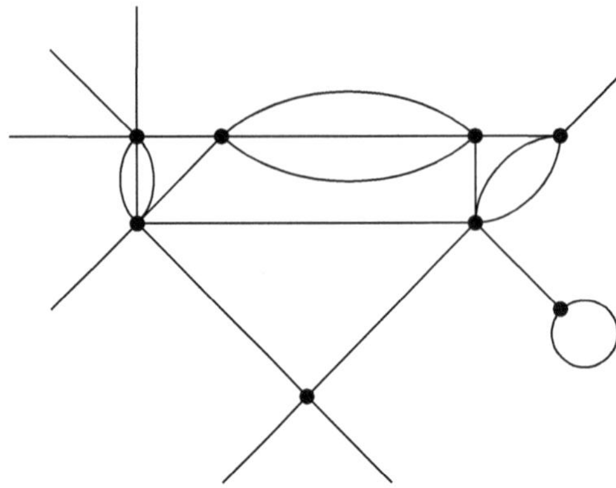

When the graph is connected, then it makes sense to speak of the distance between two points.

The main mathematical object appearing in this talk is the *Laplace operator*, which as a first definition is the second derivative of functions on the graph away from the vertices. At the vertices ambiguities arise.

Indeed, metric graphs are singular spaces, the singularities being just the vertices. And this makes such spaces interesting, as we will see both from the mathematical and the physical point of view. For the Laplace operator it turns out that things become interesting when certain boundary conditions at the vertices are introduced. As a consequence there are many Laplace operators, according to the particular choice of such boundary conditions. Here are two first examples:

The first graph is just the half-line $\mathbb{R}^+$ with a vertex at the origin.

The *Robin boundary condition* for a suitable square integrable function f on $\mathbb{R}^+$

$$\cos\phi \;\; f(0) + \sin\phi \;\; f'(0) = 0$$

is parametrized by ϕ.

The second graph is the real axis $\mathbb{R}$, again with a vertex at the origin. The boundary conditions for suitable functions g on $\mathbb{R}$

$$g(0_-) = g(0_+), \qquad g'(0_+) - g'(0_-) = \lambda g(0)$$

describe a so-called *delta-potential* of strength λ (real) at the origin.

A single vertex graph is given by an arbitrary finite set $\mathscr{E}$ of external edges, each $\cong [0, \infty)$. with one common vertex.

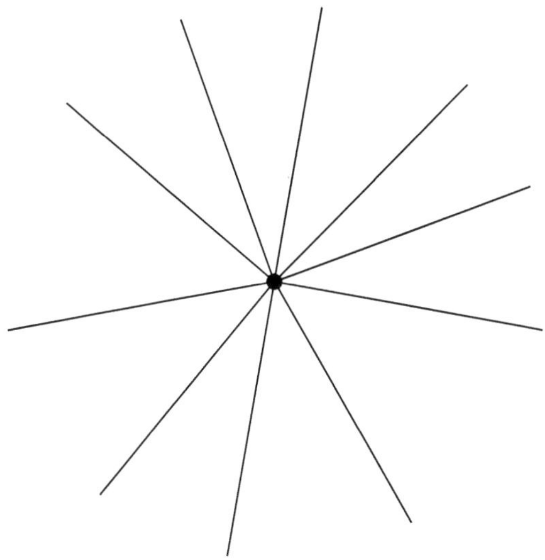

Boundary conditions, which define self-adjoint Laplace operators on the Hilbert space of all square integrable functions f on this single vertex graph are given as follows. Let f_e denote the restriction to the external edge e of a function f and set $\underline{f}(0) = \{f_e(0)\}_{e\in\mathscr{E}}, \underline{f}'(0) = \{f_e'(0)\}_{e\in\mathscr{E}}$, viewed as columns vectors of length $n = |\mathscr{E}|$, the number of edges. Let A and B be two complex $n \times n$ matrices, also labeled by the edges e, Provided the $n \times 2n$ matrix (A, B) has maximal rank $= n$, and provided $AB^{\dagger}$ is hermitian, then the boundary conditions

$$A\underline{f}(0) + B\underline{f}'(0) = 0$$

define a self-adjoint Laplace operator. The two examples given above are of this type. On any other graph boundary conditions of this form at each vertex also give rise to self-adjoint Laplace operators. Moreover, all self-adjoint Laplace operators are obtained in this way.

References: [1]. For additional references, see also the talk given by P. Exner at this conference as well as those given in the articles [1–12] (Kostrykin et al., 2008, Brownian motion on metric graphs: Feller Brownian motions on intervals revisited, unpublished, arXiv: 1008.3761; Kostrykin and Schrader, 2004, Generating functions of random walks on graphs, unpublished, arXiv: math.CO/0404467; Kostrykin and Schrader, 2006 The inverse scattering problem for metric graphs and the traveling salesman problem, with V. Kostrykin, unpublished, arXiv: math-ph/0603010).

11.2 Finite Propagation Speed for Solutions of the Wave Equation on Graphs

Light can be described as a solution of equations carrying the name of James Clerk Maxwell (1831–1879). These equations were published in 1864. The experimental verification was done by Heinrich Hertz (1857–1894) in the year 1886. Special solutions for the electric and magnetic fields in the vacuum, that is in the absence of charges and currents, are of the form

$$\mathbf{E}(\mathbf{x},t) = \mathbf{E}_0\ \mathrm{e}^{\mathrm{i}(\omega t + \mathbf{k}\cdot\mathbf{x})},$$
$$\mathbf{B}(\mathbf{x},t) = \mathbf{B}_0\ \mathrm{e}^{\mathrm{i}(\omega t + \mathbf{k}\cdot\mathbf{x})}.$$

The three vectors $\mathbf{E}_0$, $\mathbf{B}_0$ and $\mathbf{k}$ are pairwise orthogonal and $\mathbf{E}_0$ and $\mathbf{B}_0$ have equal length. Also $\mathbf{k}$ and ω are related by $\omega = \pm|\mathbf{k}|c$, where c is the speed of light. As a consequence of being solutions of the Maxwell equations they also are solutions of the wave equation

$$\Box\mathbf{E}(\mathbf{x},t) = 0,$$
$$\Box\mathbf{B}(\mathbf{x},t) = 0.$$

$\Box$ is the d'Alembert operator:

$$\Box = \frac{1}{c^2}\frac{\partial^2}{\partial t^2} - \Delta,$$

and Δ is the Laplace operator:

$$\Delta = \frac{\partial^2}{\partial x_1^2} + \frac{\partial^2}{\partial x_2^2} + \frac{\partial^2}{\partial x_3^2}.$$

An important consequence is the property of *finite propagation speed*: Any (light) signal sent out at time $t = 0$ at the point $\mathbf{x}_0$ can arrive at the point $\mathbf{x}$ at time $t > 0$ only when the distance between $\mathbf{x}_0$ and $\mathbf{x}$ is exactly ct (and not earlier). Instead of ordinary euclidean space $\mathbb{R}^3$ one may also consider a Riemannian manifold $\mathscr{M}$. In particular on $\mathscr{M}$ the notion of a distance between two points is well defined. On such a space there exists a canonical Laplace operator Δ, called the Laplace-Beltrami operator operating on functions on $\mathscr{M}$. So the d'Alembert operator is well defined. It operates on functions of $x \in \mathscr{M}$ and time t. Again one can prove finite propagation speed.

In the context of metric graphs one can find a class of boundary conditions, such that the associated Laplace operators Δ are non-positive, that is $-\Delta \leq 0$. The associated wave kernel

$$W(t) = \frac{\sin\sqrt{-\Delta}t}{\sqrt{-\Delta}}$$

allows one to solve the initial value problem: For given functions ψ_0 and $\dot{\psi}_0$ on the graph

$$\psi(t) = \partial_t W(t)\psi_0 + W(t)\dot{\psi}_0$$

satisfies $\psi(t = 0) = \psi_0$ and $\partial_t \psi(t = 0) = \dot{\psi}_0$ and is a solution of the wave equation

$$\Box\psi(t) = 0.$$

More importantly, finite propagation speed has the following formulation:

For $t \geq 0$ the support of $\psi(t)$ is contained in the set of points, that have distance at most ct from the support of ψ_0 and the support of $\dot{\psi}_0$.

We recall: The support of a function f is the smallest closed set outside of which the function vanishes identically.

References: [2, 3].

11.3 Brownian Motion on Graphs

In 1829 Robert Brown published an article in the Philosophical Magazine, where he described "rapid oscillatory motion" of pollen in water. This phenomenon now carries his name: *Brownian motion*. He made many observations on different materials, organic and inorganic.

The first to provide a full fledged theory explaining this phenomenon were Albert Einstein (1905) and Marian Smoluchowski (1906), though there were precursors like Louis Bachelier (1900) in his thesis "The theory of speculation".

Making some probabilistic assumptions Einstein derived the diffusion equation

$$\frac{\partial \rho(\mathbf{x}, t)}{\partial t} = D\Delta\rho(\mathbf{x}, t).$$

D is called the coefficient of diffusion. Thus if a Brownian particle starts (in three dimensional space) at time $t = 0$ at the point $\mathbf{y}$, that is $\rho(\mathbf{x}, t = 0) = \delta(\mathbf{x} - \mathbf{y})$, then

$$\rho(\mathbf{x}, t) = \frac{1}{(4\pi Dt)^{3/2}} \mathrm{e}^{-\frac{|\mathbf{x}-\mathbf{y}|^2}{4Dt}}$$

gives the probability of finding the particle at x at time t. For the choice $D = 1$ the right hand side is called the heat kernel, that is the integral kernel of $\exp t\Delta$. Starting from this, it is possible to construct what is now called a *Wiener process*, the mathematical description of a Brownian motion. Also in this context the Laplace operator is called the infinitesimal generator of the process.

Motivated by this we have found all *self-adjoint Laplace operators* Δ on any given graph, whose heat kernel $\exp t\Delta(p,q)$, with p and q being points on the graph, is positive. This is the essential ingredient, which allows one to construct a Wiener process on the graph having such a Laplace operator as its infinitesimal generator.

However, there exist additional processes on any given graph, whose infinitesimal generators are non-self-adjoint Laplace operators. They satisfy another set of boundary conditions, which, however, again are local: At any such vertex v of the graph the *Wentzell's boundary conditions* are of the form

$$f(v_e) = f(v) \quad \text{for all} \quad e \in \mathcal{E}$$

$$f''(v_e) = f''(v) \quad \text{for all} \quad e \in \mathcal{E}$$

$$af(v) - \sum_e b_e f'(v_e) + \frac{1}{2}c\, f''(v) = 0$$

with $a, b_e, c \geq 0$ and $a + \sum_e b_e + c = 1$.

f is a function on the graph, e labels an edge and $f'(v_e)$ denotes the outward derivative of f at v along the edge e. Also f is such all its second outward derivatives along each edge exist and are equal. They are written as $f''(v)$.

Interpretation of the parameters in these boundary conditions:

- $\beta = a/(1 - a - c)$ describes the probability on an exponential scale of the local time at the vertex that the Brownian particle will die, that is move to a cemetery, when it arrives at the vertex v.
- $\gamma = c/(1 - a - c)$ measures a certain stickiness of the vertex, given by a slow down of the Brownian motion near the vertex.
- $$0 \leq w_e = \frac{1}{1 - a - c} b_e,$$

  is the probability, that the Brownian particle, upon arrival at the vertex and unless it has died, will move from v continuously into the edge e.
- Any set of such boundary conditions at each vertex gives a so called *Feller process*.
- Conversely the infinitesimal generator of each Feller process on the graph is a Laplace operator with boundary conditions of this form.

We have constructed these processes explicitly, first by constructing them on star graphs and then by gluing them together.

Open problem (work in progress) :

The study of processes, where the Brownian particle can jump from a vertex <u>into</u> an edge or to another vertex, that is, processes subject to non-local boundary conditions.

References: [4–6] (Kostrykin et al., 2008, Brownian motion on metric graphs: Feller Brownian motions on intervals revisited, unpublished, arXiv: 1008.3761).

11.4 One Particle Quantum Dynamics on Graphs

In the year 1925 Erwin Schrödinger (1987–1961) wrote down the equation

$$i\hbar\frac{\partial}{\partial t}\psi(t) = H\psi(t),$$

which now carries his name. It is the time evolution equation for a quantum mechanical state $\psi(t)$. $\hbar$ $\psi(t)$ is an element of a complex Hilbert space and H is the Hamilton operator, a self-adjoint linear operator acting in this space. This is a very general feature:

A *dynamical quantum system* is given by the specification of the Hilbert space and of H.

An important example is the Hilbert space $L^2(\mathbb{R}^3)$ of all square integrable functions $\psi(\mathbf{x})$ on $\mathbb{R}^3$ and

$$H = -\frac{\hbar^2}{2m}\Delta + V(\mathbf{x}).$$

It provides the quantum mechanical description of a particle of mass m moving under the influence of a potential $V(\mathbf{x})$. For vanishing potential, one speaks of *free motion*.

Letting V be the Coulomb potential, Schrödinger was able to solve (with the help of Hermann Weyl) the *stationary Schrödinger equation*

$$H\psi = E\psi$$

for all *energy eigenvalues* of the hydrogen atom.

Model on the graph $\mathcal{G}$:

The Hilbert space is $L^2(\mathcal{G})$, the set of all square integrable functions on $\mathcal{G}$ (= wave packets on $\mathcal{G}$) and the Hamiltonian is

$$H = -\frac{\hbar^2}{2m}\Delta$$

where Δ is any self-adjoint Laplace operator. So this describes free motion of the quantum particle away from the vertices.

Consequence of self-adjointness:

For a given state $\psi(x)$ define the *quantum current*

$$j(x) = \frac{\mathrm{i}}{2m}\left(\overline{\psi(x)}\frac{d}{dx}\psi(x) - \left(\frac{d}{dx}\overline{\psi(x)}\right)\psi(x)\right)$$

Then at any vertex v the following quantum version of the *local Kirchhoff law* holds:

$$\sum_{e:v\in\partial(e)} j(v_e) = 0,$$

that is the sum of the currents at each "node" (=vertex) v vanishes.

So again the boundary conditions specify, how a wave-packet is split up, when it reaches a vertex.

One may discuss *scattering theory*: Imagine a wave-packet under the time evolution entering the graph through one of its external edges.

Question: What happens asymptotically for large times?

The answer is given in form of the so called on-shell S-matrix at energy E, in this case it is a matrix

$$S(E)_{e\,e'},$$

where e and e' label the external edges of the graph $\mathcal{G}$. Using linear algebra, it can be calculated explicitly in terms of the boundary conditions, which define the Laplacian.

This matrix is unitary:

$$\sum_e \overline{S(E)_{ee'}} S(E)_{ee''} = \delta_{e'e''}.$$

In particular when $e' = e''$: The total probability for a quantum particle of energy E entering the graph at any external edge e' and leaving through any edge e equals 1. In analogy to Kirchhoff's law for currents in electrical circuits, for this we coined the name

Kirchhoff's rule for quantum wires.

Given the on-shell scattering matrix for all energies, the graph and the Laplace operator (that is its boundary conditions) can be recovered. This also provides a quantum approach to solving the *traveling salesman problem.* The presence of magnetic fields can also be mimicked through so-called quantum fluxes.

References: [1, 7–12] (Kostrykin and Schrader, 2004, Generating functions of random walks on graphs, unpublished, arXiv: math.CO/0404467; Kostrykin and Schrader, 2006 The inverse scattering problem for metric graphs and the traveling salesman problem, with V. Kostrykin, unpublished, arXiv: math-ph/0603010).

References

1. Kostrykin V, Schrader R (1999) Kirchhoff's rule for quantum wires. J Phys A: Math Gen 32:595–630
2. Kostrykin V, Potthoff J, Schrader R (2012) Finite propagation speed for solutions of the wave equation on metric graphs. J Funct Anal 263:1198–1223
3. Schrader R (2009) Finite propagation speed and causal free quantum fields on networks. J Phys A: Math Gen 42:495401
4. Kostrykin V, Potthoff J, Schrader R (2012) Brownian motion on metric graphs. J Math Phys 53:095206. Special Issue in Honor of Elliott Lieb's 80th Birthday
5. Kostrykin V, Potthoff J, Schrader R (2012) Construction of the paths of Brownian motion on star graphs I. Commun Stoch Anal 6:223–245
6. Kostrykin V, Potthoff J, Schrader R (2012) Construction of the paths of Brownian motion on star graphs II. Commun Stoch Anal 6:247–261
7. Kostrykin V, Schrader R (2000) Kirchhoff's rule for quantum wires, II: the inverse problem with possible applications to quantum computers. Fortschr d Phys 48:703–716
8. Kostrykin V, Schrader R (2001) The generalized star product and the factorization of scattering matrices on graphs. J Math Phys 42:1563–1598
9. Kostrykin V, Schrader R (2003) Quantum wires with magnetic fluxes. Commun Math Phys 237:161–179
10. Kostrykin V, Schrader R (2006) Laplacians on metric graphs: eigenvalues, resolvents and semigroups. In: Berkolaiko G, Carlson R, Fulling SA, Kuchment P (eds) Quantum graphs and their applications (proceedings of joint summer research conference in the mathematical sciences on quantum graphs and their applications, Snowbird, 2005). Contemporary mathematics, vol 415. American Mathematical Society, Providence, pp 201–225
11. Kostrykin V, Potthoff J, Schrader R (2007) Heat kernels on metric graphs and a trace formula. In: Germinet F, Hislop PD (eds) Adventures in mathematical physics. Contemporary mathematics, vol 447. American Mathematical Society, Providence, pp 175–198
12. Kostrykin V, Potthoff J, Schrader R (2008) Contraction semigroups on metric graphs. In: Exner P, Keating J, Sunada T, Teplyaev A (eds) Analysis on graphs and its applications. Proceedings of symposia in pure mathematics, vol 77. American Mathematical Society, Providence, pp 423–458

Chapter 12
Resonances in Quantum Networks and Their Generalizations

Pavel Exner

Abstract We discuss resonances in quantum graphs and their more general analogs having 'edges' of different dimensions. Since the notion of resonance may mean different things, we show that the two most common definitions, scattering and resolvent resonances, are equivalent in this case. We analyze the high-energy behavior of resonances in quantum graphs and show that it may deviate from the standard Weyl law prediction; we derive a criterion which shows when such a thing happens. We also investigate influence of magnetic fields on graph resonances and show that they are field configurations which remove all 'true' resonances from such systems.

12.1 Introduction

Quantum graphs are in fact an old invention. The idea of a quantum particle the motion of which is confined to a graphs was first spelled out by Linus Pauling in the 1930s and two decades later worked out as a model of aromatic hydrocarbons [29]. After that, however, it was happily forgotten and rediscovered only in the second half of the 1980s when it found new use as a model of numerous experimentalist-made microstructures created from semiconductor or metal, later also from carbon nanotubes and other materials. Apart from its practical importance the concept appeared to be fruitful as a tool to understand properties of quantum systems with a complicated geometry and topology.

P. Exner (✉)
Doppler Institute for Mathematical Physics and Applied Mathematics, Czech Technical University, Břehová 7, 11519 Prague, Czech Republic

Nuclear Physics Institute, ASCR, 25068 Řež near Prague, Czech Republic
e-mail: exner@ujf.cas.cz

D. Matrasulov and H.E. Stanley (eds.), *Nonlinear Phenomena in Complex Systems: From Nano to Macro Scale*, NATO Science for Peace and Security Series C: Environmental Security, DOI 10.1007/978-94-017-8704-8_12,

One of the dominant attractive features of the model is in its simplicity. The motion on graph edges is described typically by a one-dimensional Schrödinger equation and the nontrivial part is the coupling of the wave function at the graph vertices. It is easy to derive how such couplings should look like, using the requirement of the probability current conservation, however, it is much more complicated to understand the meaning of these couplings. A natural idea is to start with 'fat graphs' which have only few free parameters and to look what we obtain if such a network shrinks to its graph 'skeleton'. The founding paper [29] offered a heuristic argument that such a limit would lead to the simplest coupling, usually dubbed *Kirchhoff*.

From the mathematical point of view the problem appeared to be rather difficult. After many efforts the heuristic argument was turned into a sound mathematical proof [28] which works for tubes with 'soft', or Neumann-type boundary, yielding indeed the Kirchhoff coupling. On the other hand, for tubes with 'hard', of Dirichlet boundary, the shrinking limit is completely different and its mechanism has been clarified only recently—see, e.g., [20]. Moreover, one asks whether by shrinking Neumann networks one can obtain other—or in fact all—coupling consistent with probability current conservation. An affirmative answer to this question is very fresh: it requires addition of local properly scaled potentials and magnetic fields as well as a local modification of the graph topology [16].

Since it rebirth 25 years ago the theory of quantum graphs attracted a lot of attention and a literature about the subject is vast indeed; we can refer the reader to the fresh monograph [3] for a rather exhaustive bibliography. The list of problems studied in this field is also very wide; let us mention just a few items:

- There is a spectral *duality* between this differential equation problem and a class of difference equations, originally by found by Alexander and de Gennes in the early 1980s, and mathematically established in [7, 12, 27]
- Very useful are *trace formulæ* expressing spectral properties a compact graph Hamiltonian in terms of closed orbits on the graph [4, 24]
- Attention has been paid to *inverse problems* characterized by the question 'Can one hear the shape of a graph?'—[21] and numerous subsequent papers
- Randomness on a graph can cause *Anderson localization* but some absolutely continuous spectrum may survive—see, e.g., [2, 19, 22]
- One can create gaps in a quantum graph spectrum *by decoration* [25, 30]

There are many more interesting results; for a survey we refer to the monograph [3].

The topic of this talk are *resonances* on quantum graphs and there is probably no need in explaining how important effect it is. Quantum graphs represent an environment where resonances are often observed; the reason is that they can be generated here in a purely geometric way [13]. We are going to address several questions, in the first place the meaning of a resonance and their behavior in the high-energy asymptotics, motivated by a recent surprising observation of Davies and Pushnitski [9]. Furthermore, we will investigate the influence of a magnetic field on graph resonances; we shall show that such a field can change the effective size of the graph.

In the second part of the talk we shall look into a generalization of the quantum graph model. We are going to examine a class of systems the configuration space of which is a complex, or a 'graph' whose edges may have different dimensions. We shall again show that the concept of a resonance is well defined in this context and examine the influence of a magnetic field, in particular, we will give an example of a systems in which a singular magnetic field destroys the resonances.

12.2 Resonances: Some General Observations

Quantum mechanical resonances come typically from perturbations of eigenvalues embedded in the continuum. Quantum graphs offer may examples; the reason is that for their Hamiltonians the *unique continuation principle* is often *not* valid and they can have compactly supported eigenfunction; this typically happens if zeros of an eigenfunction coincide with the vertices where external leads are attached. A caution is needed, however, since resonances are defined in different ways, in particular, as

- Poles of the analytically continued resolvent
- Singularities of the analytically continued on-shell S matrix

Consider a graph Γ consisting of families od vertices $\mathcal{V} = \{\mathcal{X}_j : j \in I\}$, finite edges $\mathcal{L} = \{\mathcal{L}_{jn} : (\mathcal{X}_j, \mathcal{X}_n) \in I_{\mathcal{L}} \subset I \times I\}$ and infinite ones $\mathcal{L}_\infty = \{\mathcal{L}_{j\infty} : \mathcal{X}_j \in I_{\mathcal{C}}\}$. The corresponding state Hilbert space is

$$\mathcal{H} = \bigoplus_{L_j \in \mathcal{L}} L^2([0, l_j]) \oplus \bigoplus_{\mathcal{L}_{j\infty} \in \mathcal{L}_\infty} L^2([0, \infty))\,;$$

its elements are columns $\Psi = (f_j : \mathcal{L}_j \in \mathcal{L}, g_j : \mathcal{L}_{j\infty} \in \mathcal{L}_\infty)^T$. The Hamiltonian acts as $-\frac{\mathrm{d}^2}{\mathrm{d}x^2}$ on each link being defined on H^2_{loc} functions satisfying the boundary conditions $(U - I)\Psi_n + i(U + I)\Psi' = 0$ at the n-th vertex, where Ψ_n, Ψ_n' are subcolumns of boundary values at the vertex and U_n is a unitary matrix. For more complicated graphs it might be cumbersome to write them; a useful trick is to replace Γ by the 'flower-like' graph putting all the vertices to a single point. Its degree is $2N + M$, where $N := \mathrm{card}\,\mathcal{L}$ and $M := \mathrm{card}\,\mathcal{L}_\infty$, and the coupling is described by a 'big', $(2N + M) \times (2N + M)$ unitary block diagonal matrix U,

$$(U - I)\Psi + i(U + I)\Psi' = 0\,; \tag{12.1}$$

the blocks U_j correspond to the vertices of Γ and in this way the structure of U encodes the original topology of the graph—cf. Fig. 12.1.

Poles of the analytically continued resolvent $(H - \lambda)^{-1}$ can be find by *exterior complex scaling* which turns H into a non-selfadjoint operator; resonances are its complex eigenvalues. Looking for them we leave the compact part of the graph

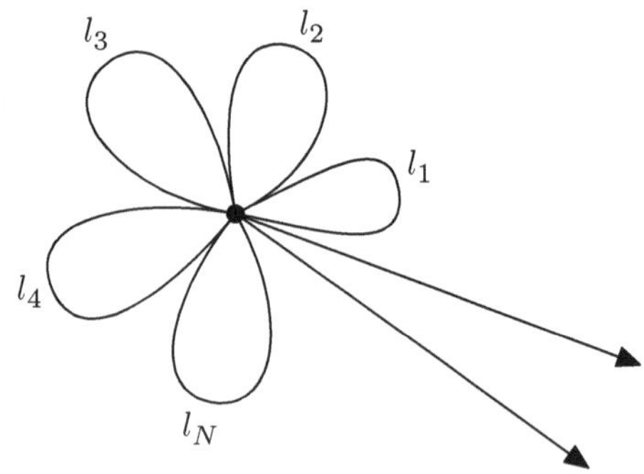

Fig. 12.1 A way to encode the graph topology into the coupling description

without change and set $f_j(x) = a_j \sin kx + b_j \cos kx$ on the j-th internal edge. On the other hand, functions on the external leads are scaled by $g_{j\theta}(x) = e^{\theta/2} g_j(xe^{\theta})$ with an imaginary θ rotating the essential spectrum into the lower complex halfplane; the 'exterior' boundary values at energy k^2 are thus to $g_j(0) = e^{-\theta/2} g_{j\theta}$ and $g'_j(0) = ike^{-\theta/2} g_{j\theta}$. Substituting them into (12.1) we get the condition

$$\det\left[(U - I)\, C_1(k) + ik(U + I)\, C_2(k)\right] = 0\,, \tag{12.2}$$

where $C_j := \operatorname{diag}\left(C_j^{(1)}(k), C_j^{(2)}(k), \ldots, C_j^{(N)}(k), i^{j-1} I_{M\times M}\right)$ with

$$C_1^{(j)}(k) = \begin{pmatrix} 0 & 1 \\ \sin kl_j & \cos kl_j \end{pmatrix}, \qquad C_2^{(j)}(k) = \begin{pmatrix} 1 & 0 \\ -\cos kl_j & \sin kl_j \end{pmatrix};$$

complex values of k that solve this condition indicate the resonance pole positions.

To deal with scattering resonances, on the other hand, we choose a combination of two planar waves, $g_j = c_j e^{-ikx} + d_j e^{ikx}$, as an Ansatz on the external edges; we ask about poles of the matrix $S = S(k)$ which maps the amplitudes of the incoming waves $c = \{c_n\}$ into amplitudes of the outgoing waves $d = \{d_n\}$ by the relation $d = Sc$. The boundary conditions then yield

$$(U - I)C_1(k) \begin{pmatrix} a_1 \\ b_1 \\ a_2 \\ \vdots \\ b_N \\ c_1 + d_1 \\ \vdots \\ c_M + d_M \end{pmatrix} + ik(U + I)C_2(k) \begin{pmatrix} a_1 \\ b_1 \\ a_2 \\ \vdots \\ b_N \\ d_1 - c_1 \\ \vdots \\ d_M - c_M \end{pmatrix} = 0$$

Since we are interested in zeros of $\det S^{-1}$, we regard the above relation as an equation for variables a_j, b_j and d_j while c_j are just parameters. Eliminating the variables a_j, b_j one derives from here a system of M equations expressing the map $S^{-1}d = c$. It is *not* solvable, $\det S^{-1} = 0$, provided

$$\det\left[(U - I)\, C_1(k) + ik(U + I)\, C_2(k)\right] = 0\,.$$

This is the same solvability condition as for the previous system of equations, hence we are able to make the following conclusion [13]:

Theorem 12.1. *The two above resonance notions, the resolvent and scattering one, are equivalent for quantum graphs.*

Before proceeding further let us show how the task can be reduced to investigation of the compact subgraph only by replacing the coupling at the vertex where external semi-infinite edges are attached by an effective one obtained by eliminating the external variables. We suppose that the matrix in (12.2) is of the form $U = \begin{pmatrix} U_1 & U_2 \\ U_3 & U_4 \end{pmatrix}$, where U_1 is the $2N \times 2N$ square matrix referring to the compact subgraph, U_4 is the $M \times M$ square matrix related to the exterior part, and U_2 and U_3 are rectangular matrices of the size $M \times 2N$ and $2N \times M$, respectively, connecting the two. If the matrix $[(1-k)U_4 - (k+1)]$ is regular, one obtains from here

$$(\tilde{U}(k) - I)(f_1, \dots, f_{2N})^{\mathrm{T}} + i(\tilde{U}(k) + I)(f_1', \dots, f_{2N}')^{\mathrm{T}} = 0\,,$$

where the effective coupling matrix

$$\tilde{U}(k) = U_1 - (1-k)U_2[(1-k)U_4 - (k+1)I]^{-1}U_3 \qquad (12.3)$$

is obviously energy-dependent and, in general, may not be unitary.

12.3 High-Energy Asymptotics

The following considerations are motivated by a serendipitous observation of Davies and Pushnitski [9] concerning the high-energy asymptotics of graph resonances. To describe it, we employ the *counting function* $N(R, F)$ defined as the number of zeros of $F(k)$ in the circle $\{k : |k| < R\}$ of given radius $R > 0$, algebraic multiplicities taken into account. If the function F comes from resonance secular equation we count in this way *number of resonances* within the given circle; let us stress that by resonances we mean here both the 'true' resonances and *embedded eigenvalues* corresponding to resolvent poles at the real axis.

Davies and Pushnitski noted that if the coupling is *Kirchhoff* and some external vertices are *balanced*, i.e. connecting the same number of internal and external edges, then the leading term in the asymptotics may be *smaller than the standard*

Weyl formula prediction. Let us look how the situation looks like for graphs in which Kirchhoff is replaced by a *more general vertex coupling*. To this aim we first rewrite the secular equation; denoting $e_j^{\pm} := \mathrm{e}^{\pm ikl_j}$ and $e^{\pm} := \Pi_{j=1}^N e_j^{\pm}$, one can cast it into the form

$$0 = \det\left\{\frac{1}{2}[(U-I)+k(U+I)]E_1(k)+\frac{1}{2}[(U-I)+k(U+I)]E_2+k(U+I)E_3\right.$$
$$\left.+(U-I)E_4+[(U-I)-k(U+I)]\,\mathrm{diag}\,(0,\dots,0,I_{M\times M})\right\},$$

where $E_i(k) = \mathrm{diag}\left(E_i^{(1)}, E_i^{(2)}, \dots, E_i^{(N)}, 0, \dots, 0\right)$, $i = 1,2,3,4$, consists of N nontrivial 2×2 blocks

$$E_1^{(j)} = \begin{pmatrix} 0 & 0 \\ -ie_j^+ & e_j^+ \end{pmatrix},\quad E_2^{(j)} = \begin{pmatrix} 0 & 0 \\ ie_j^- & e_j^- \end{pmatrix},\quad E_3^{(j)} = \begin{pmatrix} i & 0 \\ 0 & 0 \end{pmatrix},\quad E_4^{(j)} = \begin{pmatrix} 0 & 1 \\ 0 & 0 \end{pmatrix}$$

and the trivial $M \times M$ part.

Beauty of mathematics comes from the fact it is æternal, often you can take advantage of results obtained before any of the conference participants was born. In the present case, looking for zeros of the right-hand side we can employ a modification of a classical result on zeros of exponential sums [26]:

Theorem 12.2. *Let $F(k) = \sum_{r=0}^n a_r(k)\,\mathrm{e}^{ik\sigma_r}$, where $a_r(k)$ are rational functions of the complex variable k with complex coefficients, and the numbers $\sigma_r \in \mathbb{R}$ satisfy $\sigma_0 < \sigma_1 < \cdots < \sigma_n$. Suppose that $\lim_{k\to\infty} a_0(k) \neq 0$ and $\lim_{k\to\infty} a_n(k) \neq 0$. Then there are a compact $\Omega \subset \mathbb{C}$, real numbers m_r and positive K_r, $r = 1,\dots,n$, such that the zeros of $F(k)$ outside Ω lie in the logarithmic strips bounded by the curves $-\mathrm{Im}\,k + m_r \log|k| = \pm K_r$ and the counting function behaves in the limit $R \to \infty$ as*

$$N(R,F) = \frac{\sigma_n - \sigma_0}{\pi} R + \mathcal{O}(1)$$

Applying this result to our situation, where $\sigma_n = L$ and $\sigma_0 = -L$ it might seem we get the Weyl asymptotics, however, this happens only if the coefficients of $e^{\pm}$ in the resonance condition do not vanish. In order to find them, we pass to the effective boundary conditions described above,

$$0 = \det\left\{\frac{1}{2}[(\tilde U(k)-I)+k(\tilde U(k)+I)]\tilde E_1(k)\right.$$
$$+\frac{1}{2}[(\tilde U(k)-I)-k(\tilde U(k)+I)]\tilde E_2(k)+k(\tilde U(k)+I)\tilde E_3$$
$$\left.+(\tilde U(k)-I)\tilde E_4\right\},$$

where $\tilde{E}_j$ are the nontrivial $2N \times 2N$ parts of the matrices E_j and I denotes the $2N \times 2N$ unit matrix. By a direct computation we find the sought coefficient values.

Lemma 12.1. *The coefficients of senior and junior term, $e^{\pm}$ respectively, in the above equation are* $\left(\frac{i}{2}\right)^N \det[(\tilde{U}(k) - I) \pm k(\tilde{U}(k) + I)]$.

This leads us to the following conclusion [10]:

Theorem 12.3. *Consider a quantum graph (Γ, H_U) corresponding to Γ with finitely many edges and the coupling at vertices $\mathscr{X}_j$ given by unitary matrices U_j. The asymptotics of the resonance counting function as $R \to \infty$ is of the form*

$$N(R, F) = \frac{2W}{\pi} R + \mathscr{O}(1)\,,$$

where W is the effective size of the graph. One always has

$$0 \leq W \leq V := \sum_{j=1}^{N} l_j\,.$$

Moreover $W < V$—that is, graph is non-Weyl in the terminology of [9]—if and only if there exists a vertex where the corresponding energy dependent coupling matrix $\tilde{U}_j(k)$ has an eigenvalue $(1-k)/(1+k)$ *or* $(1+k)/(1-k)$.

Let us apply this result first to graphs with the coupling which is *invariant w.r.t. edge permutations*. Any such coupling in the j-th vertex is described a matrix of the form $U_j = a_j J + b_j I$, where the coefficients $a_j, b_j \in \mathbb{C}$ are such that $|b_j| = 1$ and $|b_j + a_j \deg \mathscr{X}_j| = 1$; matrix J has all the entries equal to one; note that both the δ and δ'_s [11] are particular cases of such a coupling. We need a couple of simple auxiliary statements:

Lemma 12.2. *(a) The matrix $U = aJ_{n\times n} + bI_{n\times n}$ has eigenvalue b of multiplicity $n-1$ and simple eigenvalue $na+b$. Its inverse is* $U^{-1} = -\frac{a}{b(an+b)} J_{n\times n} + \frac{1}{b} I_{n\times n}$.
(b) Let p internal and q external edges be coupled with boundary conditions given by $U = aJ_{(p+q)\times(p+q)} + bI_{(p+q)\times(p+q)}$. Then the energy-dependent effective matrix of the compact part is $\tilde{U}(k) = \frac{ab(1-k)-a(1+k)}{(aq+b)(1-k)-(k+1)} J_{p\times p} + bI_{p\times p}$.

Combining them with the above theorem we find easily that within the permutation-invariant class there are only two cases which exhibit non-Weyl asymptotics [10]:

Theorem 12.4. *Let (Γ, H_U) be a quantum graph with the coupling described by the matrices $U_j = a_j J + b_j I$. Then it has* non-Weyl asymptotics if and only if *at least one of its vertices is balanced, $p = q$, and the coupling at this vertex is either*

(a) $f_j = f_n, \quad \forall j,n \leq 2p, \quad \sum_{j=1}^{2p} f'_j = 0, \quad$ *i.e.* $U = \frac{1}{p} J_{2p\times 2p} - I_{2p\times 2p}$, *or*
(b) $f'_j = f'_n, \quad \forall j,n \leq 2p, \quad \sum_{j=1}^{2p} f_j = 0, \quad$ *i.e.* $U = -\frac{1}{p} J_{2p\times 2p} + I_{2p\times 2p}$.

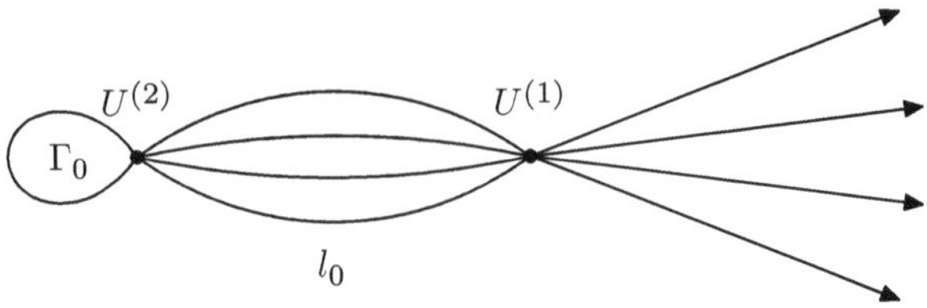

Fig. 12.2 A graph with a balanced vertex

In other words, there are only two possibilities of non-Weyl behavior in the permutation-invariant class, one of them being the one noticed in [9].

One asks naturally where the effect comes from. We are going to show that, roughly speaking, the (anti)Kirchhoff conditions at balanced vertices are 'too easy to decouple' which leads to diminishing the size of the graph. Consider graph sketched in Fig. 12.2 which has a balanced vertex $\mathcal{X}_1$ which connects p internal edges, supposed to be of the same length l_0, and p external edges with the coupling given by a unitary $U^{(1)} = aJ_{2p\times 2p} + bI_{2p\times 2p}$. The coupling to the rest of the graph, denoted as Γ_0, is described by a $q\times q$ matrix $U^{(2)}$, where $q \geq p$ holds necessarily; needless to say such a matrix can hide different topologies of this part of the graph. Note that the assumption about the same edge length is made here for convenience only; we can always satisfy it by adding 'dummy' Kirchhoff vertices.

Lemma 12.3. *Consider Γ be the coupling given by arbitrary $U^{(1)}$ and $U^{(2)}$. Let V be any unitary $p\times p$ matrix, $V^{(1)} := \mathrm{diag}\,(V, V)$ and $V^{(2)} := \mathrm{diag}\,(I_{(q-p)\times(q-p)}, V)$ be $2p\times 2p$ and $q\times q$ block diagonal matrices, respectively. Then H on Γ is* unitarily equivalent *to the Hamiltonian H_V on topologically the same graph with the coupling given by the matrices $[V^{(1)}]^{-1}U^{(1)}V^{(1)}$ and $[V^{(2)}]^{-1}U^{(2)}V^{(2)}$.*

In application to our system we choose $U^{(1)} = aJ_{2p\times 2p} + bI_{2p\times 2p}$ at $\mathcal{X}_1$, while the columns of W will be an orthonormal set of eigenvectors of the $p\times p$ block $aJ_{p\times p} + bI_{p\times p}$, the first one being $\frac{1}{\sqrt{p}}\,(1, 1, \dots, 1)^{\mathrm{T}}$. The transformed matrix $[V^{(1)}]^{-1}U^{(1)}V^{(1)}$ decouples into blocks connecting only pairs (v_j, g_j). The first one corresponding to a *symmetrization* of all the u_j's and f_j's, leads to the 2×2 matrix $U_{2\times 2} = apJ_{2\times 2} + bI_{2\times 2}$, while the other lead to *separation of the corresponding internal and external edges* described by the Robin conditions, $(b-1)v_j(0) + i(b+1)v_j'(0) = 0$ and $(b-1)g_j(0) + i(b+1)g_j'(0) = 0$ for $j = 2, \dots, p$. The 'overall' Kirchhoff/anti-Kirchhoff condition at $\mathcal{X}_1$ is transformed to the 'line' Kirchhoff/anti-Kirchhoff condition in the subspace of permutation-symmetric functions, *reducing the graph size by* l_0. In all the other cases the point interaction corresponding to the matrix $apJ_{2\times 2} + bI_{2\times 2}$ is nontrivial, and consequently, *the graph size is preserved.*

Remark 12.1. Before proceeding further, let us note that the situation is different if we leave the class of permutation symmetric couplings. Using again the trick with unitary equivalence of the involved matrices, one show that in the general case the graph can have a non-Weyl asymptotics even if none of its vertices is balanced; a simple example of Γ consisting of a line with a segment attached is analyzed in [10].

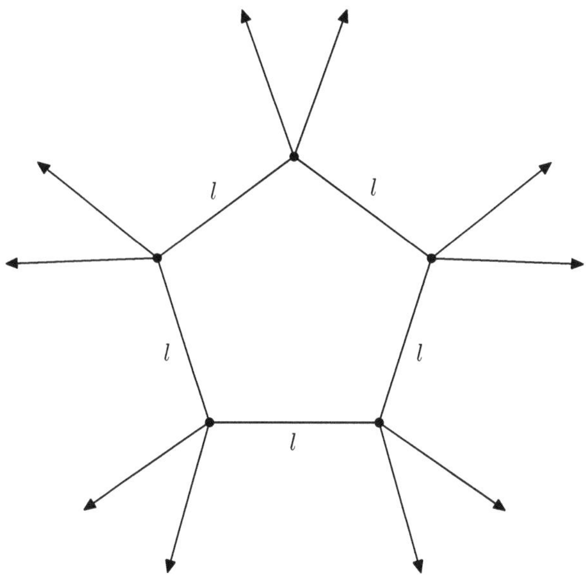

Fig. 12.3 A graph with a discrete rotational symmetry

One might have the feeling that the effect is related to the coupling in each vertex separately, but we are going to show that this is not the case—*the effective size is a global property*. To this aim, let us analyze the following simple example with Γ of the form sketched above. We fix $n \geq 3$ and consider a regular n-gon, each edge having length ℓ, and attach two semi-infinite leads to each vertex, assuming Kirchhoff coupling there. Hence each vertex is balanced and we know that the *effective size* W_n *is strictly less than* $V_n = n\ell$. We have, however, the following result [10] (Fig. 12.3).

Proposition 12.1. *The effective size of the graph* Γ_n *is given by*

$$W_n = \begin{cases} n\ell/2 & \textit{if } n \neq 0 \mod 4, \\ (n-2)\ell/2 & \textit{if } n = 0 \mod 4. \end{cases}$$

Let us *sketch the proof.* We employ Bloch-Floquet decomposition of H w.r.t. the cyclic rotation group $\mathbb{Z}_n$. It leads to analysis of one segment with 'quasimomentum' ω satisfying $\omega^n = 1$; after a short computation we find that H_ω has a resonance *iff*

$$-2(\omega^2 + 1) + 4\omega e^{-ik\ell} = 0.$$

Hence the effective size W_ω of the system of resonances of H_ω is $\ell/2$ if $\omega^2 + 1 \neq 0$ but it is zero if $\omega^2 + 1 = 0$. Now $\omega^2 + 1 = 0$ is not soluble if $\omega^n = 1$ and $n \neq 0$ mod 4, but it has two solutions if $n = 0$ mod 4. □

Let us next look what happens if we *add a magnetic field*. In that case the Hamiltonian acts as $-\mathrm{d}^2/\mathrm{d}x^2$ at the infinite leads and as $-(\mathrm{d}/\mathrm{d}x + iA_j(x))^2$ at the internal edges, where A_j is the tangent component of the vector potential. Its domain consists of functions in $W^{2,2}(\Gamma)$ satisfying

$$(U_j - I)\Psi_j + i(U_j + I)(\Psi_j' + iA_j\Psi_j) = 0\,.$$

Using the local gauge transformation $\psi_j(x) \mapsto \psi_j(x)\mathrm{e}^{-i\chi_j(x)}$ with $\chi_j(x)' = A_j(x)$ one gets unitary equivalence to *free Hamiltonian* with the coupling

$$(U_A - I)\Psi + i(U_A + I)\Psi' = 0\,, \quad U_A := \mathcal{F}U\mathcal{F}^{-1}\,,$$

where $\mathcal{F} = \mathrm{diag}\,(1, \exp{(i\Phi_1)}, \ldots, 1, \exp{(i\Phi_N)}, 1, \ldots, 1)$ containing magnetic fluxes $\Phi_j = \int_0^{l_j} A_j(x)\,\mathrm{d}x$. Using the effective coupling matrix $\tilde{U}(k)$ as in the non-magnetic case [14] we infer that

Theorem 12.5. *Let Γ be a quantum graph with N internal and M external edges and coupling given by a $(2N + M) \times (2N + M)$ unitary matrix U. Let Γ_V be obtained from Γ by replacing U by $V^{-1}UV$ where $\left(\begin{smallmatrix} V_1 & 0 \\ 0 & V_2 \end{smallmatrix}\right)$ is unitary block-diagonal matrix consisting of a $2N \times 2N$ block V_1 and an $M \times M$ block V_2. Then Γ_V has a non-Weyl resonance asymptotics* iff *Γ does.*

Corollary 12.1. *Let Γ be a quantum graph with Weyl resonance asymptotics. Then Γ_A has also the Weyl asymptotics for any profile of the magnetic field.*

The magnetic field *can* change, though, the *effective size of the graph.* To illustrate this claim, consider the example of the graph sketched below, assuming again the Kirchhoff coupling at the vertex. Since the latter is balanced, the graph is non-Weyl for $A = 0$, and thus for any A. The resonance condition is easily found to be

$$-2\cos\Phi + \mathrm{e}^{-ikl} = 0\,,$$

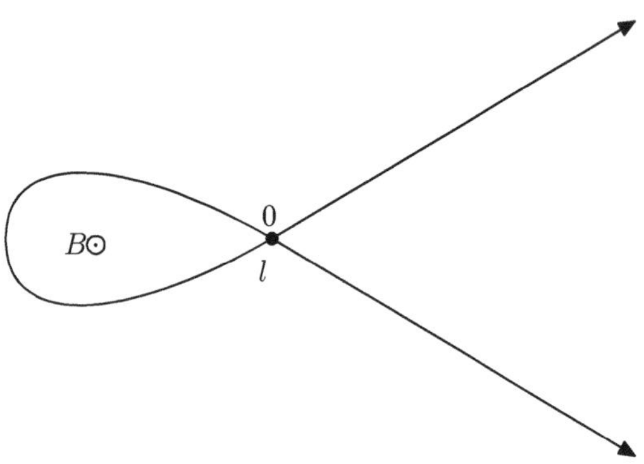

Fig. 12.4 A magnetic loop with two leads

where $\Phi = Al$ is the loop flux. The non-Weyl character is associated with the absence of the factor e^{ikl} at the left-hand side. For $\Phi = \pm\pi/2 \pmod{\pi}$, i.e. for odd multiples of a quarter of the flux quantum 2π, the l-independent term also disappears. The effective size of the graph is then zero; it is straightforward to see that in such a case there are *no resonances at all* (Fig. 12.4).

12.4 Generalized Graphs

Let us turn next to a generalization of quantum graphs to complexes with 'edges' of different dimensions. To be illustrative we shall use for configuration space of such system the term *hedgehog manifold*. An example is sketched here; for the sake of simplicity we consider the case with a single (two-dimensional) manifold part only (Fig. 12.5).

First we have to describe the formalism. The basic idea how to couple manifolds of different dimensions is old [17] and it was applied to the problem considered here, e.g., in [5, 6, 18]. Comparing to those paper we use here a slightly general approach allowing more than one lead to be attached to a given manifold point. Let thus Ω be a compact Riemannian manifold of dimension $d = 2, 3$ with metric g from which we make Γ by attaching $M = \sum_j n_j$ halflines at points $x_j \in \Omega$. Consequently, the state Hilbert space is

$$\mathscr{H} = L^2(\Omega, \mathrm{d}g) \bigoplus_{i=1}^{M} L^2(\mathbb{R}_+^{(i)}) .$$

In the manifold part, let H_0 be Laplace-Beltrami operator acting on $C_0^\infty(\Omega)$ as $-g^{-1/2}\partial_r(g^{1/2}\partial_r)$ with suitable boundary conditions if Ω has a boundary; by H_0' we denote it restriction to functions $\{f \in C_0^\infty : f(x_j) = 0\}$ which is a symmetric operator with deficiency indices (n, n).

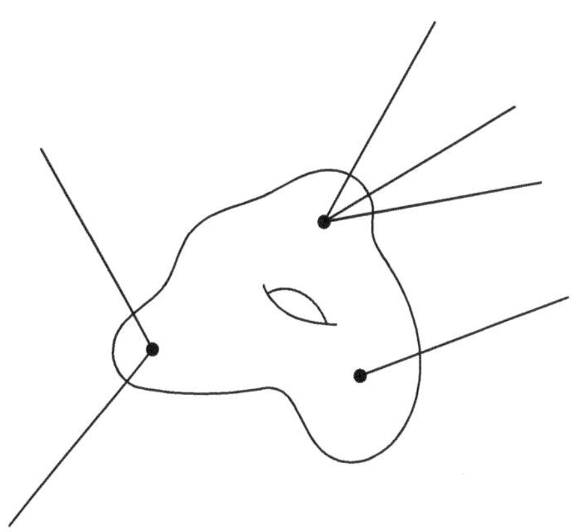

Fig. 12.5 A hedgehog manifold

Let further H_i' be restriction of the Laplacian on i-th halfline to functions from $C_0^\infty(\mathbb{R}_+^{(i)})$. The operator

$$H' = H_0' \oplus H_1' \oplus \cdots \oplus H_M'$$

has obviously deficiency indices $(n+M, n+M)$. To construct *self-adjoint extensions* of the operator H' we need (generalized) boundary values. A function $f \in D((H_0')^*)$ can be expanded near x_j as $f(x) = c_j(f)F_0(x,x_j) + d_j(f) + \mathcal{O}(r(x,x_j))$, where

$$F_0(x,x_j) = \begin{cases} -\frac{q_2(x,x_j)}{2\pi} \ln r(x,x_j) & d=2 \\ \frac{q_3(x,x_j)}{4\pi}(r(x,x_j))^{-1} & d=3 \end{cases}$$

Here q_2, q_3 are continuous functions of x with $q_i(x_j,x_j) = 1$ and $r(x,x_j)$ denotes the geodesic distance between x and x_j. Using the described expansion we write the boundary values as

$$\begin{aligned} \Psi &= (d_1(f), \dots, d_n(f), f_1(0), \dots, f_M(0))^{\mathrm{T}}, \\ \Psi' &= (c_1(f), \dots, c_n(f), f_1'(0), \dots, f_M'(0))^{\mathrm{T}}, \end{aligned}$$

and describe any self-adjoint extension of H' by the conditions

$$(U-I)\Psi + i(U+I)\Psi' = 0,$$

where U is an $(n+M)\times(n+M)$ unitary matrix. This covers all the self-adjoint extensions of H' including those allowing ‘hopping’ between vertices. We are interested, however, in *local* ones only; they are characterized by *block-diagonal* matrices U which does not couple different points x_i and x_j.

As in the ordinary quantum-graph case we can investigate resonances on hedgehog manifolds replacing external leads by an effective energy-dependent coupling at the points $x_j \in \Omega$ in the following way,

$$(\tilde{U}_j(k) - 1)d_j(f) + i(\tilde{U}_j(k)+1)c_j(f) = 0,$$

where $\tilde{U}_j(k) \in \mathbb{C}$ is easily seen to be given by

$$\tilde{U}_j(k) = U_{1j} - (1-k)U_{2j}[(1-k)U_{4j} - (k+1)I]^{-1}U_{3j} \tag{12.4}$$

and U_{1j} denotes the top-left entry of U_j, U_{2j} the rest of the first row, U_{3j} the rest of the first column and U_{4j} is $n_j \times n_j$ part corresponding to the coupling between the halflines attached to the manifold. In a sense we have replaced again the leads by *k-dependent point interactions* on the manifold Ω itself, in analogy with (12.3).

The first question to address concerns again the meaning of resonances. From the scattering point of view, it is natural to regard Ω as a geometric scatterer. To find the on-shell scattering matrix we take for solution of Schrödinger equation of energy k^2 on the j-th lead the Ansatz $a_j(k)e^{-ikx} + b_j(k)e^{ikx}$ and look for the matrix that relates the coefficient columns $\mathbf{a}(k)$ and $\mathbf{b}(k)$. If we want to compare the case of two leads with scattering on the line, there is a caveat: in our formalism all the leads are *positive* halflines, our S-matrix satisfies $S(k)^{-1} = S(-k) = S^*(\bar{k})$, where star and bar denote the Hermitian and complex conjugation, respectively.

By *scattering resonance* we mean a pole of the scattering matrix, more precisely, the (complex) energy at which some of its entries has a pole. On the other hand, by *resolvent resonance* we mean a pole in *analytical continuation* of $(H - k^2)^{-1}$; to determine it one can employ as before a *exterior complex scaling* on the edges turning resonances into eigenvalues of the *non-selfadjoint operator* $H_\theta := U_\theta H U_\theta^{-1}$. We need an auxiliary result which can be obtained, e.g., by the method used in [23].

Lemma 12.4. *Let* $H\big|_\Omega f(x,k) = kf(x,k)$ *hold for* $k^2 \notin \sigma(H_0)$*, then* f *is as a linear combination of Green's functions of* H_Ω*,*

$$f(x,k) = \sum_{j=1}^{n} c_j G(x, x_j; k) .$$

This allows us to establish again equivalence of the two resonance notions [15].

Theorem 12.6. *Consider the open lower complex halfplane of momentum,* $\mathrm{Im}\, k < 0$ *and* $k^2 \notin \mathbb{R}$*. There is a scattering resonance in* k_0 *iff there is a resolvent resonance in* k_0*, and the algebraic multiplicities of resonances defined in both ways coincide.*

Let us *sketch the proof.* Using the above lemma in combination with the Ansatz $a_j(k)e^{-ikx} + b_j(k)e^{ikx}$ on the leads one arrives at the condition

$$A(k_0)\mathbf{a} + B(k_0)\mathbf{b} + C(k_0)\mathbf{c} = 0 ,$$

where A, B are $(P+M)\times M$ matrices, C is $(P+M)\times P$ matrix, P is the number of internal parameters of the geometric scatterer and M is the number of halflines. Since $k_0^2 \notin \mathbb{R}$ the columns of $C(k_0)$ are linearly independent, otherwise k_0 would have to be an eigenvalue. A rearrangement allows us to express $\mathbf{c}$; substituting it to the remaining conditions we get

$$\tilde{A}(k_0)\mathbf{a} + \tilde{B}(k_0)\mathbf{b} = 0$$

with $\tilde{A}(k_0)$ and $\tilde{B}(k_0)$ being $M \times M$ matrices. If $\det \tilde{A}(k_0) = 0$ holds than there is a solution with $\mathbf{b} = 0$ and k_0 is an eigenvalue of H with $\mathrm{Im}\, k_0 < 0$, however, this contradicts the assumed self-adjointness of H. Hence by Cramer's rule the scattering resonances are given by the condition $\det \tilde{B}(k) = 0$.

The solution $a_j(k)\mathrm{e}^{-ikx}$ on the j-th halfline is taken by U_θ into an exponentially growing one if $\mathrm{Im}\, k_0 < 0$ and $\mathrm{Im}\,\theta > 0$, while $b_j(k)\mathrm{e}^{ikx}$ becomes square integrable. Hence solving the eigenvalue problem for $\mathbf{H}_\theta$ one needs to find solutions with $\mathbf{a} = 0$. This leads again to the condition $\det \tilde{B}(k) = 0$ proving thus the claim. □

After clarifying the meaning of resonances we can proceed to analyzing their behavior. The above lemma allows us to express the boundary values using expansion of deficiency subspace elements,

$$f(x_i,k) = c_i F_0(x_i,x_i) + \sum_{j\neq i}^{n} c_j F_0(x_i,x_j) + \sum_{j=1}^{n} c_j F_1(x_i,x_j;k) + \sum_{j=1}^{n} c_j R(x_i,x_j;k)\,,$$

where $R(x_i,x_j;k)$ is $\mathcal{O}(r(x_i,x_j))$. Let $\tilde{U}(k) = \mathrm{diag}\,(\tilde{U}_1(k),\dots,\tilde{U}_n(k),)$; the definition (12.4) of $\tilde{U}_j(k)$ shows that $\tilde{U}(k)$ diverges at most M values of k. We define

$$Q_0(k) = \begin{cases} G(x_i,x_j;k) & i \neq j \\ F_1(x_i,x_i;k) & i = j \end{cases}$$

The resonance condition then reads as follows,

$$\det\left[(\tilde{U}(k)-I)Q_0(k) + i(\tilde{U}(k)+I)\right] = 0\,,$$

reducing again a complicated differential equation problem to an algebraic one. We could use it to analyze, for instance, the high-energy behavior of resonances. It is very interesting, though, because the manifold part will dominate the asymptotics in view of it higher dimension and no 'surprises' analogous to Theorem 12.3 are expected.

Instead we are going to present an example illustrating a behavior similar to that of magnetic quantum graphs mentioned at the end of the previous section: we shall demonstrate that an appropriately chosen magnetic field can remove all 'true resonances' on a hedgehog manifold, i.e. those corresponding to poles in the open lower complex halfplane. If the reader wonders whether this does not contradict to the previous paragraph, we recall that embedded eigenvalues have been included among resonances in Theorem 12.3; in the present case the embedded eigenvalues of the system corresponding to higher partial waves with eigenfunctions vanishing at the junction will persist being just shifted. The manifold to be considered consists of a disc of radius R with a halfline lead attached at its centre, for definiteness we assume that it to be perpendicular to the disc plane, as sketched below. In the disc we use the polar coordinates r, φ, and Dirichlet boundary conditions are imposed at the boundary, $r = R$. We suppose that the system is under the influence of a

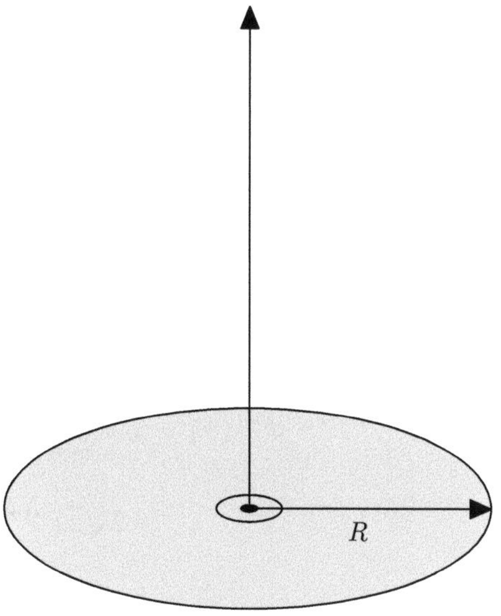

Fig. 12.6 A disc with a lead in the field of an AB flux

magnetic field in the form of an Aharonov-Bohm string which coincides in the 'upper' halfspace with the lead. The effect of an Aharonov-Bohm field piercing a surface was studied in numerous papers—see, e.g., [1, 8]—the idea is that the 'true' resonances will disappear if we manage to choose such a coupling in which the radial part of the disc wave function will match the halfline wave function in a trivial way (Fig. 12.6).

The Hilbert space of the model is $\mathscr{H} = L^2((0,R), r dr) \otimes L^2(S^1) \oplus L^2(\mathbb{R}^+)$ and admissible Hamiltonians are self-adjoint extensions of the operator $\dot{H}_\alpha$ acting as

$$\dot{H}_\alpha \begin{pmatrix} u \\ f \end{pmatrix} = \begin{pmatrix} -\frac{\partial^2 u}{\partial r^2} - \frac{1}{r}\frac{\partial u}{\partial r} + \frac{1}{r^2}\left(i\frac{\partial}{\partial \varphi} - \alpha\right)^2 u \\ -f'' \end{pmatrix}$$

on function columns $\binom{u}{f}$ with $u \in H^2_{\text{loc}}(B_R(0))$ satisfying $u(0,\varphi) = u(R,\varphi) = 0$ and $f \in H^2_{\text{loc}}(\mathbb{R}^+)$ satisfying $f(0) = f'(0) = 0$. The parameter α denotes the magnetic flux of the Aharonov-Bohm string measured in the units of the flux quantum; since the entire part of the flux can be easily removed by a gauge transformation we may restrict our attention to the values $\alpha \in (0,1)$.

Using the partial-wave decomposition together with the standard unitary transformation $(Vu)(r) = r^{1/2}u(r)$ describing the pass to the reduced radial functions we get

$$\dot{H}_\alpha = \bigoplus_{m=-\infty}^{\infty} V^{-1}\dot{h}_{\alpha,m} V \otimes I$$

where the component $\dot{h}_{\alpha,m}$ act on the upper component of $\psi = \binom{\phi}{f}$ as

$$\dot{h}_{\alpha,m}\phi = -\frac{\mathrm{d}^2\phi}{\mathrm{d}r^2} + \frac{(m+\alpha)^2 - 1/4}{r^2}\phi\,. \tag{12.5}$$

To construct the self-adjoint extensions of $\dot{H}_\alpha$ which describe the coupling between the disc and the lead the following functionals are needed,

$$\Phi_1^{-1}(\psi) = \sqrt{\pi}\lim_{r\to 0}\frac{r^{1-\alpha}}{2\pi}\int_0^{2\pi} u(r,\varphi)\mathrm{e}^{i\varphi}\mathrm{d}\varphi\,,$$

$$\Phi_2^{-1}(\psi) = \sqrt{\pi}\lim_{r\to 0}\frac{r^{-1+\alpha}}{2\pi}\Bigg[\int_0^{2\pi} u(r,\varphi)\mathrm{e}^{i\varphi}\mathrm{d}\varphi$$
$$-2\sqrt{\pi}r^{-1+\alpha}\Phi_{-1}^1(\psi)\Bigg],$$

$$\Phi_1^0(\psi) = \sqrt{\pi}\lim_{r\to 0}\frac{r^{\alpha}}{2\pi}\int_0^{2\pi} u(r,\varphi)\mathrm{d}\varphi\,,$$

$$\Phi_2^0(\psi) = \sqrt{\pi}\lim_{r\to 0}\frac{r^{-\alpha}}{2\pi}\Bigg[\int_0^{2\pi} u(r,\varphi)\mathrm{d}\varphi$$
$$-2\sqrt{\pi}r^{-\alpha}\Phi_1^0(\psi)\Bigg],$$

$$\Phi_1^{\mathrm{h}}(\psi) = f(0)\,,\quad \Phi_2^{\mathrm{h}}(\psi) = f'(0)\,.$$

The first two of them are, in analogy with [8], multiples of the coefficients of the two leading terms of asymptotics as $r \to 0$ of the wave functions from $\dot{H}_\alpha^*$ belonging to the subspace with $m = -1$, the second two correspond to the analogous quantities in the subspace with $m = 0$, and finally, the last two are the standard boundary values for the Laplacian on a halfline.

It is obvious that if the s-wave resonances should be absent, one has to get rid of the second term in the expression (12.5) for the $m = 0$ function, with this fact in mind we focus our attention on the case $\alpha = 1/2$. In analogy with the Aharonov-Bohm flux piercing a plane treated in [8] one obtains

$$(\psi_1, H\psi_2) = -\int_0^{2\pi}\int_0^R \overline{u_1}\, r^{-1/2}\frac{\mathrm{d}^2}{\mathrm{d}r^2}r^{1/2}u_2\, r\,\mathrm{d}r\,\mathrm{d}\varphi - \int_0^\infty \overline{f_1}f_2''\,\mathrm{d}x$$
$$= -\int_0^{2\pi}\int_0^R \overline{\tilde{u}_1}\tilde{u}_2''\,\mathrm{d}r\,\mathrm{d}\varphi - \int_0^\infty \overline{f_1}f_2''\,\mathrm{d}x$$
$$= -\int_0^{2\pi} \overline{\tilde{u}_1}\tilde{u}_2'\,\mathrm{d}\varphi + \int_0^{2\pi}\int_0^R \overline{\tilde{u}_1}'\tilde{u}_2'\,\mathrm{d}r\,\mathrm{d}\varphi - \overline{f_1}(0+)f_2'(0+)$$
$$+\int_0^\infty \overline{f_1}'f_2'\,\mathrm{d}x\,,$$

where $\tilde{u}_\iota = r^{1/2}u_\iota$, $\iota = 1, 2$, corresponds of the disc function component u_ι, prime denotes the derivative with respect to r, and f_ι is the corresponding halfline component. This allows us to express $(\psi_1, H\psi_2) - (H\psi_1, \psi_2)$, and using asymptotic expansion of u near $r = 0$, namely

$$\begin{aligned}\sqrt{\pi}u(r,\theta) &= (\Phi_1^{-1}(\psi)r^{-1/2} + \Phi_2^{-1}(\psi)r^{1/2})\mathrm{e}^{-i\theta} \\ &\quad + \Phi_1^0(\psi)r^{-1/2} + \Phi_2^0(\psi)r^{1/2}, \\ -2r\sqrt{\pi}u'(r,\theta) &= (\Phi_1^{-1}(\psi)r^{-1/2} - \Phi_2^{-1}(\psi)r^{1/2})\mathrm{e}^{-i\theta} \\ &\quad + \Phi_1^0(\psi)r^{-1/2} - \Phi_2^0(\psi)r^{1/2},\end{aligned}$$

one is able to write the boundary form as

$$(\psi_1, H\psi_2) - (H\psi_1, \psi_2) = \Phi_1(\psi_1)^*\Phi_2(\psi_2) - \Phi_1(\psi_2)^*\Phi_2(\psi_1),$$

where $\Phi_\iota(\psi) = (\Phi_\iota^{\mathrm{h}}, \Phi_\iota^0, \Phi_\iota^{-1})^{\mathrm{T}}$ for $\iota = 1, 2$. Consequently, to get a self-adjoint operator one has to impose coupling conditions similar to (12.1), namely

$$(U - I)\Phi_1(\psi) + i(U + I)\Phi_2(\psi) = 0 \tag{12.6}$$

with a unitary matrix U. For the purpose of our example, we choose the latter in the form

$$U = \begin{pmatrix} 0 & 1 & 0 \\ 1 & 0 & 0 \\ 0 & 0 & \mathrm{e}^{i\rho} \end{pmatrix}, \tag{12.7}$$

in other words, the nonradial part, $m = -1$, of the disc wave function is coupled to neither one of the other two, while the radial part, $m = 0$, is coupled to the halfline representing the lead by means of Kirchhoff conditions. To see that this choice removes indeed all the 'true' resonances, we choose the Ansatz

$$f(x) = a\ \sin kx + b\ \cos kx\,, \quad u(r) = r^{-1/2}(c\ \sin k(R - r))$$

which yields the boundary values

$$\Phi_1(\psi) = (b, c\sqrt{\pi}\sin kR, 0)^{\mathrm{T}}, \quad \Phi_2(\psi) = k(a, -c\sqrt{\pi}\cos kR, 0)^{\mathrm{T}}.$$

It follows now from the coupling conditions that $b = c\sqrt{\pi}\sin kR$ and $a = c\sqrt{\pi}\cos kR$ which yields $f(x) = c\sqrt{\pi}\sin k(R + x)$. Hence for any $k \notin \mathbb{R}$ and $c \neq 0$ the function f contains necessarily a nontrivial part of the wave e^{-ikx}, however, as we have argued above, a resolvent resonance can must have the asymptotics e^{ikx} only. In this way we come to the indicated conclusion [15]:

Proposition 12.2. *The disc-and-lead system has no true resonances for the coupling described by the matrix (12.7) and the magnetic flux equal to* $\alpha = \frac{1}{2}$.

Since the effect occurs for a Aharonov-Bohm string, it is also interesting to inquire what happens if the field changes, in other words, if α runs from 0 to $\frac{1}{2}$. The symmetry of the problem allows us to use the Ansatz $u(r,\varphi) = R(r)\,\mathrm{e}^{im\varphi}$; it shows that one has to solve the equation

$$-\frac{\partial^2 R(r)}{\partial r^2} - \frac{1}{r}\frac{\partial R(r)}{\partial r} + \frac{1}{r^2}(m+\alpha)^2 R(r) = k^2 R(r),$$

which can be easily transformed into Bessel equation in the variable kr with the constant $(m+\alpha)$. Hence the radial part of the wavefunction on the disc is given as a combination of Bessel functions and

$$u(r,\varphi) = \sum_m (a_{1m} J_{m+\alpha}(kr) + a_{2m} Y_{m+\alpha}(kr))\,\mathrm{e}^{im\varphi}.$$

We employ the behaviour of Bessel functions in the vicinity of zero,

$$J_\alpha(x) \approx \frac{1}{\Gamma(\alpha+1)}\left(\frac{x}{2}\right)^\alpha, \quad Y_\alpha(x) \approx -\frac{\Gamma(\alpha)}{\pi}\left(\frac{2}{x}\right)^\alpha,$$

which yields the values of the above functionals,

$$\Phi_1^0 = \sqrt{\pi}\lim_{r\to 0}\frac{r^\alpha}{2\pi}2\pi\frac{-\Gamma(\alpha)}{\pi}a_{20}\left(\frac{2}{kr}\right)^\alpha = -\frac{\Gamma(\alpha)}{\sqrt{\pi}}\left(\frac{2}{k}\right)^\alpha a_{20},$$

$$\Phi_2^0 = \sqrt{\pi}\lim_{r\to 0}\frac{r^{-\alpha}}{2\pi}2\pi a_{10}\frac{1}{\Gamma(\alpha+1)}\left(\frac{kr}{2}\right)^\alpha = \frac{\sqrt{\pi}}{\Gamma(\alpha+1)}\left(\frac{k}{2}\right)^\alpha a_{10},$$

$$\Phi_1^{-1} = \sqrt{\pi}\lim_{r\to 0}\frac{r^{1-\alpha}}{2\pi}2\pi\frac{1}{\Gamma(\alpha)}\left(\frac{kr}{2}\right)^{-1+\alpha} a_{1,-1} = \frac{\sqrt{\pi}}{\Gamma(\alpha)}\left(\frac{2}{k}\right)^{1-\alpha} a_{1,-1},$$

$$\Phi_2^{-1} = -\sqrt{\pi}\lim_{r\to 0}\frac{r^{-1+\alpha}}{2\pi}2\pi\frac{\Gamma(\alpha-1)}{\pi}\left(\frac{2}{kr}\right)^{-1+\alpha} a_{2,-1} = -\frac{\Gamma(\alpha-1)}{\sqrt{\pi}}\left(\frac{k}{2}\right)^{1-\alpha} a_{2,-1}.$$

The resonances are given by Eq. (12.6) in combination with Dirichlet condition at the disc boundary, $a_{10}J_\alpha(kR) + a_{20}Y_\alpha(kR) = 0$ and $a_{1,-1}J_{\alpha-1}(kR) + a_{2,-1}Y_{\alpha-1}(kR) = 0$. In the particular case of the coupling described by matrix (12.7) this yields the equation

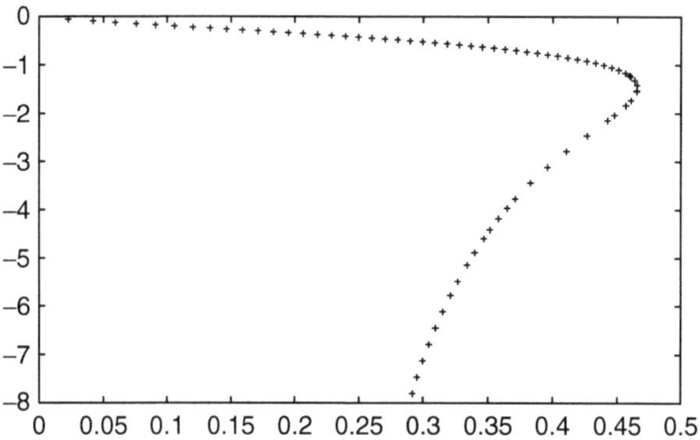

Fig. 12.7 Trajectory of a resonance for α running from 0 to $\frac{1}{2}$

$$\det\left[\begin{pmatrix} -1 & 1 \\ 1 & -1 \end{pmatrix}\begin{pmatrix} 1 & 0 \\ 0 & -\frac{\Gamma(\alpha)}{\sqrt{\pi}}\left(\frac{2}{k}\right)^{\alpha} J_{\alpha}(kR) \end{pmatrix} + i\begin{pmatrix} 1 & 1 \\ 1 & 1 \end{pmatrix}\right.$$
$$\left.\begin{pmatrix} ik & 0 \\ 0 & \frac{\sqrt{\pi}}{\Gamma(\alpha+1)}\left(\frac{k}{2}\right)^{\alpha} Y_{\alpha}(kR) \end{pmatrix}\right] = 0\,,$$

which can be rewritten as

$$\frac{i\sqrt{\pi}}{\Gamma(\alpha+1)}\left(\frac{k}{2}\right)^{\alpha} Y_{\alpha}(kR) + k\frac{\Gamma(\alpha)}{\sqrt{\pi}}\left(\frac{2}{k}\right)^{\alpha} J_{\alpha}(kR) = 0\,.$$

In the flux has a half-integer value, $\alpha = 1/2$, we get $\frac{1}{\sqrt{\pi R}}(\sin kR - i\cos kR) = 0$ which shows one more time the absence of resonances.

For other values of α the condition can be solved numerically. In Fig. 12.7 we plot the trajectory of one of the resonances as the value of α increases from 0 to $\frac{1}{2}$. The step is taken to be 0.01 for values until $\alpha = 0.49$, which corresponds roughly to the sharp bend of the curve, from this point on the linear step is replaced by a sequence of exponentially increasing density accumulating at $\alpha = \frac{1}{2}$.

Acknowledgements The research the results of which are reported here has been supported by the Czech Science Foundation within the project P203/11/0701.

References

1. Adami R, Teta A (1998) On the Aharonov-Bohm Hamiltonian. Lett Math Phys 43:43–54
2. Aizenman M, Sims R, Warzel S (2006) Absolutely continuous spectra of quantum tree graphs with weak disorder. Commun Math Phys 264:371–389
3. Berkolaiko G, Kuchment P (2013) Introduction to quantum graphs. American Mathematical Society, Providence

4. Bolte J, Endres S (2009) The trace formula for quantum graphs with general self-adjoint boundary conditions. Ann H Poincaré 10:189–223
5. Brüning J, Geyler V (2003) Scattering on compact manifolds with infinitely thin horns. J Math Phys 44:371–405
6. Brüning J, Exner P, Geyler VA (2003) Large gaps in point-coupled periodic systems of manifolds. J Phys A Math Gen 36:4875–4890
7. Cattaneo C (1997) The spectrum of the continuous Laplacian on a graph. Monatsh Math 124:215–235
8. Dabrowski L, Šťovíček P (1998) Aharonov-Bohm effect with δ-type interaction. J Math Phys 36:47–62
9. Davies EB, Pushnitski A (2011) Non-Weyl resonance asymptotics for quantum graphs. Anal PDE 4:729–756
10. Davies EB, Exner P, Lipovský J (2010) Non-Weyl asymptotics for quantum graphs with general coupling conditions. J Phys A Math Theor 43:474013
11. Exner P (1996) Contact interactions on graph superlattices. J Phys A Math Gen 29:87–102
12. Exner P (1997) A duality between Schrödinger operators on graphs and certain Jacobi matrices. Ann Inst H Poincaré Phys Théor 66:359–371
13. Exner P, Lipovský J (2010) Resonances from perturbations of quantum graphs with rationally related edges. J Phys A Math Theor 43:105301
14. Exner P, Lipovský J (2011) Non-Weyl resonance asymptotics for quantum graphs in a magnetic field. Phys Lett A 375:805–807
15. Exner P, Lipovský J (2013, to appear) Resonances on hedgehog manifolds. Acta Polytech. arXiv:1302.5269
16. Exner P, Post O (2013, to appear) Approximation of quantum graph vertex couplings by scaled Schrödinger operators on thin branched manifolds. Commun Math Phys. arXiv:1205.5129
17. Exner P, Šeba P (1987) Quantum motion on a halfline connected to a plane. J Math Phys 28:386–391; erratum p 2254
18. Exner P, Tater M, Vaněk D (2001) A single-mode quantum transport in serial-structure geometric scatterers. J Math Phys 42:4050–4078
19. Exner P, Helm M, Stollmann P (2007) Localization on a quantum graph with a random potential on the edges. Rev Math Phys 19:923–939
20. Grieser D (2008) Spectra of graph neighborhoods and scattering. Proc Lond Math Soc 97:718–752
21. Gutkin B, Smilansky U (2001) Can one hear the shape of a graph? J Phys A Math Gen 34:6061–6068
22. Hislop P, Post O (2009) Anderson localization for radial tree-like random quantum graphs. Wave Random Media 19:216–261
23. Kiselev A (1997) Some examples in one-dimensional 'geometric' scattering on manifolds. J Math Anal Appl 212:263–280
24. Kottos T, Smilansky U (1999) Periodic orbit theory and spectral statistics for quantum graphs. Ann Phys 274:76–124
25. Kuchment P (2004) Quantum graphs: I. Some basic structures. Waves Random Media 14:S107–S128
26. Langer RE (1931) On the zeros of exponential sums and integrals. Bull Am Math Soc 37:213–239
27. Pankrashkin K (2012) Unitary dimension reduction for a class of self-adjoint extensions with applications to graph-like structures. J Math Anal Appl 396:640–655
28. Post O (2011) Spectral analysis on graph-like spaces. Lecture notes in mathematics, vol 2039. Springer, Berlin
29. Ruedenberg K, Scherr CW (1953) Free-electron network model for conjugated systems, I. Theory J Chem Phys 21:1565–1581
30. Schenker JH, Aizenman M (2000) The creation of spectral gaps by graph decoration. Lett Math Phys 53:253–262

Chapter 13
Quantum Graph and Quantum Filter

Taksu Cheon

Abstract We show that the quantum star graphs, with added potentials on some of graph lines and with suitable choice of the connection condition at the graph node, can be made into spectral branching filters whose properties is controllable through the tuning of the strength of the potentials.

13.1 Introduction

The quantum graphs can be regarded as mathematical models of quantum single-electron devices [1,2]. The quantum star graph, which is the "atoms" of all quantum graphs, serve particularly well for the purpose. They allow to design devices, that are both simple and rich enough thanks to their large parameter spaces. One of the first applications of quantum star graphs emerged in the spectral filtering. An $n = 2$ star graph with the δ-interaction in its center is already usable as a high-pass filter, and similarly, a graph with the δ'-interaction works as a low-pass filter. Besides of these two simple designs, the existence of an $n = 3$ branching filter, functioning as a high-pass/low-pass junction, has been proved [3]. Such a system, in principle, can be controlled by a variation of the vertex parameters. However such realisation is difficult in practice since it requires real-time adjustments of a nanoscale object. It would be highly desirable if the control is achieved through an external field applied onto one of the lines, preferably on lines other than those along which we want the quantum particles to propagate.

In this article, we show that a quantum filter controllable by an external potential can be indeed designed. Besides the filter, we construct one more similar device, namely a quantum "sluice-gate" which allows to increase and decrease the quantum

T. Cheon (✉)
Kochi University of Technology, Kami, Kochi Prefecture, Japan
e-mail: taksu.cheon@kochi-tech.ac.jp

D. Matrasulov and H.E. Stanley (eds.), *Nonlinear Phenomena in Complex Systems: From Nano to Macro Scale*, NATO Science for Peace and Security Series C: Environmental Security, DOI 10.1007/978-94-017-8704-8_13,

flux from one line to another by adjusting the external potential applied to another line. Our constructions are based on very simple star graphs with $n = 3$ and $n = 4$, respectively. The presented result may serve also as a starting point in a search for other controllable quantum device models based on quantum graphs.

13.2 Scattering Matrix for Quantum Graph with Potentials on Lines

Let us consider a quantum particle with energy E residing on a star graph. When it comes in the vertex from the j-th line, it is scattered at the vertex into all the lines. The i-th component of the final-state wave function equals

$$\psi_{ij}(x) = \begin{cases} \frac{1}{\sqrt{k_j}} \mathrm{e}^{-\mathrm{i}k_j x} + \mathscr{S}_{jj} \frac{1}{\sqrt{k_j}} \mathrm{e}^{\mathrm{i}k_j x} & \text{for } i = j\,, \\ \mathscr{S}_{ij} \frac{1}{\sqrt{k_i}} \mathrm{e}^{\mathrm{i}k_i x} & \text{for } i \neq j\,, \end{cases} \tag{13.1}$$

where $\mathscr{S}_{ij}$ are scattering amplitudes, k_i are momenta on the corresponding lines, and coefficients $1/\sqrt{k_i}$ are involved for proper normalization. For any i, the momentum k_i is equal to $k_i = \sqrt{E - U_i}$, where U_i is the potential on the i-th line. The matrix $\mathscr{S} = \{\mathscr{S}_{ij}\}$ is the *scattering matrix* of the graph. For a normalized wave function coming in from the j-th line, $\mathscr{S}_{ij}$ is interpreted as the complex amplitude of transmission into the i-th line (for $i \neq j$), whereas $\mathscr{S}_{jj}$ represents the complex amplitude of reflection. The matrix $\mathscr{S}$ depends, besides the internal properties of the vertex, on E and $U_1, U_2, \ldots, U_n$.

Define two matrices M and M' by

$$M = \{\psi_{ij}(0)\}, \quad M' = \{\psi'_{ij}(0)\}. \tag{13.2}$$

With regard to (13.1), it holds

$$M = K^{-1} + K^{-1}\mathscr{S}, \quad M' = \mathrm{i}K(-K^{-1} + K^{-1}\mathscr{S}), \tag{13.3}$$

where $K = \{\sqrt{k_i}\delta_{ij}\}$. Any wave function $\Psi(x) = (\psi_1(x), \ldots, \psi_n(x))^T$ (the superscript T denotes the transposition) on the graph obeys the *boundary condition* determining the vertex, which is usually written in the form $A\Psi(0) + B\Psi'(0) = 0$ for certain fixed $A, B \in \mathbb{C}^{n,n}$, cf. [4]. In particular, b. c. must be satisfied by the final-state wave function $\Psi_j(x) = (\psi_{1j}(x), \ldots, \psi_{nj}(x))^T$ determined in (13.1) for all j, hence $AM + BM' = 0$, which together with (13.3) leads to the sought expression for $\mathscr{S}$:

$$\mathscr{S} = -(AK^{-1} + \mathrm{i}BK)^{-1}(AK^{-1} - \mathrm{i}BK)\,. \tag{13.4}$$

Squared moduli of the elements of $\mathscr{S}$ have the following interpretation: $|S_{ij}|^2$ for $j \neq i$ represents the probability of transmission from the i-th to the j-th line, $|S_{jj}|^2$ is the probability of reflection on the j-th line.

13.3 Threshold Resonance in Star Graph with External Potential

We now consider an $n = 3$ star graph with a Fülöp-Tsutsui (also called "scale invariant") singular coupling (cf. [5–7]) in its vertex. For the sake of convenience, the coupling will be described by a boundary condition written in the so-called ST-form ($B\Psi' = -A\Psi$ with specially structured A, B, see [8] and [9]) with explicit notation

$$\begin{pmatrix} 1 & a & b \\ 0 & 0 & 0 \\ 0 & 0 & 0 \end{pmatrix} \begin{pmatrix} \psi_1'(0) \\ \psi_2'(0) \\ \psi_3'(0) \end{pmatrix} = \begin{pmatrix} 0 & 0 & 0 \\ -a & 1 & 0 \\ -b & 0 & 1 \end{pmatrix} \begin{pmatrix} \psi_1(0) \\ \psi_2(0) \\ \psi_3(0) \end{pmatrix} . \tag{13.5}$$

The graph is schematically illustrated in Fig. 13.1. The roles of individual lines are the following:

- Line 1 is *input*. Particles of various energies are coming in the vertex along this line.
- Line 2 is *output*. Particles passed through the vertex are gathered on this line.
- Line 3 is *controling line*. We assume that this line is subjected to a constant (but adjustable) external potential U.

A quantum particle with energy $E = k^2$ coming in the vertex from the input line 1 is scattered at the vertex into all the lines. The scattering amplitudes can be calculated by substituting the matrices A, B from the b. c. (13.5) together with

$$k_1 = k_2 = k, \quad k_3 = \sqrt{k^2 - U} \tag{13.6}$$

into Eq. (13.4). We obtain:

$$\mathscr{S}_{21}(k; U) = \frac{2a}{1 + a^2 + b^2\sqrt{1 - \frac{U}{k^2}}}, \tag{13.7}$$

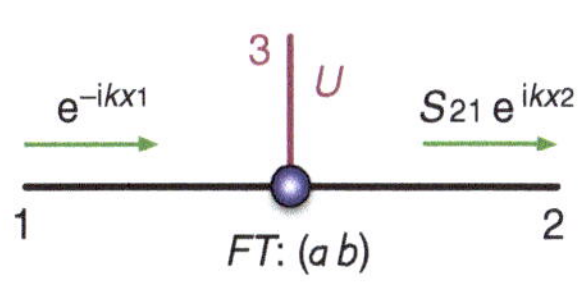

Fig. 13.1 Schematic depiction of the $n = 3$ star graph with an external potential U on the line 3

$$\mathscr{S}_{11}(k;U) = \frac{1 - a^2 - b^2\sqrt{1 - \frac{U}{k^2}}}{1 + a^2 + b^2\sqrt{1 - \frac{U}{k^2}}}, \tag{13.8}$$

$$\mathscr{S}_{31}(k;U) = \frac{2b\left(1 - \frac{U}{k^2}\right)^{\frac{1}{4}} \Theta(k - \sqrt{U})}{1 + a^2 + b^2\sqrt{1 - \frac{U}{k^2}}}. \tag{13.9}$$

The Heaviside step function $\Theta(k - \sqrt{U})$ is added in Eq. (13.9) to make the expression valid for all energies k^2, including $k^2 < U$. It represents asymptotically no transmission to the line 3 below the threshold momentum $k_{\text{th}} = \sqrt{U}$.

We are interested in the probability of transmission into the output line 2, which we denote by $\mathscr{P}(k;U)$, and particularly in its k-dependence. Since $\mathscr{P}(k;U) = |\mathscr{S}_{21}(k;U)|^2$, we have from (13.7)

$$\mathscr{P}(k;U) = \begin{cases} \frac{4a^2}{\left(1+a^2+b^2\sqrt{1-U/k^2}\right)^2} & \text{for } k \geq \sqrt{U}, \\ \\ \frac{4a^2}{(1+a^2)^2+b^4(U/k^2-1)} & \text{for } k \leq \sqrt{U}. \end{cases} \tag{13.10}$$

We observe that for a given constant potential on the line 3, $\mathscr{P}(k;U)$ as a function of k grows in the interval $(0, \sqrt{U})$, attains its maximum at $k = \sqrt{U}$, and decreases in the interval $(\sqrt{U}, \infty)$. In particular, it holds:

$$\lim_{k \to 0} \mathscr{P}(k;U) = 0, \tag{13.11}$$

$$\mathscr{P}(\sqrt{U};U) = \left(\frac{2a}{1 + a^2}\right)^2, \tag{13.12}$$

$$\lim_{k \to \infty} \mathscr{P}(k;U) = \left(\frac{2a}{1 + a^2 + b^2}\right)^2. \tag{13.13}$$

If the parameters a, b satisfy $b \gg a \geq 1$, the function $\mathscr{P}(k;U)$ has a sharp peak at $k = \sqrt{U}$. Equation (13.12) implies that the peak is highest possible (attaining 1) for $a = 1$. We conclude: If $b \gg a = 1$, the system has high input→output transmission probability for particles having momenta $k \approx \sqrt{U}$ and the transmission is perfect for $k = \sqrt{U}$, whereas there is just a very small transmission probability for other values of k. The situation is numerically illustrated in Fig. 13.2. The quantum graph schematically depicted in Fig. 13.1 can be therefore used as an adjustable spectral filter, controllable by the potential put on the controlling line 3. The bandwidth W of the filter, i.e., the width of the interval of energies k^2 for which $\mathscr{P}(k;U) > 1/2$, depends on U and b, and for $b \gg 1$ it is approximately given as $W \approx 4.7U/b^4$. Let us remark that the resonance at the threshold momentum $k_{\text{th}} = \sqrt{U}$ is related to the pole of the scattering matrix which is located on the positive real axis at

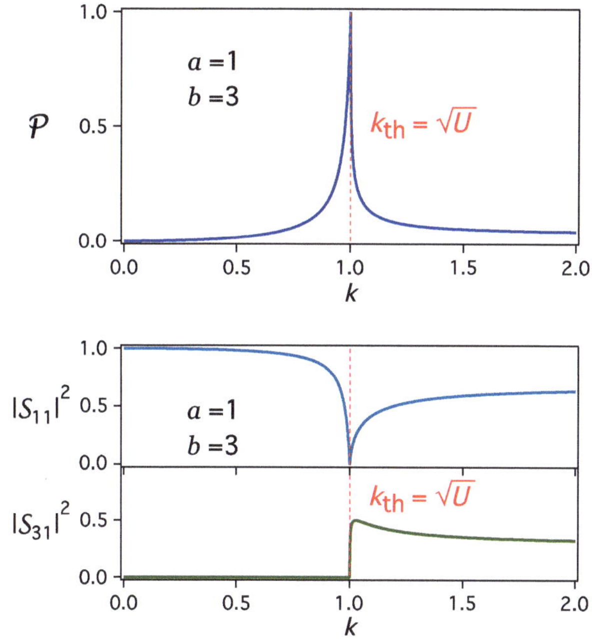

Fig. 13.2 Scattering characteristics of the graph from Fig. 13.1 with parameters $a = 1$, $b = 3$. The transmission probability $\mathscr{P}(k;U)$ as a function of k with the value of the potential set to $U = 1$ is plotted in the *top figure*. The *lower figure* shows reflection probability $|\mathscr{S}_{11}(k;U)|^2$ and the probability of transmission to the controlling line $|\mathscr{S}_{31}(k;U)|^2$

$k_{\text{pol}} = \frac{b^2}{\sqrt{b^4-(1+a^2)^2}}\sqrt{U}$ on the unphysical Riemann surface which is connected to the physical Riemann surface at the cut that runs between $k = \pm\sqrt{U}$.

13.4 Flux Control and Quantum Sluice-Gate

Let us now consider an $n = 4$ star graph as schematically illustrated in Fig. 13.3. The meaning of the first three lines will be the same as in the previous model: 1 = input, 2 = output, 3 = controlling line subjected to a constant external potential U. The line No. 4 is a *drain* and is included in the model for technical reasons: our considerations showed that the device we wish to construct is mathematically infeasible using a vertex of degree $n = 3$. The vertex coupling is again assumed to be of the Fülöp-Tsutsui type, and its properties are determined by the boundary condition written for convenience in the ST-form

$$\begin{pmatrix} 1 & 0 & a & a \\ 0 & 1 & a & -a \\ 0 & 0 & 0 & 0 \\ 0 & 0 & 0 & 0 \end{pmatrix} \begin{pmatrix} \psi_1'(0) \\ \psi_2'(0) \\ \psi_3'(0) \\ \psi_4'(0) \end{pmatrix} = \begin{pmatrix} 0 & 0 & 0 & 0 \\ 0 & 0 & 0 & 0 \\ -a & -a & 1 & 0 \\ -a & a & 0 & 1 \end{pmatrix} \begin{pmatrix} \psi_1(0) \\ \psi_2(0) \\ \psi_3(0) \\ \psi_4(0) \end{pmatrix} \tag{13.14}$$

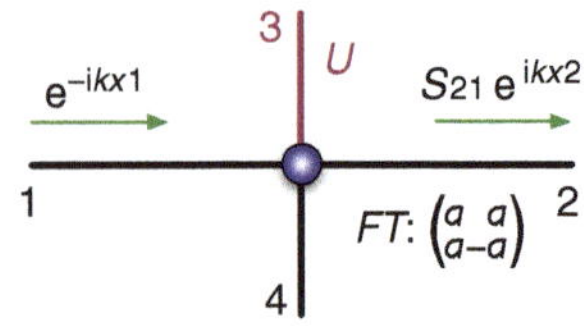

Fig. 13.3 Schematic depiction of the $n = 4$ star graph with an external potential U on the line No. 3

with $a \in$ R. The block $\begin{pmatrix} a & a \\ a & -a \end{pmatrix}$ is a special choice that ensued from our analysis; generally, the ST-form admits $\begin{pmatrix} a & b \\ c & d \end{pmatrix}$ for any $a, b, c, d \in$ C, cf. [8] or [9].
For a particle with energy $E = k^2$ coming in the vertex from the input line 1, we have

$$k_1 = k_2 = k, \quad k_3 = \sqrt{k^2 - U}, \quad k_4 = k, \tag{13.15}$$

and the scattering amplitudes can be calculated as

$$S_{21}(k;U) = \frac{2a^2\left(1 - \sqrt{1 - \frac{U}{k^2}}\right)}{(1 + 2a^2) + 2a^2(1 + 2a^2)\sqrt{1 - \frac{U}{k^2}}} \tag{13.16}$$

and

$$S_{11}(k;U) = \frac{1 - 4a^4\sqrt{1 - \frac{U}{k^2}}}{(1 + 2a^2) + 2a^2(1 + 2a^2)\sqrt{1 - \frac{U}{k^2}}}, \tag{13.17}$$

$$S_{31}(k;U) = \frac{2a(1 + 2a^2)\left(1 - \frac{U}{k^2}\right)^{\frac{1}{4}} \Theta(k - \sqrt{U})}{(1 + 2a^2) + 2a^2(1 + 2a^2)\sqrt{1 - \frac{U}{k^2}}}, \tag{13.18}$$

$$S_{41}(k;U) = \frac{2a + 4a^3\sqrt{1 - \frac{U}{k^2}}}{(1 + 2a^2) + 2a^2(1 + 2a^2)\sqrt{1 - \frac{U}{k^2}}}. \tag{13.19}$$

We again denote the transmission probability input→output by $\mathscr{P}(k;U) = |\mathscr{S}_{21}(k;U)|^2$. It holds

$$\mathscr{P}(k;U) = \begin{cases} \dfrac{4a^4U/k^2}{(1+2a^2)^2\left(1-4a^4+4a^4U/k^2\right)} & \text{for } k \le \sqrt{U}, \\[2ex] \dfrac{4a^4\left(1-\sqrt{1-U/k^2}\right)^2}{(1+2a^2)^2\left(1+2a^2\sqrt{1-U/k^2}\right)^2} & \text{for } k \ge \sqrt{U}, \end{cases} \tag{13.20}$$

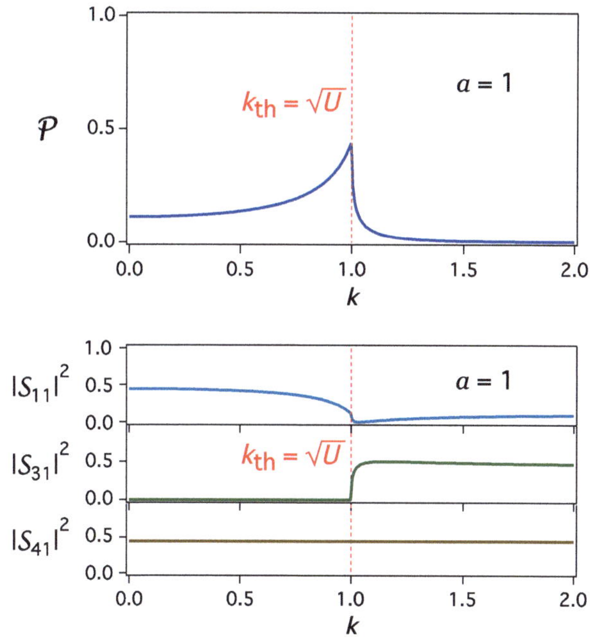

Fig. 13.4 Scattering characteristics of the graph from Fig. 13.3 with parameter $a = 1$. The transmission probability $\mathscr{P}(k;U)$ as a function of k with the value of the potential set to $U = 1$ is plotted in the *top figure*. The *lower figure* shows the reflection probability $|S_{11}(k;U)|^2$ and the probabilities of transmission to the controlling line $|S_{31}(k;U)|^2$ and to the drain line $|S_{41}(k;U)|^2$

hence

$$\lim_{k\to 0} \mathscr{P}(k;U) = \frac{1}{(1+2a^2)^2}, \tag{13.21}$$

$$\mathscr{P}(\sqrt{U};U) = \frac{4a^4}{(1+2a^2)^2}, \tag{13.22}$$

$$\lim_{k\to\infty} \mathscr{P}(k;U) = 0. \tag{13.23}$$

If U is fixed, $\mathscr{P}(k;U)$ as a function of k quickly falls off to zero at $k > \sqrt{U}$. A typical behaviour is illustrated in a numerical example in Fig. 13.4. The peak at the threshold momentum $k_{\text{th}} = \sqrt{U}$, appearing for $a > 1/\sqrt{2}$, is again related to the pole in the unphysical Riemann plane at $k_{\text{pol}} = \frac{2a^2}{\sqrt{(4a^4-1)}}\sqrt{U}$.

There is a value of the parameter a that deserves a particular attention, namely $a = 1/\sqrt{2}$. For this choice of a the peak disappears and the function $\mathscr{P}(k;U)$ becomes constant in the whole interval $(0, \sqrt{U})$:

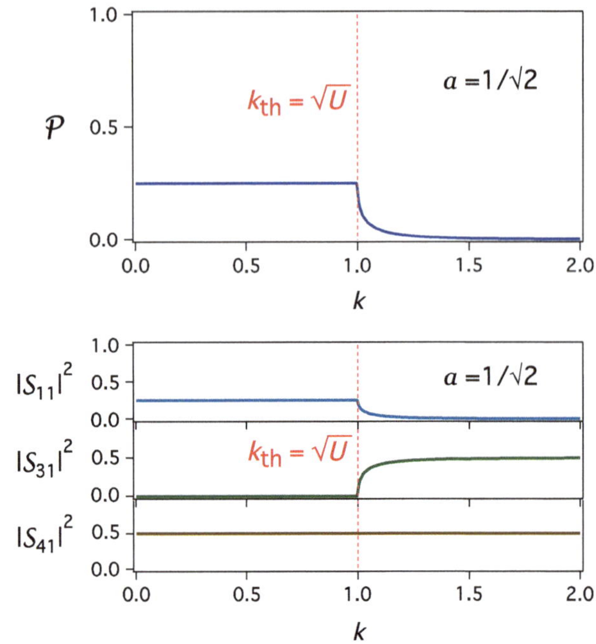

Fig. 13.5 Characteristics of the flat spectral filter obtained from the graph on Fig. 13.3 for $a = 1/\sqrt{2}$. The transmission probability $\mathscr{P}(k;U)$ as a function of k with the value of the potential set to $U = 1$ is plotted in the *top figure*. The *lower figure* shows the reflection probability $|S_{11}(k;U)|^2$ and the probabilities of transmission to the controlling line $|S_{31}(k;U)|^2$ and to the drain line $|S_{41}(k;U)|^2$

$$\mathscr{P}(k;U) = \begin{cases} \frac{1}{4} & \text{for } k \leq \sqrt{U}, \\ \frac{1}{4}\cdot\left(\frac{1-\sqrt{1-U/k^2}}{1+\sqrt{1-U/k^2}}\right)^2 & \text{for } k > \sqrt{U}, \end{cases} \tag{13.24}$$

see Fig. 13.5. The device then behaves as a spectral filter with a flat passband that transmits (with the probability of $1/4$) quantum particles with momenta $k \in [0, \sqrt{U}]$ to the output, whereas particles with higher momenta are diverted to other lines, mainly to 3 and 4. The process is directly controlled by the external potential U. Since increasing U opens the channel $1 \to 2$ for more particles, the device can be regarded as a quantum sluice-gate, applicable as a quantum flux controller. When there are many particles described by the momentum distribution $\rho(k)$ on the line 1, the flux J to the line 2 is given by $J(U) = \int dk \rho(k) k \mathscr{P}(k;U)$. Assuming the Fermi distribution with Fermi momentum k_F larger than our range of operation of $\sqrt{U}$, we can set $\rho(k) = \rho = \text{const}$. With the approximation $\mathscr{P}(k;U) \approx \frac{1}{4}\Theta(\sqrt{U} - k)$, we obtain $J(U) = \frac{1}{8}\rho U$, which indicates the linear flux control.

The sluice-gate built from an $n = 4$ star graph has one more operation mode. If the line No. 4 (the drain) is subjected to another external field V, $0 < V < U$, the channel 1→2 opens for particles with $k \in [\sqrt{V}, \sqrt{U}]$ and mostly closes for particles with k outside this interval. The gate then works as a fully tunable band spectral filter. However, in contrast to the standard $V = 0$ operation mode, the filter with $V > 0$ does not have a flat passband.

13.5 Realization of Exotic Node with Delta Graphs

We have to emphasize that the studied controllable filter devices using the threshold resonance became possible only with "exotic" Fülöp-Tsutsui-type couplings in the vertices. Standard vertex couplings (the free and the δ-coupling) would not work this way. It is therefore essential, for the proposed designs to be viable, that the required Fülöp-Tsutsui vertices can be created using standard couplings, which themselves have a simple physical interpretation [10]. This problem has been addressed in [8] and [9], where it was proved that any Fülöp-Tsutsui coupling given by b. c. with real matrices A, B can be approximately constructed by assembling a few δ-couplings. The solution for our case is shown in Fig. 13.6.

This research has been supported by the Japan Ministry of Education, Culture, Sports, Science and Technology under the Grant number 24540412.

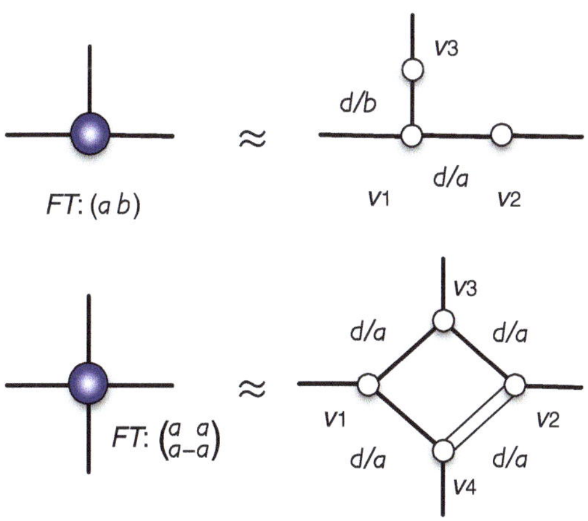

Fig. 13.6 Finite constructions of the Fülöp-Tsutsui couplings used. The design, based on [9], utilizes the δ-couplings connected by short lines. The small size limit $d \to 0$ with the δ-coupling strengths scaled with d effectively produces the required F-T vertex coupling. For the $n = 3$ case (*top*), the δ-coupling strengths are given by $v_1 = [a(a-1) + b(b-1)]/d$, $v_2 = (1-a)/d$ and $v_3 = (1-b)/d$. For the $n = 4$ case (*bottom*), the strengths are $v_1 = v_2 = 2a(a-1)/d$, $v_3 = v_4 = (1-2a)/d$. The double line represents a line with a "magnetic" vector potential, which can be alternatively replaced by a line carrying the δ-coupling of strength $v_5 = -8a/d$ in its center, together with changing v_2 and v_4 to $v_2 = 2a(a-2)/d$, $v_4 = (1-4a)/d$

References

1. Albeverio S, Gesztesy F, Hoegh-Krohn R, Holden H (2005) Solvable models in quantum mechanics, 2nd edn., with appendix by P. Exner. AMS Chelsea, Providence
2. Exner P, Keating JP, Kuchment P, Sunada T, Teplyaev A (eds) (2008) Analysis on graphs and applications. Proceedings of symposia in pure mathematics, vol 77. AMS, Providence. *and references therein*
3. Cheon T, Exner P, Turek O (2009) Spectral filtering in quantum Y-junction. J Phys Soc Jpn 78:124004. (7pp)
4. Kostrykin V, Schrader R (1999) Kirchhoff's rule for quantum wires. J Phys A Math Gen 32:595–630
5. Fülöp T, Tsutsui I (2000) A free particle on a circle with point interaction. Phys Lett A 264:366–374
6. Naimark K, Solomyak M (2000) Eigenvalue estimates for the weighted Laplacian on metric trees. Proc Lond Math Soc 80:690–724
7. Sobolev AV, Solomyak M (2002) Schrödinger operator on homogeneous metric trees: spectrum in gaps. Rev Math Phys 14:421–467
8. Cheon T, Exner P, Turek O (2010) Approximation of a general singular vertex coupling in quantum graphs. Ann Phys (NY) 325:548–578
9. Cheon T, Turek O (2010) Fulop-Tsutsui interactions on quantum graphs. Phys Lett A 374:4212–4221
10. Exner P (1996) Weakly coupled states on branching graphs. Lett Math Phys 38:313–320

Chapter 14
From Continuous-Time Random Walks to Continuous-Time Quantum Walks: Disordered Networks

Oliver Mülken and Alexander Blumen

Abstract Recent years have seen a growing interest in dynamical quantum processes; thus it was found that the electronic energy transfer through photosynthetic antennae displays quantum features, aspects also known from the dynamics of charge carriers along polymer backbones. Hence, in modeling energy transfer one has to extend the classical, master-equation-type formalism and incorporate quantum-mechanical aspects, while still aiming to describe complex networks of molecules over which the transport takes place. The continuous time random walk (CTRW) scheme is widely employed in modeling transport in random environments (Sokolov et al, Phys Today 55:48, 2002) and is mathematically akin to quantum-mechanical Hamiltonians of tight-binding type (Mülken and Blumen, Phys Rep 502:37, 2011; Mülken and Blumen, Phys Rev E 73:066117, 2006); a simple way to see it is to focus on the time-evolution operators in statistical and in quantum mechanics: The transition to the quantal domain leads then to continuous-time quantum walks (CTQW). In this way the CTQW problem stays linear, and thus many results obtained in solving CTRW (such as eigenvalues and eigenfunctions) can be readily reutilized for CTQW. However, the physically relevant properties of the two models differ vastly: In the absence of traps CTQW are time-inversion symmetric and no energy equipartition takes place at long times. Also, the quantum system keeps memory of the initial conditions, a fact exemplified by the occurrence of quasi-revivals (Mülken and Blumen, Phys Rep 502:37, 2011). Here we will exemplify the vastly different behaviors of CTQW and CTRW on disordered networks, namely on small-world networks (Mülken et al, Phys Rev E 76:051125, 2007) and on star-graphs with randomly added bonds (Anishchenko et al, Quantum Inf Process 11:1273, 2012).

O. Mülken • A. Blumen (✉)
Physikalisches Institut, Universität Freiburg, Hermann-Herder-Str. 3, 79104 Freiburg, Germany
e-mail: muelken@physik.uni-freiburg.de; blumen@physik.uni-freiburg.de

D. Matrasulov and H.E. Stanley (eds.), *Nonlinear Phenomena in Complex Systems: From Nano to Macro Scale*, NATO Science for Peace and Security Series C: Environmental Security, DOI 10.1007/978-94-017-8704-8_14,

14.1 Introduction

The recent, growing interest in quantum mechanical problems ranging from quantum computation [18] and ultra-cold atomic assemblies [24] up to the very efficient energy transfer in biological systems at room temperature [6, 8] calls for appropriate theoretical models. Now, quantum transport over regular lattices [7, 17, 23], over branched [12], and over fractal structures [1, 2] differs from its classical counterpart. Accounting for quantum aspects of the transport can be based on several schemes that start from a classical picture. Here, one distinguishes between discrete-time quantum walks which require coin operators [11] and continuous-time quantum walks (CTQW) which do not need additional operators [11, 15]. Recently, we have presented for several types of graphs an overview of CTQW under different conditions [15]. In the present contribution, we will follow Ref. [15] and will recall the basic features of continuous-time random walks (CTRW) and of the related CTQW. As applications we choose to focus on disordered graphs. Now, many investigations in solid state physics start from regular networks, such as lattices. In these, even for finite structures, the majority of the nodes of the network have the same number of outgoing bonds, i.e., the same functionality. However, there exist also other types of networks such as star graphs or complete graphs in which the majority of their nodes has the same functionality. Star graphs and complete graphs are of great interest for quantum computation [15, 19, 20].

We will study now disordered systems built from regular structures, by adding in a random fashion bonds to the original graph, but allowing any pair of nodes to be connected by a single bond at most. Starting from a ring one first obtains random, so-called small-world networks (SWN); starting from a star one is led to other types of random structures. In both cases, for an extremely high density of additional bonds one gets close to having complete graphs. The insertion of additional bonds changes the topology of the original network and thus the spectrum of eigenvalues of the corresponding Hamiltonian. Now, one may envisage that the spreading of the excitations over the network may get enhanced by the additional bonds. We observe indeed that for stars additional bonds may lead – depending on the number of additional bonds – to an increase of the spreading of CTQW, see [4]. This is in contrast to the case of small-world networks (SWN), where starting to randomly add bonds to a ring of nodes slows down the CTQW, as shown in Ref. [16].

14.2 CTRW and CTQW on Networks

In general terms, we study the transport of excitations over networks; these are built from N nodes connected by bonds. The information about a network's topological structure is reflected in the $N \times N$ connectivity matrix $\mathbf{A}$, whose elements A_{kj} are:

$$A_{kj} = \begin{cases} f_j & \text{if } k = j \\ -1 & \text{if } k \text{ and } j \text{ are connected by one bond} \\ 0 & \text{otherwise.} \end{cases} \tag{14.1}$$

Here f_j is the functionality of node j. If the network is simply connected, **A** has a single vanishing eigenvalue $E_{\min} = 0$ [15]. We model the dynamics of the purely coherent and of the purely incoherent (diffusive) transport through CTQW and through CTRW, respectively [15]. In both cases, the dynamics is strongly influenced by **A**. An excitation localized at node j is associated with the state $|j\rangle$; the $|j\rangle$ states form an orthonormal basis set $\{|j\rangle\}$ with $j = 1, \ldots, N$.

For CTRW, the spreading is governed by a master equation in which the transfer matrix **T** is in the simplest case given by **A**, see [9, 15, 21, 22]. For CTQW, the states $|j\rangle$ span the whole accessible Hilbert space and the time evolution of an excitation initially placed at node $|j\rangle$ is determined by the system's Hamiltonian **H** which can be identified with **A** [9, 15].

The classical and the quantum probabilities to be in state $|k\rangle$ at time t when starting at $t = 0$ from state $|j\rangle$ are then [15]:

$$p_{k,j}(t) = \langle k|\exp(\mathbf{T}t)|j\rangle \qquad \text{for CTRW} \tag{14.2}$$

and

$$\pi_{k,j}(t) = |\langle k|\exp(-i\mathbf{H}t)|j\rangle|^2 \qquad \text{for CTQW,} \tag{14.3}$$

respectively. Here, we set $\hbar = 1$ and take as initial condition at $t = 0$ that $p_{k,j}(0) = \pi_{k,j}(0) = \delta_{k,j}$, where $\delta_{k,j}$ is Kronecker's delta. Given that both **T** and **H** are identified with **A**, yields that the eigenvalues and eigenvectors of both **T** and **H** are the same, while the difference in the dynamics of CTQW and of CTRW originates from the different functional forms of Eqs. (14.2) and (14.3), see Refs. [9] and [15].

14.2.1 Efficiency of CTRW and CTQW

While the eigenvectors are needed for calculating the exact transition probabilities $p_{k,j}(t)$ and $\pi_{k,j}(t)$, there are quantities (such as the lower bounds for $\pi_{k,j}(t)$, see Ref. [15]) which only depend on the eigenvalues. This allows to analyse several features which depend on the network's topology. For CTRW one such aspect is the probability to be (return or remain) at the initially excited node j, averaged over all j [3, 5]:

$$\bar{p}(t) = \frac{1}{N}\sum_j p_{j,j}(t) = \frac{1}{N}\sum_{n=1}^{N} e^{-E_n t}. \tag{14.4}$$

Therefore, $\bar{p}(t)$ depends only on the eigenvalues E_n of **A**, but not on the eigenvectors $|\phi_n\rangle$.

For CTQW $\bar{\pi}(t) \equiv \frac{1}{N}\sum_{n=1}^{N} \pi_{j,j}(t)$ still depends on the $|\phi_n\rangle$. Nevertheless, it is possible to define the quantity

$$\bar{\alpha}(t) \equiv \frac{1}{N}\sum_{j=1}^{N} \alpha_{j,j}(t) = \frac{1}{N}\sum_{n=1}^{N} e^{-iE_n t}, \tag{14.5}$$

which also depends only on the eigenvalues E_n and which is a lower bound to $\bar{\pi}(t)$, see Refs. [12] and [15]:

$$\bar{\pi}(t) = \frac{1}{N}\sum_{j} \left|\alpha_{j,j}(t)\right|^2 \geq \left|\bar{\alpha}(t)\right|^2. \tag{14.6}$$

Equations (14.4) and (14.5) allow to assess the spreading [14]: A quick decrease of $\bar{p}(t)$ reflects a quick increase in the probability of finding the excitation anywhere but at the initial node. Thus, the spreading is faster when $\bar{p}(t)$ decreases more quickly. For CTQW the unitary time evolution leads to $\bar{\pi}(t)$ and $|\bar{\alpha}(t)|^2$ showing oscillations, but the overall decay of $\bar{p}(t)$ and of $|\bar{\alpha}(t)|^2$ can be used to make statements about the spreading, say, by considering the decay of the envelope of $|\bar{\alpha}(t)|^2$ [14].

For regular d-dimensional networks, quantum walks appear to be faster than the classical ones [14]: the envelope of $|\bar{\alpha}(t)|^2$ decays as t^{-d}, while $\bar{p}(t)$ decays as $t^{-d/2}$. For other networks, which may behave differently, it is convenient to use the long-time averages [4]

$$P_{\mathrm{RW}} \equiv \lim_{T\to\infty} \frac{1}{T}\int_0^T dt\ \bar{p}(t) \qquad \text{and} \qquad P_{\mathrm{QW}} \equiv \lim_{T\to\infty} \frac{1}{T}\int_0^T dt\ |\bar{\alpha}(t)|^2, \tag{14.7}$$

in order to extract a global measure for the spreading.

For CTRW, due to the eigenvalue $E_1 = E_{\min} = 0$, $\bar{p}(t)$ will eventually drop to the equipartition value $1/N$. Thus $\lim_{T\to\infty} \bar{p}(t) = 1/N$ and also $P_{\mathrm{RW}} = 1/N$. This means that in the incoherent case at long times both quantities do not depend on the topology of the considered network [4]. For CTQW, $|\bar{\alpha}(t)|^2$ oscillates and one obtains [4]:

$$P_{\mathrm{QW}} = \lim_{T\to\infty} \frac{1}{T}\int_0^T dt \frac{1}{N^2}\sum_{n,n'} e^{-i(E_n - E_{n'})t} = \frac{1}{N^2}\sum_{n,n'} \delta(E_n - E_{n'}),$$

where $\delta(E_n - E_{n'})$ is unity if $E_n = E_{n'}$ and vanishes otherwise. Thus, the DOS fully determines the long-time average of $|\bar{\alpha}(t)|^2$. For highly degenerate eigenvalues, $|\bar{\alpha}(t)|^2$ will not decay to zero, but it will oscillate around a finite value [4]. Thus, for CTQW there is a large probability to remain at or to return to the initially excited node. Consequently, this slows down the average spreading for CTQW on networks with highly degenerate eigenvalues.

We now turn to the special disordered lattices which we want to consider here. Let us focus on situations which follow from randomly adding bonds to a regular lattice and calculate ensemble averaged quantities. Formally we set:

$$\langle \ldots \rangle_R \equiv \frac{1}{R}\sum_{r=1}^{R}[\ldots]_r, \tag{14.8}$$

where r runs over the particular realizations. In this way we can determine the ensemble-averaged probabilities $\langle \bar{p}(t)\rangle_R$ and $\langle |\bar{\alpha}(t)|^2\rangle_R$, along with the long-time average $\langle P_{\mathrm{QW}}\rangle_R$. For the long-time average of $\langle \overline{\pi}(t)\rangle_R$ we obtain (see also Eq. (17) of Ref. [17])

$$\langle \overline{\chi}\rangle_R \equiv \left\langle \lim_{T\to\infty}\frac{1}{T}\int_0^T dt\, \overline{\pi}(t)\right\rangle_R = \frac{1}{RN}\sum_{r,j,n,n'} \delta(E_{n,r}-E_{n',r})\big|\langle j|\Phi_{n,r}\rangle\langle j|\Phi_{n',r}\rangle\big|^2. \tag{14.9}$$

For the long-time average of $\langle |\overline{\alpha}(t)|^2\rangle_R$ one finds [4]:

$$\langle P_{\mathrm{QW}}\rangle_R \equiv \left\langle \lim_{T\to\infty}\frac{1}{T}\int_0^T dt\, |\overline{\alpha}(t)|^2\right\rangle_R = \frac{1}{RN^2}\sum_{r,n,n'} \delta(E_{n,r}-E_{n',r}), \tag{14.10}$$

which is a lower bound to $\langle \overline{\chi}\rangle_R$.

14.3 Small World Networks

The translational invariance of regular networks can be destroyed by randomly adding B bonds to them. In doing so, one creates "shortcuts", i.e., shorter paths, between many pairs of nodes. The obtained structures are called small-world networks (SWN). Here, we obtain SWN starting from a one-dimensional ring, see Ref. [16].

We note that CTQW on SWN never loose the information of the initial node j. Thus, the averaged probabilities to return to j are a good measure to quantify the efficiency of the transport on such networks [14]. For CTRW the decay is depicted, for instance, in Fig. 5(a) of Ref. [16]. The initial decay of $\langle \overline{p}(t)\rangle_R$ occurs faster when the numbers of additional bonds B is larger. At intermediate times the decay follows a power-law, namely $t^{-1/2}$, for the ring and changes to a decay of stretched exponential-type when B increases [10]. Therefore, CTRW will quickly explore the SWN until equipartition is reached, which happens after a relatively short time.

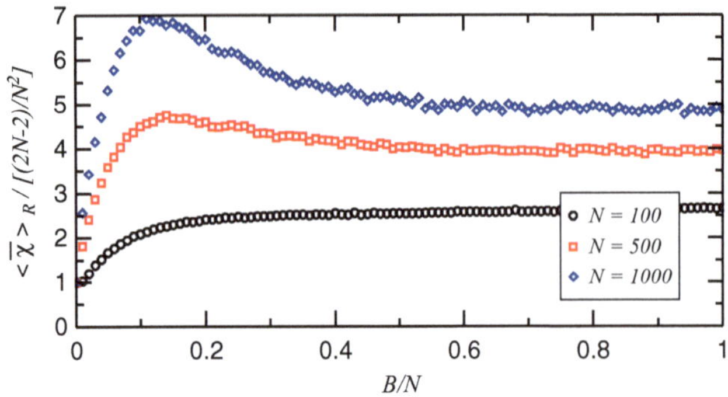

Fig. 14.1 (Color online) The long-time average of $\langle\overline{\pi}(t)\rangle_R$, $\langle\overline{\chi}\rangle_R$, for SWN with $N = 100$, 500, and 1,000 nodes as a function of B/N

The CTQW situation is more subtle and is depicted in Fig. 5(b) of Ref. [16]. The ensemble average $\langle\overline{\pi}(t)\rangle_R$ for a ring of N nodes and for times smaller than $N/2$ shows an oscillating behavior, whose maxima decay as t^{-1}. At longer times interference sets in and results in an irregular behavior, see also [13]. The oscillatory behavior remains intact as long as B is significantly less than N. With increasing B the curves for $\langle\overline{\pi}(t)\rangle_R$ are smoothed out, such that the heights of the first maxima as well as the depths of the minima decrease. At longer times the curves get flat and $\langle\overline{\pi}(t)\rangle_R$ approaches a limiting value. The lower bound $\langle|\overline{\alpha}(t)|^2\rangle_R$ behaves in a similar fashion as $\langle\overline{\pi}(t)\rangle_R$.

Figure 14.1 shows $\langle\overline{\chi}\rangle_R$, calculated based on Eq. (14.9), for SWN for several N, namely for $N = 100$, 500, and 1,000, as a function of B/N. For $B = 0$ there is only one realization and for N even one has [16]:

$$\langle\overline{\chi}_{\text{ring}}\rangle_R \equiv \overline{\chi} = \frac{1}{N}\sum_j \chi_{jj} = \frac{2N-2}{N^2}. \tag{14.11}$$

Increasing B up to N, $\langle\overline{\chi}\rangle_R$ is sometimes not monotonic; for B/N comparable to unity it reaches a plateau which increases with N. Therefore, increasing the number of bonds results to a less efficient transport from the initial node to all others when compared to the situation for $B = 0$, see also [16].

14.4 Transition from Star Graphs to Complete Graphs

In this section we study the transition from star graphs to complete graphs, achieved by adding new bonds to the original graph. Here, the addition of bonds leads to a domain in which the CTQW transport is more efficient: this is in contrast to the findings of the previous section devoted to SWN. We start from a star graph of size N with one central node and $N - 1$ peripheral nodes connected to the central one.

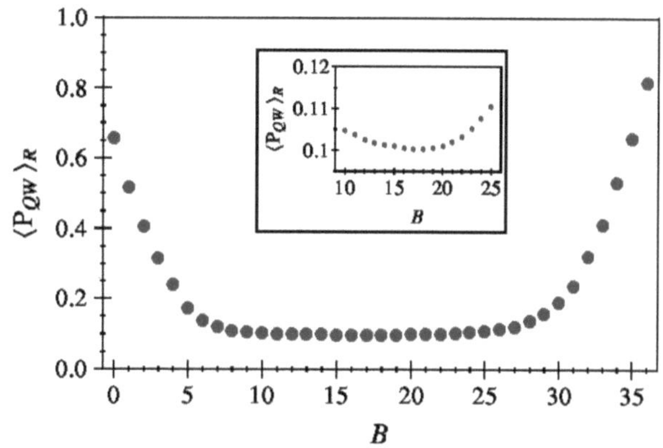

Fig. 14.2 $\langle P_{\mathrm{QW}} \rangle_R$ for networks obtained from a star graph with $N = 10$ nodes. The inset shows a magnification of the vertical scale in the region close to $B_{\max}/2$

To this graph we add B bonds. The maximal number of bonds that can be added is $B_{\max}(N) = [(N-1)(N-2)]/2$, where we do not allow more than one bond between any pair of nodes. When $B = B_{\max}(N)$ we obtain the complete graph, in which every node is connected to every other node [4]. The star graph and the complete graph have very distinct topologies: The star graph has a central node which has functionality $N - 1$, while each of the peripheral nodes has functionality 1. On the other hand, the complete graph has only nodes of functionality $N-1$. Both networks are "regular", in the sense that they are invariant under the "exchange" symmetry: Exchanging the positions of any pair of nodes (except for the central node of the star graph) leaves the networks unchanged, see Ref. [4]. The eigenvalue sets for these graphs can be obtained analytically: The star graph has only three distinct eigenvalues [14], namely $E_1 = 0$, $E_2 = \cdots = E_{N-1} = 1$, and $E_N = N$, with degeneracies $D(0) = 1$, $D(1) = N-2$, and $D(N) = 1$, respectively. The complete graph has only two distinct eigenvalues [25], $E_1 = 0$ and $E_2 = \cdots = E_N = N$, with degeneracies $D(0) = 1$ and $D(N) = N - 1$.

Randomly adding bonds to the star graph leads to networks with distinct topologies and with different eigenvalue sets: For B very close to 1 or to N, the number of graphs which are topologically distinct is small. For $B \ll 1$ and $N - B \ll 1$, this number is large, see Ref. [4]. The influence of the additional bonds on the transport efficiency gets reflected in the average probabilities $\bar{p}(t)$ for CTRW and in the lower bound $|\bar{\alpha}(t)|^2$ for CTQW, see Ref. [4]. The long-time CTQW behavior is given by $\langle P_{\mathrm{QW}} \rangle_R$. Figure 14.2 shows $\langle P_{\mathrm{QW}} \rangle_R$ as a function of B in which the number of realizations was taken to be $R = 10{,}000$. As B increases from zero one observes a decrease of $\langle P_{\mathrm{QW}} \rangle_R$ from the star graph-value, $P_{\mathrm{QW}}^{\mathrm{SG}} = (N^2 - 4N + 6)/N^2 \geq 1/N$. This decrease continues until a broad plateau centered around $B = B_{\max}/2$ is reached. Increasing B further results in an increase of $\langle P_{\mathrm{QW}} \rangle_R$ until the P_{QW} value for the complete graph, $P_{\mathrm{QW}}^{\mathrm{CG}} = (N^2 - 2N + 2)/N^2 \geq 1/N$, is attained. In the intermediate (plateau) domain $\langle P_{\mathrm{QW}} \rangle_R$ is bounded from below by $1/N$ and becomes comparable to the CTRW value. Therefore, we infer that by adding bonds to the star graph (until $B \approx B_{\max}/2$) one enhances – in the ensemble average – the efficiency of the CTQW.

14.5 Conclusions

In this article we have demonstrated differences in transport over given graphs in cases in which the transport is purely classical and given by CTRW and in cases when it is quantum mechanical in nature and described by CTQW. While under simple assumptions the corresponding factors are, for each graph, based on exactly the same set of eigenvalues and eigenfunctions, the implementation of the CTRW and of the CTQW leads to vastly different behaviors. We have exemplified this by focusing on two types of disordered structures, namely on SWN that start from a ring and on random structures intermediate between star- and complete graphs. We have also shown that the intuitive notion that adding bonds may lead to a quicker transport is not always fulfilled in the quantum case. In summary, we find that the physical properties of the transport are much more rich in the quantum than in the classical case.

Acknowledgements We thank Anastasiia Anishchenko for stimulating discussions. Support from the Deutsche Forschungsgemeinschaft (DFG Grant No. MU2925/1-1), from the Fonds der Chemischen Industrie, from the Deutscher Akademischer Austauschdienst (DAAD Grant No. 56266206), and from the Marie Curie International Research Staff Exchange Science Fellowship within the 7th European Community Framework Program SPIDER (Grant No. PIRSES-GA-2011-295302) is gratefully acknowledged.

References

1. Agliari E, Blumen A, Mülken O (2008) Dynamics of continuous-time quantum walks on restricted geometries. J Phys A 41:445301
2. Agliari E, Blumen A, Mülken O (2010) Quantum walk approach to search on fractal structures. Phys Rev A 82:012305
3. Alexander S, Orbach R (1982) Density of states on fractals: fractons. J Phys (Paris) Lett 43:625
4. Anishchenko A, Blumen A, Mülken O (2012) Enhancing the spreading of quantum walks on star graphs by additional bonds. Quantum Inf Process 11:1273
5. Bray A, Rodgers G (1988) Diffusion in a sparsely connected space: a model for glass relaxation. Phys Rev B 38:11461
6. Collini E, Wong CY, Wilk KE, Curmi PMG, Brumer P, Scholes GD (2010) Coherently wired light-harvesting in photosynthetic marine algae at ambient temperature. Nature 463:644
7. Darázs Z, Kiss T (2010) Pólya number of the continuous-time quantum walks. Phys Rev A 81:062319
8. Engel GS, Calhoun TR, Read EL, Ahn T-K, Manal T, Cheng Y-C, Blankenship RE, Fleming GR (2007) Evidence for wavelike energy transfer through quantum coherence in photosynthetic systems. Nature 446:782
9. Farhi E, Gutmann S (1998) Quantum computation and decision trees. Phys Rev A 58:915
10. Jespersen S, Sokolov IM, Blumen A (2000) Relaxation properties of small-world networks. Phys Rev E 62:4405
11. Kempe J (2003) Quantum random walks-an introductory overview. Contemp Phys 44:307
12. Mülken O, Bierbaum V, Blumen A (2006) Coherent exciton transport in dendrimers and continuous-time quantum walks. J Chem Phys 124:124905

13. Mülken O, Blumen A (2005) Spacetime structures of continuous-time quantum walks. Phys Rev E 71:036128
14. Mülken O, Blumen A (2006) Efficiency of quantum and classical transport on graphs. Phys Rev E 73:066117
15. Mülken O, Blumen A (2011) Continuous-time quantum walks: models for coherent transport on complex networks. Phys Rep 502:37
16. Mülken O, Pernice V, Blumen A (2007) Quantum transport on small-world networks: a continuous-time quantum walk approach. Phys Rev E 76:051125
17. Mülken O, Volta A, Blumen A (2005) Asymmetries in symmetric quantum walks on two-dimensional networks. Phys Rev A 72:042334
18. Nielsen MA, Chuang IL (2000) Quantum computation and quantum information. Cambridge University Press, Cambridge
19. Reitzner D, Hillery M, Feldman E, Bužek V (2009) Quantum searches on highly symmetric graphs. Phys Rev A 79:012323
20. Salimi S (2009) Continuous-time quantum walks on star graphs. Ann Phys 324:1185
21. Sokolov IM, Klafter J, Blumen A (2002) Fractional kinetics. Phys Today 55:48
22. van Kampen N (1992) Stochastic processes in physics and chemistry. North Holland, Amsterdam
23. Volta A, Mülken O, Blumen A (2006) Quantum transport on two-dimensional regular graphs. J Phys A 39:14997
24. Westermann S, Amthor T, de Oliveira AL, Deiglmayr J, Reetz-Lamour M, Weidemüller M (2006) Dynamics of resonant energy transfer in a cold Rydberg gas. Eur Phys J D 40:37
25. Xu X-P (2009) Exact analytical results for quantum walks on star graph. J Phys A 42:115205

Chapter 15
Excitations Transfer and Random Walks on Dynamic Contacts Networks

Raffaella Burioni, Elena Agliari, and Davide Cassi

Abstract Diffusion and spreading processes are strongly influenced by the topology of the substrate. An interesting example is provided by exchange reactions in the diffusion-limited regime, that model the spreading of an excitation among a population of randomly moving agents. In this case, the excitation is diffusing on an evolving dynamical graph, created by the istantaneous contacts of the moving agents. In recent works, the excitation random walk on the contacts graph generated by a set of random walkers moving on restricted geometries has been considered. We review here the properties of the process and we extend our results to the case of multiple excitations transfer with fermionic and bosonic statistics.

15.1 Introduction

Networks are one of the most flexible representations of a set of elements connected pairwise by a relation, and they are widely used to model complex phenomena in many different disciplines, ranging from physics to chemistry, to biology and social sciences [11, 19]. The elements connected by the relation are represented by sites, and relations or interaction are described by links, connecting two sites. When a dynamical or a statistical model is defined on the network, with variables living on the sites for example, the topology of the interactions has a strong influence on the dynamics or the equilibrium properties. Understanding this influence has been one of the central problems in networks research in the last decades.

In a large variety of systems however, the relations connecting two elements are not static and can change with time, leading to an effective dynamical network that

---

R. Burioni (✉) • E. Agliari • D. Cassi
Dipartimento di Fisica e Scienza della Terra and INFN, Universita' di Parma, Viale G.P.Usberti 7/A 43124 Parma, Italy
e-mail: raffaella.burioni@unipr.it; elena.agliari@unipr.it; davide.cassi@unipr.it

D. Matrasulov and H.E. Stanley (eds.), *Nonlinear Phenomena in Complex Systems: From Nano to Macro Scale*, NATO Science for Peace and Security Series C: Environmental Security, DOI 10.1007/978-94-017-8704-8_15,

evolves on the same time scale of the dynamical system itself. This is the rule rather than an exception in social relations, person to person communications, molecular interactions in biological systems, metabolic, ecologic and neural networks [9, 10, 12, 15, 17, 20, 25], and this observation has recently led to a new line of research on time varying and temporal networks [21, 23]. Interestingly, the dynamic nature of time varying networks has been shown to play a crucial role in equilibrium and non-equilibrium phenomena, still to be unraveled.

A particularly interesting class of dynamical network is that generated by interactions of moving agents in physical proximity, i.e. in contact. Contacts networks are observed in reaction diffusion processes in chemical species, in movements of humans and animals, and they come into play when a "reaction" occurs in case two agents are in contact, being the reaction a contagion process, an information spreading or word-of-mouth process, an excitation transfer or a chemical reaction: the excitation or the wave front moves on an effective network created by the instantaneous contacts of agents. In all these situations, the topological, metric and spectral properties of the dynamic networks evolve with time, and the time dependent adjacency matrix of the network can be determined once the rule for the contact process, or the explicit data for the contacts, are given [17].

We will consider here a basic yet interesting model of spreading on a dynamic network, generated by the contacts of N agents performing a simple random walk on a static substrate [6]. Each agent diffusing on the substrate represents a particle that can be either in an excited (A^*) or in an unexcited (A) state; when an excited particle meets an unexcited one, they react [13] according to the scheme

$$A^* + A \rightarrow A + A^*. \tag{15.1}$$

This reaction mechanism is known as *homogeneous energy transfer* and it takes place from a so-called donor (A^*) to a so-called acceptor (A), which occur to be sufficiently close. In this way, the excitation moving from one particle to another one can be looked at as a "second-level" random walker moving on the contact graph generated by the set of N agents, i.e. the set of particles A and A^*. Even if apparently simple, this model features interesting non linear many-body effects, that can be detected in more realistic dynamic contacts networks, induced by physical proximity. Moreover, the model can be extended to describe spreading of information and contagion processes on the dynamic contacts network [3, 5, 7]. We review here our previous results on a single excitation transfer and we extend our analysis to the case of transfer of multiple excitations with different exclusion properties, i.e. following either Fermi-Dirac or Bose-Einstein statistics.

As for the underlying static substrates, we devote a particular attention to a class of interesting experimental frameworks, the so called restricted geometries [4, 14]. This expression refers to two, possibly concurrent, situations: low dimensionality and small spatial extent. In the low dimensional case, we expect strong effects due to multiple encounters between random walkers. In finite volumes and mesoscopic samples, when the process occurs on a spatial scale too small to consider the infinite volume limit, an explicit and non trivial dependence on the number of walkers N and on the volume V, not only through the density N/V, can arise [7].

The paper is organized as follows. In Sect. 15.2 we introduce our model, and in Sect. 15.3 we review the main results on single excitation transfer. In Sect. 15.4 we discuss the case of multiple excitations with bosonic and fermionic statistics. Finally, Sect. 15.5 contains our conclusions and perspectives.

15.2 The Model: Excitations Transfer and Random Walks on the Dynamic Contacts Network

We consider N simple random walkers, moving on a finite structure of volume V (henceforth, the substrate), and we indicate the position of the i-th walker at time t with $x_i(t)$, being $i = 1, \ldots, N$. At time $t = 0$, the walkers are randomly distributed on the substrate and one of them carries an excitation. When two walkers i and j meet at time t, i.e their distance on the substrate is smaller than R, a time dependent link is established between them and if walker i is carrying the excitation, the excitation is transferred to walker j. We fix the collision radius $R = 1$ to avoid parity effects. If more than one walker meet the carrier (multiple hit), the excitation jumps on one of them, chosen randomly. The model described is a discrete-time stochastic process $\mathscr{X}(t)$, where the state space of the system is composed by the set of the random walkers [6]. At time t, the system is in state i if the excitation is on walker i. Then, the transition (or jump) probabilities, are themselves a stochastic process. In particular, at time t the transition probability from state i ($\mathscr{X}(t) = i$) to state j ($\mathscr{X}(t+1) = j$) is a function of the positions $x_i(t)$ and $x_j(t)$ of the two agents i and j, hence a function of two stochastic processes. In Figs. 15.1 and 15.2 we depict the random walk of the excitation among moving walkers.

For the moving agents, the interesting quantities are the probability $P_0(r,t)$ of being at 0, the origin, at time t, being r the initial position at $t = 0$ and the probability $F_0(r,t)$ of being at 0 for the first time at time t, again, being r the initial position at $t = 0$. The diffusion of the excitation on the contact network can also be looked as a random walk and this will be described through the following quantities:

- $\mathscr{J}(t)$, the average number of jumps performed by the excitation at time t.
- $\mathscr{S}(t)$, the average number of different states visited at time t; the probability $\mathscr{S}(t,k)$ that k different states have been visited by the system at time t, $\mathscr{S}(t) = \sum_{k=1}^{N} k\mathscr{S}(t,k)$.
- The *cover time* τ, defined as the average time required to the excitation to visit all the N walkers (an interesting quantity, analogous to the lattice-covering time for random walks [28]). We also define the *cover walk-length* as the average number of jumps required to visit all the states, and we denote it as π; of course, $\pi \leq \tau$.

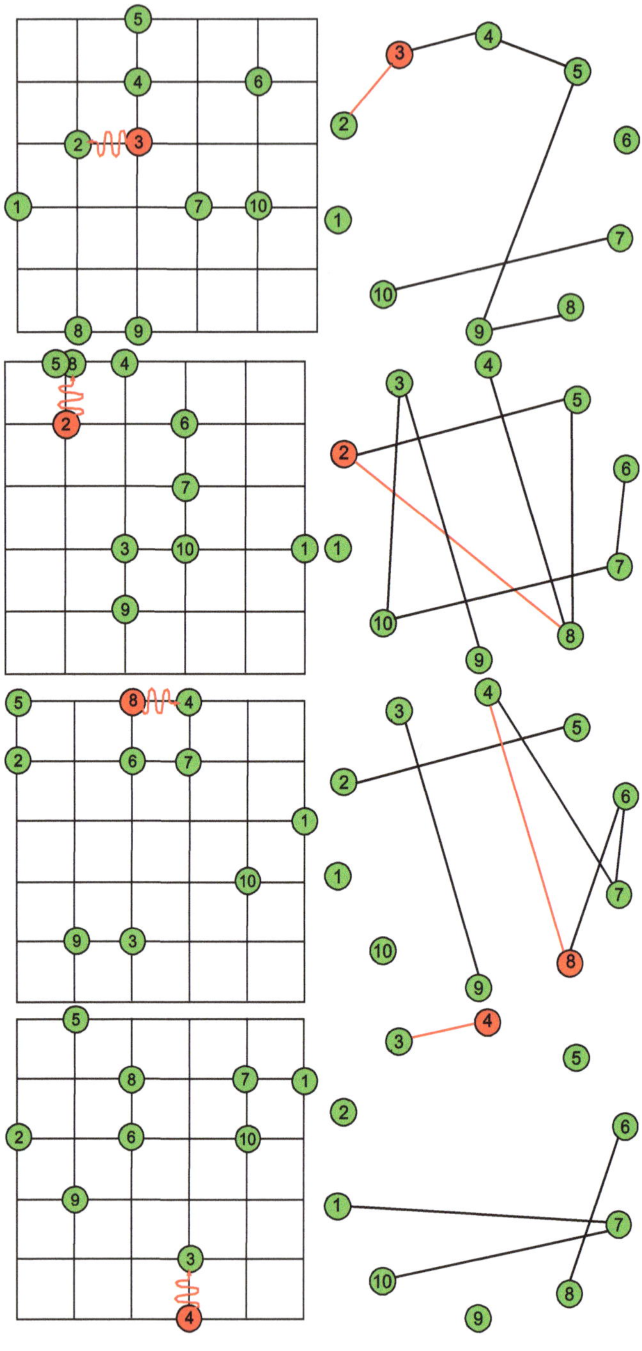

Fig. 15.1 A system made up of $N = 10$ particles on a square toroidal lattice sized $L = 5$ is depicted from two different perspectives: particles react via exchange (*left panels*); the excitation performs a random walk on the related dynamic graph (*right panels*). Time runs from top to bottom

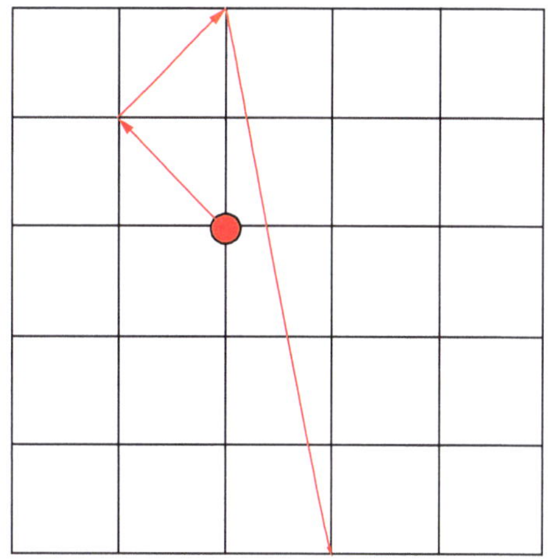

Fig. 15.2 This figure refers to Fig. 15.1 and it shows the motion of the excitation on the static substrate. Notice that, differently by a simple random walk, at each time step t, the excitation can either stop or move and jumps can possibly be towards next-nearest-neighbours

We consider as substrates hypercubic lattices of linear size L and volume $V = L^d$ (with $d = 1, 2, 3$), with periodic boundary conditions. To take into account the effects of topology, we also investigate the case of self-similar lattices, with given fractal and spectral dimensions, d_F and d_s respectively [16].

15.3 Single Excitation Random Walk

An analytical description of the system can be obtained in the limit of low density of walkers [6], i.e. $N/V \ll 1$. In fact, under this condition, multiple hits can be neglected and the model can be mapped into two equivalent problems. For example, the process can be seen in the reference frame of the excitation. In this frame, the walker carrying the excitation is stuck at the origin, and the remaining $N-1$ agents perform a simple random walk, with 2 jumps on each time step. The jump of the excitation from walker i to walker j corresponds to the following process: walker j hits the origin and gets stuck, while walker i gets free and starts performing its own walk. In this picture, the process is a double-state random-walk process [28], as each walker can exist in two different states: either stuck at the origin or free. In a second picture, the problem is mapped into a label switching process: the walker originally carrying the excitation stays fixed at the origin, and when a walker crosses the origin they exchange their label. This second picture has the advantage of mapping all the relevant random walk quantities into trapping problems for the free particles [26]. Both pictures are rigorously valid only on translation invariant lattices, but the validity can be extended, in the large time limit, to self similar structures [6].

First, the number of jumps at time t is easily calculated in the label switching picture. In the low density regime, the number of jumps performed at time t is the number of passages through the origin made by $N-1$ random walkers at time t, that

Finally, let us consider the probability distribution $\mathscr{S}(k,t)$ for the k distinct agents visited at time t. This quantity corresponds to the probability that the number of walkers absorbed into a trap at the origin is k. Recalling that $U(t)$ is the probability that a given walker has survived up to t, we have

$$\mathscr{S}(k,t) = U(t)^{N-k}(1-U(t))^{k-1}\binom{N-1}{k-1}, \quad 1 \le k \le N,$$

that is, recalling that for Euclidean lattices $d_s \equiv d$,

$$\mathscr{S}(k,t) = e^{-(N-1)\lambda_{d_s}t/g_{d_s}(V)}\left(e^{\lambda_{d_s}t/g_{d_s}(V)} - 1\right)^{k-1}\binom{N-1}{k-1}. \tag{15.9}$$

Notice that, in the thermodynamic limit, Eq. 15.9 becomes a Poissonian distribution with average $\mu = \lambda_{d_s}(N-1)t/g_{d_s}(V)$.

The time $t_{\text{peak}}(k)$, each distribution is peaked at, can be directly derived from Eq. 15.9 as

$$t_{\text{peak}}(k) = \frac{V}{\lambda_d}\log\left(\frac{N-1}{N-k}\right). \tag{15.10}$$

An important feature concerning $\mathscr{S}(k, t_{\text{peak}}(k))$ is that it exhibits a minimum at $k = \tilde{k} = (N+1)/2$, as can be deduced from Eqs. 15.9 and 15.10.

It is as well possible to calculate the average time τ_{N-k} spent by the system in state corresponding to *exactly* k distinct states visited:

$$\tau_{N-k} = \sum_{t=0}^{\infty}\mathscr{S}(k,t) \sim \frac{V}{\lambda_{d_s}(N-k)}, \tag{15.11}$$

where the last relation was derived in the continuum limit for t.

All the previous calculations are very well verified numerically, as shown by the plots in Figs. 15.3 and 15.4.

Finally, we mention that the same techniques presented here, based on first passage quantities, can be used to investigate different types of contact reactions. A particular interesting one is, using the same chemical language, the so called autocatalytic reaction:

$$A^* + A \rightarrow A^* + A^*, \tag{15.12}$$

so that after the contact both particles are in the excited state. Clearly, this reaction is able to describes contagion processes, information spreading and word-of-mouth diffusion [3, 5, 7] among moving agents. In particular, in the case of information spreading in presence of progressive degradation, an interesting effect due to

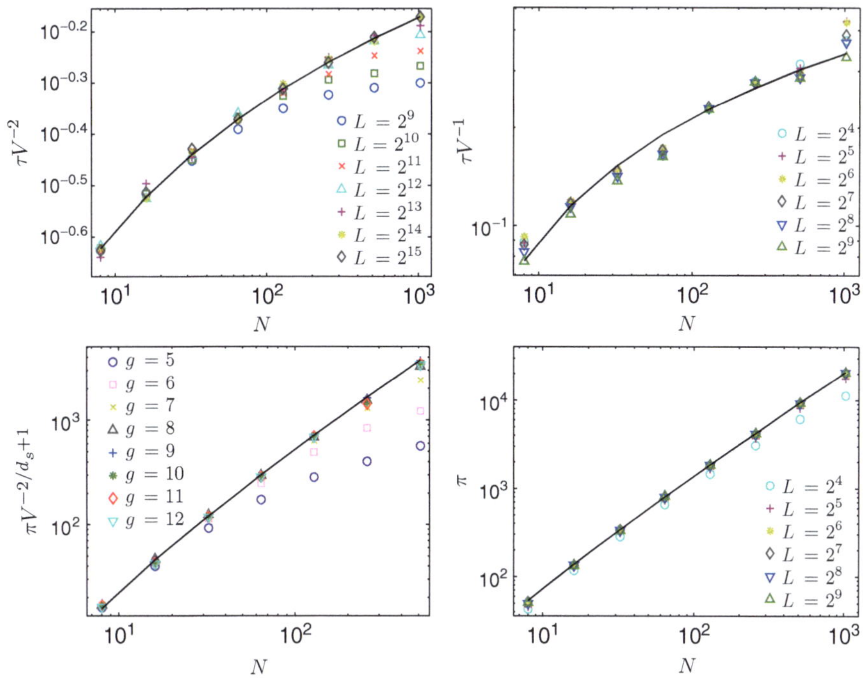

Fig. 15.3 *Upper panels*: Average cover time $\tau(N, V)$, properly rescaled, versus the number N of particles making up the system and diffusing on a periodic chain (*left panel*) and on a cubic lattice (*right panel*). Equation 15.4 provides the best fit when reactants concentration is small. *Lower panels*: Average cover walk-length $\pi(N, V)$ versus N for a system of particles diffusing and reacting on a Sierpinski gasket (*left panel*) and in a cubic lattice with periodic boundary conditions (*right panel*). Different sizes are depicted, as shown by the legend. Equation 15.5 provides the best fit when reactants concentration is small. Notice that in the latter case $\pi(N, V)$ is independent of V

crossing from high density to low density regimes has been identified, leading to non trivial effects in the dependence on N and V separately [7]. Interestingly, this kind of effects are signatures of contact networks dynamics.

15.4 Multiple Excitations Random Walk with Bosonic and Fermionic Statistics

In this section we consider a system where the number of excitations N_e diffusing on the dynamic graph is larger than one. In this case, we can distinguish between two alternative scenarios, according to possible constraints on the number ϵ_i of excitations which can be hosted by the same particle i:

- $\epsilon_i \in \{0, 1\}, i = 1, \ldots, N,$
- $\epsilon_i \in \{0, 1, \ldots, N_e\}, i = 1, \ldots, N.$

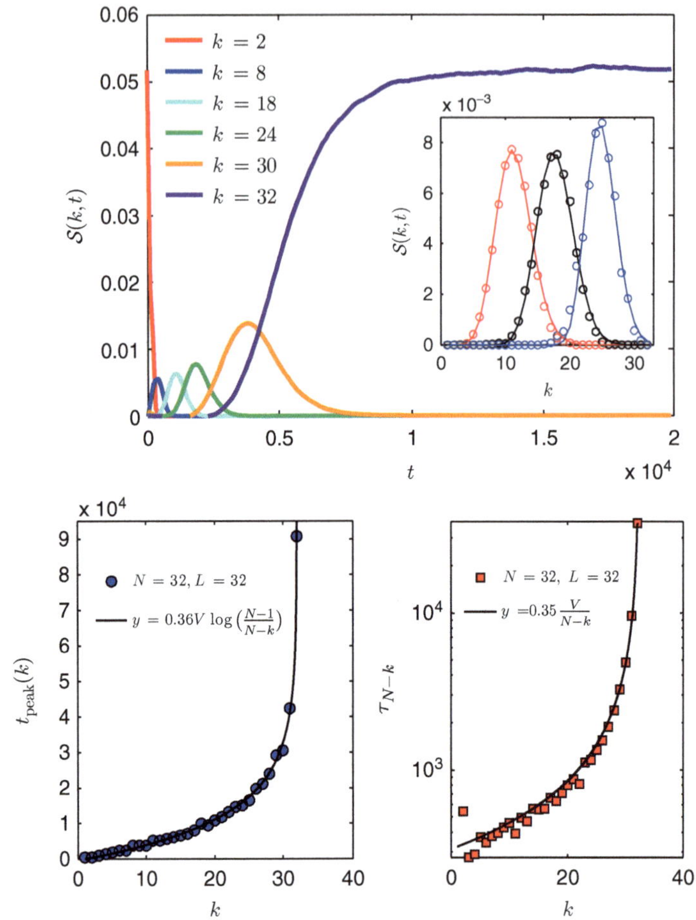

Fig. 15.4 *Upper panel*: Probability distribution $\mathscr{S}(k,t)$ versus time t for a system of $N = 32$ particles diffusing on a cubic lattice sized $L = 16$ with periodic boundary conditions. Each curve represents a different value of k; as explained by the legend, only a subset of curve is depicted seeking for readability. The *inset* shows $\mathscr{S}(k,t)$ versus the number of visited random walkers k for three different instants of time; from left to right: $t = 6 \cdot 10^2, 1.1 \cdot 10^3, 2 \cdot 10^3$. Data points ($\circ$) are fitted by a Poissonian distribution with average $\mu = \lambda_d t N/V$ in agreement with Eqs. 15.4 and 15.9. *Lower panels*: $t_{\text{peak}}(k)$ (*left*) and τ_{N-k} (*right*) as a function of k for a periodic cubic lattice. The *solid lines* (whose equations are reported) represent the best fits in agreement with Eqs. 15.10 and 15.11, respectively. The only free parameter is λ_d and we get $\lambda_d^{fit} = 2.79 \pm 0.07$

With obvious meaning, we refer to such distinct cases as *fermionic* (hereafter labeled as F) and *bosonic* (hereafter labeled as B), respectively.

The latter case is the easiest to treat since each excitation performs an independent walk on the dynamic graph, yet one has to take into account the fact that the

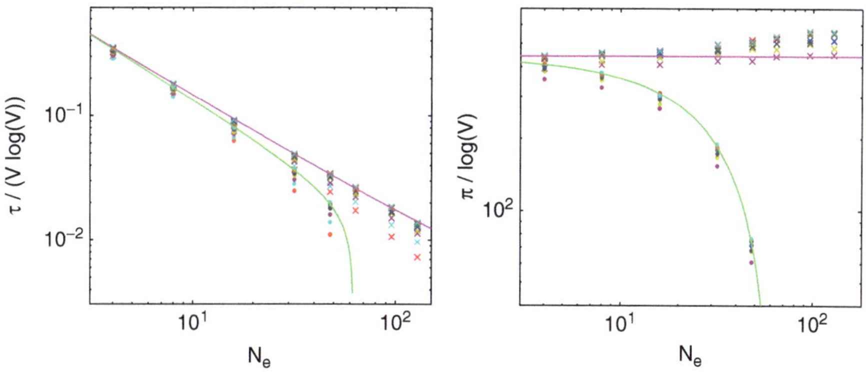

Fig. 15.5 Average final time $\tau(N, N_e, V)$ (*left*) and walk length $\pi(N, N_e, V)$ (*right*), as a function of N_e, for fermionic (•) and bosonic (×) systems. Data points refers to $N = 128$ particles reacting and diffusing on a toroidal square lattice with size ranging from $L = 2^8$ to $L = 2^{12}$. The lower the concentration the best the agreement with Eqs. 15.13 and 15.14, respectively

number of donors is a stochastic variable. Of course, when excitation concentration $\rho_e = N_e/N$ is small enough, we expect the two cases to asymptotically converge.

Following arguments similar to those described above, we can properly extend previous results allowing for $N_e \neq 1$; here we will explicitly consider only the case of homogeneous substrates, but the analysis can be straightforwardly extended also to self-similar structures like fractals.

Let us start with the average cover time, for which we find

$$\tau^B(N, N_e, V) = \frac{\tau(N, V)}{N_e} \sim g_d(V)\frac{\log N}{N_e},$$
$$\tau^F(N, N_e, V) \sim g_d(V)\frac{\log(N - N_e)}{N_e}, \tag{15.13}$$

while the average cover walk-length is given by

$$\pi^B(N, N_e, V) \sim \frac{g_d(V)}{V}\begin{cases} N \log N & \text{for } N_e \ll N, \\ N_e \log N & \text{for } N_e \approx N, \end{cases}$$
$$\pi^F(N, N_e, V) \sim \frac{g_d(V)}{V}(N - N_e)\log(N - N_e); \tag{15.14}$$

where we simply provided the leading term. Notice that the latter quantity best evidences the difference between the two systems: $\pi^B(N, N_e, V)$ is a (non-strictly) monotonically increasing function of N_e, conversely, just due to the constraints on ϵ_i, $\pi^F(N, N_e, V)$ decreases with N_e.
The analytical results for $\tau^{B,F}$ and $\pi^{B,F}$ are checked numerically in Fig. 15.5.

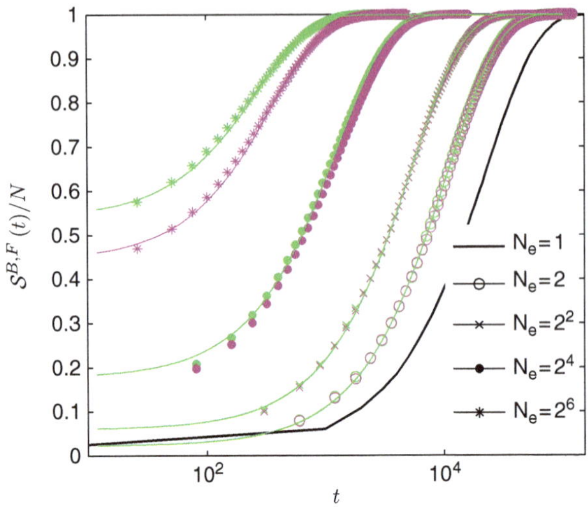

Fig. 15.6 Average percentage of distinct particles visited by N_e excitations for a system made up of $N = 128$ particles diffusing on a toroidal square lattice sized $L = 128$; different choices of N_e correspond to different symbols as explained by the legend. Notice that the difference between bosonic (*pink*) and fermionic (*green*) system is negligible when $N_e/N \ll 1$. The curves represent the best fit according to Eq. 15.15 and its extension for the bosonic case. The survival probability $U(N_e/N, t)$ considered is pure exponential; despite this approximation, fit coefficients agree, within the error (about 5 %), with predictions

Let us now consider the average number of distinct sites visited by the whole set of excitations diffusing on the dynamic graph. For the fermionic system the problem can again be solved by exploiting known results about the survival probability $U(\rho_{\text{trap}}, t)$ for a random walk with trap concentration $\rho_{\text{trap}} = N_e/V$:

$$\mathscr{S}^F(t) = N_e + (N - N_e)\left[1 - U\left(\frac{N_e}{N}, t\right)\right]. \tag{15.15}$$

In the bosonic case we must take into account that the number of donors follows a probability distribution and, for a given realization, it is given by $\sum_{i=0}^{N}(1 - \delta_{\epsilon_i,0})$. When N_e is large, we can replace the quantity N_e/N appearing in Eq. 15.15 with $1 - (1 - 1/N)^{N_e}$. Consequently, and as can be verified from Fig. 15.6, $\mathscr{S}^B(t) \leq \mathscr{S}^F(t)$ and $\left[\mathscr{S}^B(t) - \mathscr{S}^F(t)\right]$ goes to zero as N_e is lowered.

The "efficiency" of the fermionic system can also be evinced from another point of view: in Fig. 15.7 (left panel) we show numerical results for the probability distributions $\mathscr{S}^B(k, t)$ and $\mathscr{S}^F(k, t)$: the former are more spread and peaked at larger times. Such a difference can possibly be reduced by setting the same (fermionic) initialization for both systems (see the right panel in Fig. 15.7). This means that, at $t = 0$, also the bosonic system displays exactly N_e donors. With this new arrangement the difference between $\mathscr{S}^B(k, t)$ and $\mathscr{S}^F(k, t)$ is reduced, though the trend described above emerges more and more as time goes by. In any case,

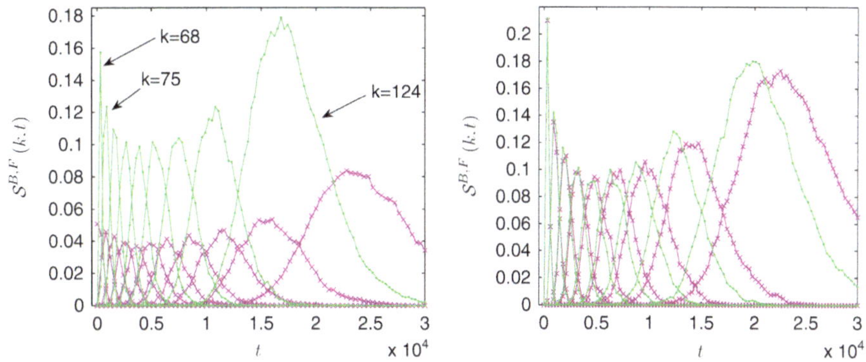

Fig. 15.7 Probability distributions $\mathscr{S}^B(k,t)$ and $\mathscr{S}^F(k,t)$ for a bosonic system ($\times$) and a fermionic system ($\bullet$) whose parameters are $N = 128$, $N_e = 64$; the substrate is a square lattice sized $L = 512$, with periodic boundary conditions. Several different values for k are depicted; lines are guide to the eye. *Left*: systems are initialized differently according to their own nature; *Right*: both systems are initialized in such a way that, at $t = 0$, the number of donors is N_e

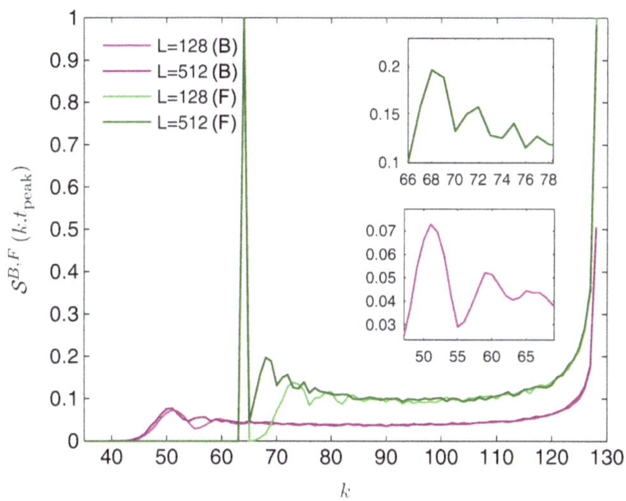

Fig. 15.8 $\mathscr{S}^B(k,t_{\text{peak}})$ *(pink curves)* and $\mathscr{S}^F(k,t_{\text{peak}})$ *(green curves)* for a two dimensional square lattice with $N = 128$, $N_e = 64$; two different sizes are considered, as shown by the legend. *Insets* show a magnification of the region where oscillations emerge

$\mathscr{S}(k,t_{\text{peak}})$ exhibits a minimum at $k = \tilde{k}$ such that $\mathscr{S}(\tilde{k},t_{\text{peak}}) \approx \frac{1}{2}[N + \mathscr{S}(0)]$, whose origin lies, again, in the maximum extent for the configurational space pertaining to that time.

There is also a new feature with respect to the case $N_e = 1$ which is, indeed, a very effect of excitations collective behaviour. For small values of k, both distributions $\mathscr{S}^B(k,t_{\text{peak}})$ and $\mathscr{S}^F(k,t_{\text{peak}})$, represented in Fig. 15.8, exhibit an oscillatory profile. The period of such oscillations depends, essentially, on $\frac{N-\mathscr{S}(t)}{V}$, and it stems from the peaked distribution governing the number of excitations jumping at each time. As k grows such distribution is more and more spread and, consequently, oscillations vanish.

15.5 Conclusions

Time varying networks generated by contacts of moving agents are one of the most interesting and intriguing topics in network sciences, nowadays. Their relevance covers topics in chemical kinetics, biology and ecology, social sciences and communications. Here we have considered a random walk process mimicking the diffusion of excitations among a population of agents moving on a restricted geometry. In particular, we have considered the case of a single excitation transfer and of multiple excitations with fermionic and bosonic statistics, evidencing the main properties of this more complex case that features interesting collective behavior. Analytical estimates of first passage related quantities have been calculated in the low density limit on lattices and self similar substrates, which are in excellent agreement with numerical simulations. These quantities feature a strong dependence on the topology of the substrate in low dimensions, as well as interesting non trivial dependence on the number of agents and on the volume of the substrate.

References

1. Agliari E (2008) Exact mean first-passage time on the T-graph. Phys Rev E 77:011128
2. Agliari E, Burioni R (2009) Random walks on deterministic scale-free networks: exact results. Phys Rev E 80:031125
3. Agliari E, Burioni R, Cassi D, Neri FM (2006) Efficiency of information spreading in a population of diffusing agents. Phys Rev E 73:046138
4. Agliari E, Burioni R, Cassi D, Neri FM (2007) Autocatalytic reaction-diffusion processes in restricted geometries. Diffus Fundam 7:1.1–1.8
5. Agliari E, Burioni R, Cassi D, Neri FM (2007) Universal features of information spreading efficiency on d-dimensional lattices. Phys Rev E 75:021119
6. Agliari E, Burioni R, Cassi D, Neri FM (2008) Random walk on a population on random walkers. J Phys A 41:015001–015013
7. Agliari E, Burioni R, Cassi D, Neri FM (2010) Word-of-mouth and dynamical inhomogeneous markets: efficiency measure and optimal sampling policies for the pre-launch stage. IMA J Manag Math 21:67
8. Agliari E, Burioni R, Manzotti A (2010) Effective target arrangement in a deterministic scale-free graph. Phys Rev E 82:011118
9. Bajardi P, Barrat A, Natale F, Savini L, Colizza V (2011) Dynamical patterns of cattle trade movements. PLoS One 6:e19869
10. Bansal S, Read J, Pourbohloul B, Meyers LA (2010) The dynamic nature of contact networks in infectious disease epidemiology. J Biol Dyn 4:478

11. Barrat A, Barthélemy, Vespignani A (2008) Dynamical processes on complex networks. Cambridge University Press, Cambridge
12. Bassett DS, Wymbs NF, Porter MA, Mucha PJ, Carlson JM, Grafton ST (2011) Dynamic reconfiguration of human brain networks during learning. Proc Natl Acad Sci USA 108:7641–7646
13. ben-Avraham D, Havlin S (2000) Diffusion and reactions in fractals and disordered systems. Cambridge University Press, Cambridge
14. Bénichou O, Chevalier C, Klafter J, Meyer B, Voituriez R (2010) Geometry-controlled kinetics. Nat Chem 2:472
15. Bullmore E, Sporns O (2009) Complex brain networks: graph theoretical analysis of structural and functional systems. Nat Rev Neurosci 10:186
16. Burioni R, Cassi D (2005) Random walks on graphs: ideas, techniques and results. J Phys A 38:R45–R78
17. Cattuto C, van den Broeck W, Barrat A, Colizza V, Pinton JF, Vespignani A (2010) Dynamics of person-to-person interactions from distributed RFID sensor networks. PLoS ONE 5:e11596
18. Condamin S, Bénichou O, Voituriez R, Klafter J (2007) First passage times in complex scale invariant media. Nature 450:77–80
19. Durrett R (2007) Random graphs dynamics. Cambridge University Press, Cambridge
20. Harary F, Gupta G (1997) Dynamic graph models. Math Comput Model 25:79
21. Holme P, Saramaki J (2012) Temporal networks. Phys Rep 519:97–125
22. Kozak JJ, Balakrishnan V (2002) Mean first-passage time for random walks in general graphs with a deep trap. Phys Rev E 64:021105
23. Kuhn F, Oshman R (2011) Dynamic networks: models and algorithms. ACM SIGACT News 42:82–96
24. Montroll EW, Weiss GH (1965) Random walks on lattices. J Math Phys 6:167
25. Perra N, Baronchelli A, Mocanu D, Goncalves B, Pastor-Satorras R, Vespignani A (2012) Random walks and search in time-varying networks. Phys Rev Lett 109:238701
26. Redner S (2001) A guide to first-passage processes. Cambridge University Press, Cambridge
27. Tejedor V, Bénichou O, Voituriez R (2009) Global mean first-passage times of random walks on complex networks. Phys Rev E 80:065140
28. Weiss GH (1994) Aspects and applications of the random walk. North-Holland, Amsterdam
29. Yuste SB, Acedo L, Lindenberg K (2001) Order statistics for d-dimensional diffusion processes. Phys Rev E 64:052102

Chapter 16
Ballistic Soliton Transport in Networks

Zarif A. Sobirov, K.K. Sabirov, Davron Matrasulov, A.A. Saidov, and K. Nakamura

Abstract We treat the problem reflectionless (ballistic) soliton transport on simplest networks and their combinations via solving nonlinear Schrödinger equation on simplest graphs. In particular, the relation between the nonlinearity coefficients of different bonds describing conditions for ballistic transport is derived for star graph. It is shown that the method can be extended to different simplest graph topologies and their combinations.

16.1 Introduction

Particle and wave transport in discrete structures and networks is of considerable fundamental and practical interest for many problems arising in mesoscopic physics. The problem of nonlinear waves and solitons in networks described by stationary and time-dependent Schrodinger equation on graphs has been attacked by few groups during last few years.

In particular, soliton solutions and connection formulae are derived for simple graphs in the Ref. [12]. The problem of fast solitons on star graphs is treated in the Ref. [5] where the estimates for the transmission and reflection coefficients are obtained in the limit of very high velocities. The problem of soliton transmission and reflection is studied in [9] by solving numerically the stationary NLSE on graphs. Dispersion relations for linear and nonlinear Schrödinger equations on graphs are

Z.A. Sobirov (✉)
Tashkent State Financial Institute, Tashkent, Uzbekistan
e-mail: sobirovzar@gmail.com

K.K. Sabirov • D. Matrasulov • A.A. Saidov • K. Nakamura
Turin Polytechnic University in Tashkent, 17, Niyazov Str., 100095 Tashkent, Uzbekistan

D. Matrasulov and H.E. Stanley (eds.), *Nonlinear Phenomena in Complex Systems: From Nano to Macro Scale*, NATO Science for Peace and Security Series C: Environmental Security, DOI 10.1007/978-94-017-8704-8_16,

discussed in [8]. The problem of scattering from nonlinear networks described by stationary NLSE is studied in the Ref. [10].

The stationary NLSE with power focusing nonlinearity on star graphs was studied in recent papers [5–7], where existence of the nonlinear stationary states are shown for δ–type boundary conditions. In particular, the authors of [5] considered a star graph with N semi-infinite bonds, for which they obtain the exact solutions for the boundary conditions with $\alpha \neq 0$. Exact analytical solutions of stationary NLSE on graphs are obtained in [11] for simplest topologies.

In this work we treat the problem of reflectionless (ballistic) transport of solitons on simple networks. Nonlinearity coefficients for different bonds are considered to be different. The relation between the nonlinearity coefficients for which ballistics transport is possible is derived. Soliton solution for such case is obtained analytically.

16.2 NLSE on Network: Conservation Rules

Let us first consider an elementary branched chain or a primary star graph (PSG) where the vertex site is now taken as origin O. Space coordinates on individual bonds are here defined as follows: $b_1 \sim (-\infty, 0), b_2 \sim (0, +\infty)$ and $b_3 \sim (0, +\infty)$. On each bond we have the nonlinear Schrödinger equation (NLSE)

$$i\frac{\partial \Psi_k}{\partial t} + \frac{\partial^2 \Psi_k}{\partial x_k^2} + \beta_k |\Psi_k|^2 \Psi_k = 0, \quad k = 1, 2, 3, \tag{16.1}$$

with x_k defined on $-\infty < x_1 < 0$, $0 < x_2, x_3 < \infty$. It should be noted that the strength of nonlinearity $\beta_k (> 0)$ may be different among bonds. We will treat the solution of NLSE on PSG which satisfies the following conditions at infinity: $\Psi_1(x_1) \to 0$ at $x_1 \to -\infty$, $\Psi_k(x_k) \to 0$ at $x_k \to \infty$ for $k = 2, 3$. One of the physically important conditions for the solution in PSG is the norm conservation. The norm is defined as

$$N = \|\Psi\|^2 = \int_{-\infty}^{0} |\Psi_1(x,t)|^2 dx + \int_{0}^{\infty} |\Psi_2(x,t)|^2 dx + \int_{0}^{\infty} |\Psi_3(x,t)|^2 dx. \tag{16.2}$$

Let us find conditions for which the norm is conservative. For this purpose we calculate its time-derivative:

$$\frac{d}{dt}N = \int_{-\infty}^{0} \frac{\partial |\Psi_1(x,t)|^2}{\partial t} dx + \int_{0}^{\infty} \frac{\partial |\Psi_2(x,t)|^2}{\partial t} dx + \int_{0}^{\infty} \frac{\partial |\Psi_3(x,t)|^2}{\partial t} dx.$$

From Eq. (16.1) we have the continuity equation,

$$\begin{aligned}\frac{\partial\,|\Psi_k(x,t)|^2}{\partial t} &= -\frac{\partial}{\partial x} j_k(x,t)\\ &\equiv -2\frac{\partial}{\partial x}\mathrm{Im}\left[\Psi_k^*(x,t)\frac{\partial \Psi_k(x,t)}{\partial x}\right],\end{aligned} \tag{16.3}$$

where $j_k(x,t)$ means the current density.

Using Eq. (16.3), we find that the norm is conservative only when the following connection formula at the vertex is satisfied:

$$\mathrm{Im}\left[\Psi_1^*\frac{\partial \Psi_1}{\partial x}\right]\bigg|_{x=0} = \mathrm{Im}\left[\Psi_2^*\frac{\partial \Psi_2}{\partial x}\right]\bigg|_{x=0} + \mathrm{Im}\left[\Psi_3^*\frac{\partial \Psi_3}{\partial x}\right]\bigg|_{x=0}. \tag{16.4}$$

In Eq. (16.4) we prescribe $[\cdots]|_{x=0}$ to $\lim_{x_1\to-0}[\cdots]$ for variables on bond b_1 and to $\lim_{x_1\to+0}[\cdots]$ for variables on bond $b_{1,2}$. Hereafter the same prescription as above will be employed.

The equality in Eq. (16.4) implies the local current conservation condition at the vertex O,

$$j_1(0,t) = j_2(0,t) + j_3(0,t). \tag{16.5}$$

16.2.1 Energy Conservation Rule

The second important condition for the solution on PSG is the energy conservation. In PSG, the energy is defined as

$$E = E_1 + E_2 + E_3, \tag{16.6}$$

where

$$E_k = \int_{b_k}\left(\left|\frac{\partial \Psi_k}{\partial x}\right|^2 - \frac{\beta_k}{2}|\Psi_k|^4\right)dx. \tag{16.7}$$

Evaluating time derivative of $E(t)$ using Eq. (16.1) we obtain

$$\frac{d}{dt}E = \mathrm{Re}\left[\frac{\partial \Psi_1^*}{\partial x}\frac{\partial \Psi_1}{\partial t}\right]\bigg|_{x=0} - \mathrm{Re}\left[\frac{\partial \Psi_2^*}{\partial x}\frac{\partial \Psi_2}{\partial t}\right]\bigg|_{x=0} + \mathrm{Re}\left[\frac{\partial \Psi_3^*}{\partial x}\frac{\partial \Psi_3}{\partial t}\right]\bigg|_{x=0}. \tag{16.8}$$

Consequently, the energy is conserved if the following rule is satisfied:

$$\mathrm{Re}\left[\frac{\partial \Psi_1^*}{\partial x}\frac{\partial \Psi_1}{\partial t}\right]\bigg|_{x=0}$$
$$= \mathrm{Re}\left[\frac{\partial \Psi_2^*}{\partial x}\frac{\partial \Psi_2}{\partial t}\right]\bigg|_{x=0} + \mathrm{Re}\left[\frac{\partial \Psi_3^*}{\partial x}\frac{\partial \Psi_3}{\partial t}\right]\bigg|_{x=0}, \tag{16.9}$$

which is another connection formula at the vertex O.

16.2.2 Boundary Condition at the Vertex and Sum Rule for Nonlinearity Coefficients

The norm and energy are conserved, provided the (nonlinear) boundary conditions in Eqs. (16.4) and (16.9) at the vertex are valid. These boundary conditions are found to be satisfied by employing either one of the following linear connection formulas at the vertex O,

$$\alpha_1 \Psi_1|_{x=0} = \alpha_2 \Psi_2|_{x=0} = \alpha_3 \Psi_3|_{x=0};$$
$$\frac{1}{\alpha_1}\frac{\partial \Psi_1}{\partial x}\bigg|_{x=0} = \frac{1}{\alpha_2}\frac{\partial \Psi_2}{\partial x}\bigg|_{x=0} + \frac{1}{\alpha_3}\frac{\partial \Psi_3}{\partial x}\bigg|_{x=0}, \tag{16.10}$$

or

$$\alpha_1 \frac{\partial \Psi_1}{\partial x}\bigg|_{x=0} = \alpha_2 \frac{\partial \Psi_2}{\partial x}\bigg|_{x=0} = \alpha_3 \frac{\partial \Psi_3}{\partial x}\bigg|_{x=0};$$
$$\frac{1}{\alpha_1}\Psi_1|_{x=0} = \frac{1}{\alpha_2}\Psi_2|_{x=0} + \frac{1}{\alpha_3}\Psi_3|_{x=0}, \tag{16.11}$$

where α_1, α_2 and α_3 are arbitrary real constants.

Among many possible choices of α_1, α_2 and α_3, there is one special case in which an infinite number of constants of motion can be found and NLSE on PSG becomes completely integrable. We shall now consider this case by finding suitable values for α_1, α_2 and α_3.

Let us assume that there exists a bond-independent universal function $g(x,t)$ underlying PSG, which satisfies

$$\alpha_k \Psi_k|_{x=0} = g(0,t),$$
$$\alpha_k \frac{\partial \Psi_k}{\partial x}\bigg|_{x=0} = \frac{\partial g(x,t)}{\partial x}\bigg|_{x=0} \tag{16.12}$$

for $k = 1, 2$, and 3. The upper half of Eq. (16.10) is identical to the upper half of Eq. (16.12). With use of Eq. (16.12) the lower half of Eq. (16.10) can also be satisfied under the constraint,

$$\frac{1}{\alpha_1^2} = \frac{1}{\alpha_2^2} + \frac{1}{\alpha_3^2}. \tag{16.13}$$

Similarly, using Eqs. (16.12) and (16.13), we find that Eq. (16.11) can be satisfied as well.

The question arising after the above treatment can be formulated as following: Is there any bond-independent universal function which should satisfy Eq. (16.12)? The answer is yes. In fact, the general soliton solution of the integrable nonlinear Schrödinger equation (NLSE) in 1-d chain,

$$i\frac{\partial \Psi}{\partial t} + \frac{\partial^2 \Psi}{\partial x^2} + \beta|\Psi|^2\Psi = 0, \tag{16.14}$$

takes the form (see Zakharov-Shabat [13])

$$\Psi(x,t) = \sqrt{\frac{2}{\beta}} iq(x,t), \tag{16.15}$$

where $q(x,t)$ stands for the β-independent universal solution which satisfies the completely-integrable NLSE with $\beta = 2$,

$$iq_t + q_{xx} + 2q|q|^2 = 0, \qquad -\infty < x < +\infty. \tag{16.16}$$

Now we can introduce the solution of NLSE in Eq. (16.1) on PSG, which, on each bond, is composed of the universal soliton solution $q(x,t)$ on a simple 1-d chain but multiplied by the inverse of square root of bond-dependent nonlinearity β_k:

$$\Psi_k(x_k,t) = \sqrt{\frac{2}{\beta_k}} iq(x_k,t), \tag{16.17}$$

where the functions $q(x_1,t)$ and $q(x_{2,3},t)$ satisfy Eq. (16.16) and are defined on $(-\infty; 0]$ and $[0, +\infty)$, respectively.

Then, noting the above fact and choosing

$$\alpha_k = \sqrt{\beta_k} \qquad (k = 1,2,3), \tag{16.18}$$

Eq. (16.12) at the vertex is reduced to

$$\begin{aligned} \sqrt{\beta_k}\Psi_k|_{x=0} &= \sqrt{2}iq(0,t); \\ \sqrt{\beta_k}\frac{\partial \Psi_k}{\partial x}\bigg|_{x=0} &= \sqrt{2}i\frac{\partial q(x_k,t)}{\partial x}\bigg|_{x=0} \end{aligned} \tag{16.19}$$

with $k = 1, 2$, and 3. Here $\sqrt{2}iq(x,t)$ corresponds to $g(x,t)$ in Eq. (16.12). At the same time, the constraint in Eq. (16.13) becomes

$$\frac{1}{\beta_1} = \frac{1}{\beta_2} + \frac{1}{\beta_3}, \tag{16.20}$$

which means the sum rule for strengths of nonlinearity around the vertex. Thus the general solution in Eq. (16.17) has proved to satisfy the boundary condition that guarantees the norm and energy conservation rules for PSG.

Summarizing, the norm and energy conservation rules are satisfied by the connection formula in Eqs. (16.10) or (16.11) at the vertex. Among many possible choices of α_k, an interesting integrable case occurs when α_k takes the value in Eq. (16.18) and strengths of nonlinearity around the vertex satisfy the sum rule in Eq. (16.20). In this case the general soliton solution on PSG is given by Eq. (16.17). Equation (16.20) plays a crucial role: unless $\beta_1 \neq \beta_2, \beta_3$, no interesting bifurcation of a soliton propagation occurs at the vertex.

16.2.3 An Infinite Number of Conservation Rules

In the previous section, we showed that the norm and energy conservation rules can be satisfied by the general solution in Eq. (16.17) which is composed of Zakharov-Shabat solution of the integrable NLSE with $\beta = 2$ in Eq. (16.16). Below we shall show that, so long as the general solution on PSG is described by parts of the universal scaled function $q(x,t)$, all the conservation laws for 1-d chain should hold for PSG under the sum rule Eq. (16.20).

Applying Zakharov-Shabat's theorem [13] to each of three bonds, we now investigate the following quantity:

$$Q_n(t) \equiv \sum_{k=1}^{3} \beta_k^{-1} \int_{b_k} f_n(q(x_k,t))dx_k, \tag{16.21}$$

where $q(x_k,t)$ is the solution of Eq. (16.16) in the bond b_k and $f_n(q(x,t))$ obeys the recursion relation (see Eq. (35) of [13]):

$$f_{n+1} = q\frac{\partial}{\partial x}\left(\frac{1}{q}f_n\right) + \sum_{j+l=n} f_j f_l,$$
$$f_1 = |q|^2. \tag{16.22}$$

In fact, with use of Eq. (16.20), the r.h.s. of Eq. (16.21) turns out:

$$\beta_1^{-1}\int_{-\infty}^{0} f_n(q(x,t))dx + (\beta_2^{-1}+\beta_3^{-1})\int_{0}^{+\infty} f_n(q(x,t))dx$$
$$= \beta_1^{-1}\int_{-\infty}^{+\infty} f_n(q(x,t))dx = \beta_1^{-1}(2i)^n C_n, \tag{16.23}$$

where the second equality is due to the conservation rule for the 1-d chain [13] and C_n is constant. Hence Q_n has proved to be a constant of motion.

It is easy to see that f_n is the $2n$-th order polynomial of q and its derivatives with respect to x, written in the following form

$$f_n = \sum_{s=1}^{n} b_s P_{n,2s}(q, q_x, q_{xx}, \cdots), \tag{16.24}$$

where $P_{n,2s} = q^{k_1}(q^*)^{k_2} q_x^{k_3}(q_x^*)^{k_4}\cdots$ with $k_1+k_2+k_3+\cdots = 2s$.

Noting Eq. (16.17), one can obtain an infinite number of conservation laws in PSG,

$$(2i)^n C_n \beta_1^{-1} = \frac{1}{2}\sum_{k=1}^{3}\int_{b_k}\sum_{s=1}^{n} b_s \left(\frac{\beta_k}{2}\right)^{s-1} P_{n,2s}(\Psi_k, \Psi_{k,x}, \cdots)dx_k. \tag{16.25}$$

In Eq. (16.25), the cases $n = 1$ and 3 give the norm and energy conservation rules in Eqs. (16.2) and (16.6) with (16.7), respectively. The current conservation rule is now given by

$$(2i)^2 C_2 \beta_1^{-1} = \frac{1}{2}\sum_k \int_{b_k}\left(\Psi_k^* \frac{\partial \Psi_k}{\partial x_k}\right)(x_k,t)dx_k. \tag{16.26}$$

Some higher-order conservation rules are as follows:

$$(2i)^4 C_4 \beta_1^{-1} = \frac{1}{2}\sum_k \int_{b_k}\left(\Psi_k \frac{\partial^3 \Psi_k^*}{\partial x_k^3} + \frac{3\beta_k}{2}\Psi_k \frac{\partial \Psi_k^*}{\partial x_k}|\Psi_k|^2\right)(x_k,t)dx_k, \tag{16.27}$$

$$(2i)^5 C_5 \beta_1^{-1} = \frac{1}{2}\sum_k \int_{b_k}\left[\left|\frac{\partial^2 \Psi_k}{\partial x_k^2}\right|^2 + \frac{\beta_k^2}{2}|\Psi_k|^6 - \frac{\beta_k}{2}\left(\frac{\partial}{\partial x_k}|\Psi_k|^2\right)^2 \right.$$
$$\left. -3\beta_k \left|\frac{\partial \Psi_k}{\partial x_k}\right|^2 |\Psi_k|^2\right](x_k,t)dx_k. \tag{16.28}$$

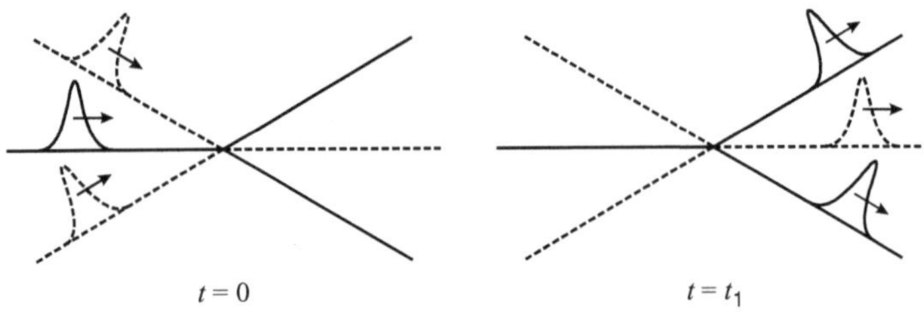

Fig. 16.1 Splitting of soliton. $t_1 > \tau$. Broken curves represent ghost solitons

The above results are also true for more general star graphs consisting of M semi-infinite bonds connected at a single vertex. In such cases, the initial soliton at an incoming bond splits into $M-1$ solitons in the remaining outgoing bonds. In this case, on the r.h.s. of the lower halves of Eqs. (16.10) and (16.11), the summation is taken over all the outgoing $M-1$ bonds. Correspondingly, the extended version of Eqs. (16.20) is given by

$$\frac{1}{\beta_1} = \sum_{j=1}^{M-1} \frac{1}{\beta_j}. \tag{16.29}$$

Another example of the graph for which the soliton solution of NLSE can be obtained analytically is a directed tree graph. The above results can be easily extended to this case.

16.2.4 Injection of a Single Soliton and Transmission Probabilities at $t \to +\infty$

Here we calculate transmission probabilities for a single soliton which is incoming through an semi-infinite bond b_0 and outgoing through the semi-infinite bonds $\beta_{\Gamma l}$.

A single (bright) soliton on a graph, which takes the general form in Eq. (16.17), is described with use of parts of Zakharov-Shabat's soliton with $\beta = 2$ [13]: Ψ_γ lying on individual bonds b_γ is given by

$$\Psi_\gamma(x_\gamma, t) = \frac{a\sqrt{2}}{\sqrt{\beta_\gamma}} \cdot \frac{\exp\left[i\frac{v}{2}x_\gamma - i\left(\frac{v^2}{4} - a^2\right)t\right]}{\cosh\left[a(x_\gamma + l - vt)\right]}, \tag{16.30}$$

where $v, -l$ and a are bond-independent parameters characterizing velocity, initial center of mass and amplitude of a soliton, respectively. In the simplest graph (: PSG) in Fig. 16.1, the soliton at bond b_1 splits into two parts and appears in both of b_2 and b_3. This is a novel feature of the soliton propagation through a branched chain and networks in general. Precisely speaking, the soliton dynamics here is governed by a single characteristic time $\tau \equiv \frac{l}{v}$. While for $0 \leq t \leq \tau$ the soliton at b_1 is a real one and those at b_2 and b_3 are ghosts, for $\tau \leq t$ the soliton at b_1 is a ghost and those at b_2 and b_3 are real. The incoming real soliton on b_1 and outgoing ghost solitons at b_2 and b_3 arrive at the vertex O. At $t = 0$ with $l \gg 1$, the soliton lying on the bond b_1 is exclusively responsible for the norm N. On the other hand, at $t \gg 1$, the solitons running through the bonds b_2 and b_3 are exclusively responsible for the norm. Therefore we can naturally define transmission probabilities at $t \to +\infty$.

Let us consider the general graph with incoming semi-infinite bond b_0 $(-\infty, 0)$ and n outgoing semi-infinite bonds $b_{\Gamma l}$ $(0, +\infty)$, $l = 1, 2, \ldots, n$. According to a combination of sum rules for nonlinearity coefficients we have

$$\frac{1}{\beta_0} = \sum_{l=1}^{n} \frac{1}{\beta_{\Gamma l}}. \tag{16.31}$$

From this rules it follows that the limit at $t \to +\infty$, transmission coefficients vanish on the part of graph of intermediate part (between incoming and outgoing bonds).

Transmission probability for arbitrary bond $b_{\Gamma l}$ are defined as

$$T_{\Gamma l} \equiv \frac{1}{N} \int_0^{+\infty} |\Psi_{\Gamma l}(x,t)|^2 dx. \tag{16.32}$$

In the case of a single soliton solution with $v > 0$ we have

$$\begin{aligned} T_{\Gamma l} &= \frac{1}{N} \frac{2a^2}{\beta_k} \int_0^{+\infty} \frac{dx}{\cosh^2(a(x + l - s_{\Gamma l} - vt))} \\ &= \frac{1}{N} \frac{2a^2}{\beta_k} \int_{l - s_{\Gamma l} - vt}^{+\infty} \frac{dx}{\cosh^2(ax)}. \end{aligned} \tag{16.33}$$

At the limit $t \to +\infty$, we have

$$T_{\Gamma l} \to \frac{1}{N} \frac{2a^2}{\beta_k} \int_{-\infty}^{+\infty} \frac{dx}{\cosh^2(ax)} = \frac{\beta_1}{\beta_{\Gamma l}}. \tag{16.34}$$

We should recognize the reflection probability at the bond b_1 is vanishing:

$$\begin{aligned} R_0 &= \frac{1}{N} \frac{2a^2}{\beta_0} \int_{-\infty}^{0} \frac{dx}{\cosh^2(a(x + l - vt))} \\ &= \frac{1}{N} \frac{2a^2}{\beta_0} \int_{-\infty}^{l - s_{\Gamma l} - vt} \frac{dx}{\cosh^2(ax)} = 0. \end{aligned} \tag{16.35}$$

The last equality is justified at the limit $t \to +\infty$.

According to the sum rule (16.31) one can see the unitarity to be satisfied.

$$\sum_l T_{\Gamma l} = 1. \tag{16.36}$$

The result in Eq. (16.34) provides new analytic expressions for the transmission probability for open networks with incoming and outgoing semi-infinite bonds.

16.3 Ablowitz-Ladik(AL) Equation on Networks

Let us consider an elementary branched chain (see Fig. 16.2), namely, a primary star graph (PSG) consisting of three semi-infinite bonds connected at the vertex O.

We denote individual lattice sites as (k, n), where $k = 1, 2$, and 3 mean the bond's number and n corresponds to the position on each bond. For the first bond $k = 1$, n is numbered as $n \in B_1 = \{0, -1, -2, \ldots\}$, where $(1, 0)$ means the branching point, i.e., the vertex. For the second ($k = 2$) and third ($k = 3$) bonds, n varies as $n \in B_k = \{1, 2, 3, \ldots\}$. (2, 1) and (3, 1) stands for the points nearest to the vertex.

Discrete nonlinear Schroödinger equation (DNLS) *a la* Ablowitz-Ladik (AL) is defined on each bond except for the vicinity of the vertex as

$$i\dot{\psi}_{k,n} + (\psi_{k,n+1} + \psi_{k,n-1})\left(1 + \gamma_k|\psi_{k,n}|^2\right) = 0, \tag{16.37}$$

where $(k, n) \notin \{(1, 0), (2, 1), (3, 1)\}$. It should be noted that γ_k may be different among bonds. There is an ambiguity about the interaction around the vertex, which is resolved as follows: Let's first introduce Hamiltonian for PSG as

$$H = -\sum_{n=0}^{-\infty}(\psi_{1,n}^*\psi_{1,n+1} + c.c.) - \sum_{k=2}^{3}\sum_{n=1}^{+\infty}(\psi_{k,n}^*\psi_{k,n+1} + c.c.), \tag{16.38}$$

where at the virtual site $(1, 1)$ we assume $\psi_{1,1} = s_2\psi_{2,1} + s_3\psi_{3,1}$ with appropriate coefficients s_2 and s_3. Then Eq. (16.37) can be obtained by the equation of motion

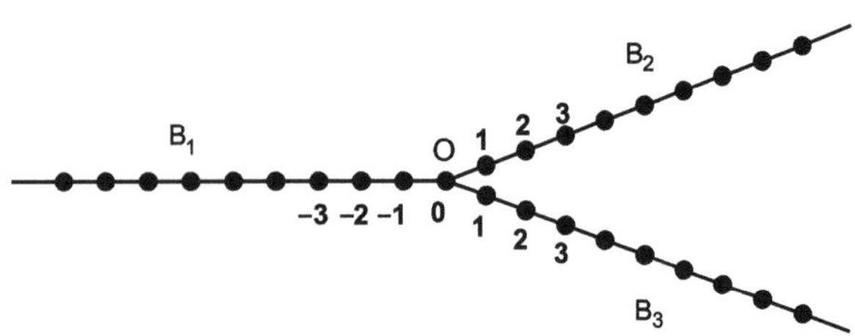

Fig. 16.2 Primary star graph. Three semi-infinite chains B_1, B_2 and B_3 connected at a vertex O

$$i\dot{\psi}_{k,n} = \{H, \psi_{k,n}\} \tag{16.39}$$

at $(k,n) \notin \{(1,0),(2,1),(3,1)\}$, with use of non-standard Poisson brackets

$$\{\psi_{k,m}, \psi^*_{k',n}\} = i(1+\gamma|\psi_{k,m}|^2)\delta_{kk'}\delta_{mn}, \{\psi_{k,m}, \psi_{k',n}\} = \{\psi^*_{k,m}, \psi^*_{k',n}\} = 0.$$

On the same footing as above, the equation of motions in Eq. (16.39) at (1,0), (2,1) and (3,1) are given, respectively, as

$$i\dot{\psi}_{1,0} + (\psi_{1,-1} + s_2\psi_{2,1} + s_3\psi_{3,1})\left(1+\gamma_1|\psi_{1,0}|^2\right) = 0, \tag{16.40}$$

$$i\dot{\psi}_{k,1} + (s_k\psi_{1,0} + \psi_{k,2})\left(1+\gamma_k|\psi_{k,1}|^2\right) = 0, \ \ k = 2,3. \tag{16.41}$$

The solution is assumed to satisfy the following conditions at infinity: $\psi_{1,n} \to 0$ at $n \to -\infty$ and $\psi_{k,n} \to 0$ at $n \to +\infty$ for $k = 2$ and 3.

16.3.1 Norm Conservation

It is known that the norm conservation is one of the most important physical conditions in conservative systems. Since Eqs. (16.37), (16.40) and (16.41) are available from Hamilton's equation of motion with non-standard Poisson brackets, the norm and energy conservations seem obvious. Below, however, we observe them explicitly. Extending the definition in the case of 1-d chain [4], the norm for PSG is given as

$$N = \|\psi\|^2 = \sum_{k=1}^{3} \frac{1}{\gamma_k} \sum_{n\in B_k} \ln\left(1+\gamma_k|\psi_{k,n}|^2\right). \tag{16.42}$$

Its time derivative is given by

$$\frac{d}{dt}N = \sum_{k=1}^{3}\sum_{n\in B_k} A_{k,n} \tag{16.43}$$

with

$$A_{k,n} = \frac{1}{1+|\gamma_{k,n}|^2}\left(\psi^*_{k,n}\dot{\psi}_{k,n} + \dot{\psi}^*_{k,n}\psi_{k,n}\right). \tag{16.44}$$

For $(k,n) \notin \{(1,0),(2,1),(3,1)\}$ with use of Eq. (16.37) we have

$$A_{k,n} = \frac{1}{i}\left(\psi_{k,n}\psi^*_{k,n+1} - \psi^*_{k,n}\psi_{k,n+1}\right)$$
$$-\frac{1}{i}\left(\psi_{k,n-1}\psi^*_{k,n} - \psi^*_{k,n-1}\psi_{k,n}\right)$$
$$\equiv j_{k,n} - j_{k,n-1}, \tag{16.45}$$

where

$$j_{k,n} \equiv \frac{1}{i}\left(\psi_{k,n}\psi^*_{k,n+1} - \psi^*_{k,n}\psi_{k,n+1}\right) \tag{16.46}$$

implies a local current. Carrying out the summation in Eq. (16.43), we have

$$\sum_k {\sum_n}' A_{k,n} = j_{1,0} - j_{2,1} - j_{3,1}, \tag{16.47}$$

where $\sum_k {\sum_n}'$ means the summation over all sites on PSG except for the points $(1,0),(2,1),(3,1)$.

Similarly, for $(k,n) = (1,0),(2,1),(3,1)$ with use of Eqs. (16.40), (16.41) we obtain

$$A_{1,0} = s_2\frac{1}{i}\left(\psi_{1,0}\psi^*_{2,1} - \psi^*_{1,0}\psi_{2,1}\right)$$
$$+s_3\frac{1}{i}\left(\psi_{1,0}\psi^*_{3,1} - \psi^*_{1,0}\psi_{3,1}\right) - j_{1,0} \tag{16.48}$$

and

$$A_{k,1} = j_{k,1} - s_k\frac{1}{i}\left(\psi_{1,0}\psi^*_{k,1} - \psi^*_{1,0}\psi_{k,1}\right) \tag{16.49}$$

for $k = 2,3$. Then we can see

$$\frac{d}{dt}N = \sum_{k=1}^{3}\sum_{n\in B_k} A_{k,n} \equiv 0, \tag{16.50}$$

i.e., norm conservation. Therefore, for any choice of values s_2 and s_3 the norm conservation turns out to hold well.

16.3.2 Energy Conservation

Energy for PSG is expressed in a symmetrical form as

$$E = -2\mathrm{Re}\Big[\sum_{n=-1}^{-\infty}\psi_{1,n}^{*}\psi_{1,n+1} + \sum_{k=2}^{3}\sum_{n=1}^{+\infty}\psi_{k,n}^{*}\psi_{k,n+1} + \psi_{1,0}^{*}(s_2\psi_{2,1} + s_3\psi_{3,1})\Big]. \tag{16.51}$$

To show that the energy is conservative, we see its time derivative

$$\begin{aligned}\frac{d}{dt}E = &-2\mathrm{Re}\sum_{n=-1}^{-\infty}\left(\psi_{1,n}^{*}\dot{\psi}_{1,n+1} + \dot{\psi}_{1,n}^{*}\psi_{1,n+1}\right)\\ &-2\mathrm{Re}\sum_{k=2}^{3}\sum_{n=1}^{+\infty}\left(\psi_{k,n}^{*}\dot{\psi}_{k,n+1} + \dot{\psi}_{k,n}^{*}\psi_{k,n+1}\right)\\ &-2\mathrm{Re}\left[\psi_{1,0}^{*}\left(s_2\dot{\psi}_{2,1} + s_3\dot{\psi}_{3,1}\right) + \dot{\psi}_{1,0}^{*}\left(s_2\psi_{2,1} + s_3\psi_{3,1}\right)\right].\end{aligned} \tag{16.52}$$

With use of Eq. (16.37) we have

$$\begin{aligned}&-\sum_{n=-1}^{-\infty}\left(\psi_{1,n}^{*}\dot{\psi}_{1,n+1} + \dot{\psi}_{1,n}^{*}\psi_{1,n+1}\right)\\ &= \frac{1}{i}\sum_{n=-1}^{-\infty}\left[|\psi_{1,n-1}|^2 - |\psi_{1,n+1}|^2\right]\left(1 + \gamma_1|\psi_{1,n}|^2\right) - \psi_{1,-1}^{*}\dot{\psi}_{1,0},\end{aligned} \tag{16.53}$$

$$\begin{aligned}&-\sum_{n=1}^{\infty}\left(\psi_{k,n}^{*}\dot{\psi}_{k,n+1} + \dot{\psi}_{k,n}^{*}\psi_{k,n+1}\right)\\ &= \frac{1}{i}\sum_{n=2}^{\infty}\left[|\psi_{k,n-1}|^2 - |\psi_{k,n+1}|^2\right]\left(1 + \gamma_1|\psi_{k,n}|^2\right) - \dot{\psi}_{k,1}^{*}\psi_{k,2}.\end{aligned} \tag{16.54}$$

Substituting Eqs. (16.53) and (16.54) into Eq. (16.52) and using Eqs. (16.40) and (16.41), we find:

$$\begin{aligned}\frac{d}{dt}E &= -2\mathrm{Re}\left[\psi_{1,0}^{*}\left(s_2\dot{\psi}_{2,1} + s_3\dot{\psi}_{3,1}\right)\right.\\ &\qquad\left. + \dot{\psi}_{1,0}^{*}\left(s_2\psi_{2,1} + s_3\psi_{3,1} + \psi_{1,-1}^{*}\dot{\psi}_{1,0}\right) + \dot{\psi}_{2,1}^{*}\psi_{2,2} + \dot{\psi}_{3,1}^{*}\psi_{3,2}\right]\\ &= 2\mathrm{Re}\Bigg[\frac{1}{i}(1 + \gamma_1|\psi_{1,0}|^2)\left(|\psi_{1,-1}|^2 - |s_2\psi_{2,1} + s_3\psi_{3,1}|^2\right)\\ &\qquad + \frac{1}{i}\sum_{k=1}^{3}(1 + \gamma_k|\psi_k, 1|^2)\left(s_k^2|\psi_{1,0}|^2 - |\psi_{k,2}|^2\right)\Bigg] \equiv 0.\end{aligned} \tag{16.55}$$

The last equality comes from the pure imaginary nature of the expression in $[\cdots]$. Equation (16.55) is nothing but the energy conservation.

As one can see from Eqs. (16.50) and (16.55) the norm and energy are conserved for any choice of the parameters s_2 and s_3. But in general one cannot guarantee other conservation laws. In the next sections we consider a special case with appropriate choice of s_2 and s_3 which guarantees an infinite number of conservation laws.

16.3.3 Completely Integrable Case: Dynamics Near Branching Point and Sum Rule

Among many possible choices of s_2 and s_3, there is one special case in which an infinite number of constants of motion can be found and DNLSE on PSG becomes completely integrable. We shall now consider this case by finding suitable values for s_1 and s_2.

For this purpose, we shall first add to each bond B_k $(k = 1, 2, 3)$ a ghost-bond counterpart B_k' so that $B_k + B_k'$ constitutes an ideal 1-d chain. Then we suppose that a solution of AL equation on PSG is given by

$$\psi_{k,n}(t) = \frac{1}{\sqrt{\gamma_k}} q_{k,n}(t), \quad k = 1, 2, 3 \tag{16.56}$$

where $q_{k,n}(t)$ are solutions of DNLSE on the ideal 1-d chain ([1–4]):

$$i\dot{q}_n + (q_{n+1} + q_{n-1})(1 + |q_n|^2) = 0, \tag{16.57}$$

with n being integers in $(-\infty, +\infty)$. The solutions of Eq. (16.57) may be different among three fictitious chains $B_k + B_k'$ $(k = 1, 2, 3)$.

Comparing Eqs. (16.40), (16.41) and (16.57), one can find at the vertex the following equalities:

$$\frac{1}{\sqrt{\gamma_1}} q_{1,1}(t) = \frac{s_2}{\sqrt{\gamma_2}} q_{2,1}(t) + \frac{s_3}{\sqrt{\gamma_3}} q_{3,1}(t), \tag{16.58}$$

$$\frac{1}{\sqrt{\gamma_k}} q_{k,0}(t) = \frac{s_k}{\sqrt{\gamma_1}} q_{1,0}(t), \quad k = 2, 3. \tag{16.59}$$

Noting the spatio-temporal behaviors of soliton solutions and to guarantee Eq. (16.59) for any t, $q_{k,n}(t) = s_k\sqrt{\frac{\gamma_k}{\gamma_1}} q_{1,n}(t)$ with $k = 2, 3$ should be satisfied for any integer n. From the latter equality we obtain

$$s_k \sqrt{\frac{\gamma_k}{\gamma_1}} = 1 \quad \text{or } s_k = \sqrt{\frac{\gamma_1}{\gamma_k}} \quad (k = 2, 3) \tag{16.60}$$

and $q_{k,n}(t) \equiv q_n(t)$, i.e., the solution $q_{k,n}(t)$ is bond independent.

Now combining this results with Eq. (16.58) we have the sum rule among nonlinearity coefficients:

$$\frac{1}{\gamma_1} = \frac{1}{\gamma_2} + \frac{1}{\gamma_3} \tag{16.61}$$

Thus one can see that the solution on PSG is defined as a common (bond-independent) soliton solution of Eq. (16.57) multiplied by square root of the inverse nonlinearity coefficient. This issue is guaranteed, so long as the sum rule for nonlinearity coefficients γ_1, γ_2 and γ_3 in Eq. (16.61) is satisfied simultaneously.

In the case of another choice of parameters s_2 and s_3 we shall see a completely different nonlinear dynamics of solitons such as reflection of a soliton at the vertex. The initial value problem for such parameters will be treated elsewhere.

We also notice that the parameters s_2 and s_3 correspond to the $\frac{\alpha_2}{\alpha_1}$ and $\frac{\alpha_3}{\alpha_1}$, respectively, in the case of continuous NLSE considered in the preceding work [12].

16.3.4 An Infinite Number of Constants of Motion

It is well known that Ablowitz-Ladik (AL) equation on the 1-d chain has an infinite number of constants of motion. Now we shall proceed to obtain an infinite number of constants of motion for general solutions of AL equation on PSG. First of all, it should be noted that the solution on PSG can now be written as

$$\psi_{k,n}(t) = \frac{1}{\sqrt{\gamma_k}} q_n(t), \quad n \in B_k, k = 1, 2, 3, \tag{16.62}$$

where $q_n(t)$ stands for a general solution of AL equation (16.57) and is restricted to each bonds B_k $(k = 1, 2, 3)$.

While we already proved the conservation of energy, we can generalize it to the general cases: Without taking the complex conjugate, Eq. (16.38) can be explicitly written as

$$\begin{aligned} Z = &- \sum_{n=-1}^{-\infty} \psi_{1,n}^* \psi_{1,n+1} - \sum_{k=2}^{3} \sum_{n=1}^{+\infty} \psi_{k,n}^* \psi_{k,n+1} \\ &- \psi_{1,0}^* (s_2 \psi_{2,1} + s_3 \psi_{3,1}). \end{aligned} \tag{16.63}$$

Substituting Eq. (16.62) into Eq. (16.63), Z is rewritten as

$$\begin{aligned} Z = &- \frac{1}{\gamma_1} \sum_{n=0}^{-\infty} q_n^* q_{n+1} - \sum_{k=2}^{3} \frac{1}{\gamma_k} \sum_{n=1}^{+\infty} q_n^* q_{n+1} \\ &+ \frac{1}{\gamma_1} q_0^* q_1 - \sum_{k=2}^{3} \frac{s_k}{\sqrt{\gamma_1 \gamma_k}} q_0^* q_1. \end{aligned} \tag{16.64}$$

Using the value s_k in Eq. (16.60) and the sum rule in Eqs. (16.61), (16.64) reduces to the constant for the case of 1-d chain [1,2]:

$$Z = -\frac{1}{\gamma_1}\sum_{-\infty}^{+\infty} q_n^* q_{n+1}. \tag{16.65}$$

Therefore Z in Eq. (16.63) is a constant of motion, and its real and imaginary parts imply energy and current, respectively.

For other higher-order conservation rules, we can write them as

$$\frac{1}{\gamma_1}C_m = \frac{1}{\gamma_1}\sum_{n=0}^{-\infty} f_m(q_{1,n}) + \sum_{k=2}^{3}\frac{1}{\gamma_k}\sum_{n=1}^{+\infty} f_m(q_{k,n}), \tag{16.66}$$

where f_m is defined as a expansion coefficient of the expression (see [2])

$$\sum_{n=-\infty}^{+\infty} \log(g_n^{(0)} + g_n^{(1)}z^2 + g_n^{(2)}z^4 + \ldots) = f_1 z^2 + f_2 z^4 + \ldots\,, \tag{16.67}$$

with

$$g_n^{(0)} = 1,\;\; g_n^{(1)} = R_{n-1}Q_{n-2},$$

$$g_n^{(m)} = \frac{R_{n-1}}{R_{n-2}} g_{n-1}^{(m-1)} - \sum_{l=1}^{m-1} g_{n-1}^{(m-l)} g_n^{(l)},\;\; m = 2, 3, 4, \ldots, \tag{16.68}$$

$$R_n = q_{n+2}^*,\;\; Q_n = -q_{n+2}. \tag{16.69}$$

The conservation laws (16.66) include some undefined terms that must be defined as

$$\psi_{1,n} = s_2\psi_{2,n} + s_3\psi_{3,n} \quad \text{with} \quad n \ge 1, \tag{16.70}$$

$$\psi_{k,n} = s_k\psi_{k,n},\;\; k = 2;3 \quad \text{with} \quad n \le 0. \tag{16.71}$$

The conservation laws in Eq. (16.66) follows from the nature of solution (16.62) and the sum rule for nonlinearity coefficients (16.61).

For $m = 1$ we obtain current and energy conservation laws. At $m \ge 2$ we obtain higher order conservation laws. Some of higher-order constants of motion are as follows:

$$\frac{1}{\gamma_1}C_2 = -\sum_{k=1}^{3}\sum_{n\in B_k}\Big(\psi_{k,n+1}^*\psi_{k,n-1}(1+\gamma_k|\psi_{k,n}|^2)$$

$$+\frac{\gamma_k}{2}\psi_{k,n}^2(\psi_{k,n+1}^*)^2\Big), \tag{16.72}$$

$$\frac{1}{\gamma_1} C_3 = -\sum_{k=1}^{3} \sum_{n \in B_k} \Big[\Big(\psi_{k,n+2}^* \psi_{k,n-1} (1 + \gamma_k |\psi_{k,n+1}|^2) + \gamma_k \psi_{k,n}^* \psi_{k,n+1}^* \psi_{k,n-1}^2$$
$$+ (\psi_{k,n+1}^*)^2 \psi_{k,n} \psi_{k,n-1} \Big) (1 + \gamma_k |\psi_{k,n}|^2) + \frac{\gamma_k^2}{3} \psi_{k,n+1}^* \psi_{k,n} \Big]. \quad (16.73)$$

16.4 Conclusions

Thus we studied the problem of reflectionless soliton transport in simple networks considering the case when the nonlinear Schrodinger equation on simplest graphs becomes completely integrable. The strength of cubic nonlinearity is assumed to be different from bond to bond, and the networks are assumed to have at least two semi-infinite bonds with one of them working as an incoming bond. The connection formula at vertices are derived obtained from norm and energy conservation rules. Under these conditions, we also showed an infinite number of constants of motion. It is shown that the results can be extended to other simplest graph topologies, i.e., general star graphs, tree graphs, loop graph and their combinations. Using the ballistic transport of Zakharov-Shabats soliton through networks we have derived expressions for the transmission probabilities on the outgoing bonds, which are inversely proportional to the bond-dependent strength of nonlinearity.

References

1. Ablowitz MJ, Ladik JF (1975) J Math Phys 16:598
2. Ablowitz MJ, Ladik JF (1976) J Math Phys 17:1011
3. Ablowitz MJ, Segur H (1975) J Math Phys 16:1054
4. Ablowitz MJ, Prinari B, Trubatch AD (2004) Discrete and continuous nonlinear Schrödinger systems. University Press, Cambridge
5. Adami R, Cacciapuoti C, Finco D, Noja D (2011) Rev Math Phys 23:4
6. Adami R, Cacciapuoti C, Finco D, Noja D (2012) J Phys A Math Gen 45:192001
7. Adami R, Noja D, Ortoleva C (2013) J Math Phys 54:013501
8. Banica V, Ignat L (2011) J Phys A Math Gen 52:083703
9. Cascaval RC, Hunter CT (2010) Libertas Math 30:85
10. Gnutzmann S, Smilansky U, Derevyanko S (2011) Phys Rev A 83:033831
11. Sabirov KK, Sobirov Z, Babajanov D, Matrasulov D (2013) Phys Lett A 377:860
12. Sobirov Z, Matrasulov D, Sabirov K, Sawada S, Nakamura K (2010) Phys Rev E 81:066602
13. Zakharov VB, Shabat AB (1972) Sov Phys JETP 34:62

Part V
Complexity in Nanoscale Systems

Chapter 17
Symmetry Breaking in Open Quantum Nonlinear Systems

Almas F. Sadreev, Evgeny N. Bulgakov, Dmitrii N. Maksimov, and Konstantin N. Pichugin

Abstract We consider symmetry breaking in the simplest open quantum nonlinear systems such as dimer and plaquette of four nonlinear sites coupled with linear tight-binding wires. If the solution is stationary, the total Hilbert space can be projected into the inner states of the dimer or plaquette by the Feshbach procedure. That derives a nonlinear analogue of the Lippmann-Schwinger equation with injected wave as a source. By neglecting radiation shifts the Lippmann-Scwinger equation limits to the coupled mode theory equations widely used in optics. We show three scenarios for the transmission through the nonlinear quantum systems. The first one inherits the linear case and preserves the symmetry. In the second scenario the symmetry is broken because of different intensities at the dimer sites. In the third scenario the intensities at the sites are equaled but phases of complex wave function are different. That results in a vortical power flow between the nonlinear sites similar to the DC Josephson current. We show how the phenomenon of symmetry breaking can used for switching of outputs symmetrically coupled to the quantum dimer. Also we reveal a domain in the parameter space where none of stationary solutions exist. As a result injection of a monochromatic symmetric wave gives rise to emission of nonsymmetric satellite waves with energies different from the energy of the incident wave. Thus, the response exhibits non monochromatic behavior.

17.1 Introduction

To the best our knowledge symmetry breaking (SB) in the nonlinear systems was first predicted by Akhmediev who considered a composite structure of a single linear layer between two symmetrically positioned nonlinear layers [1]. One could

A.F. Sadreev (✉) • E.N. Bulgakov • D.N. Maksimov • K.N. Pichugin
Institute of Physics, 660036, Krasnoyarsk, Russia
e-mail: almas@tnp.krasn.ru; ben@tnp.krasn.ru; mdn@tnp.krasn.ru; knp@tnp.krasn.ru

D. Matrasulov and H.E. Stanley (eds.), *Nonlinear Phenomena in Complex Systems: From Nano to Macro Scale*, NATO Science for Peace and Security Series C: Environmental Security, DOI 10.1007/978-94-017-8704-8_17,

easily see that if the wave length is larger than the thickness of the nonlinear layers then Akhmediev's model could be reduced to a dimer governed by the nonlinear Shrödinger equation

$$\begin{pmatrix} \lambda|\psi_1|^2 & -u \\ -u & \lambda|\psi_2|^2 \end{pmatrix}\begin{pmatrix} \psi_1 \\ \psi_2 \end{pmatrix} = E\begin{pmatrix} \psi_1 \\ \psi_2 \end{pmatrix}. \tag{17.1}$$

Independently the SB was discovered for the discrete nonlinear Schrödinger equation with a finite number of coupled cites (nonlinear dimer, trimer, etc.) [2–8]. For example, in the case of the nonlinear quantum dimer Eilbeck et al. found two different families of stationary solutions of NSE (17.1) [2]. The first family is symmetric (antisymmetric) ($|\phi_1| = |\phi_2|$) while the second is nonsymmetric ($|\phi_1| \neq |\phi_2|$). This consideration was later extended to a nonlinear dimer embedded into an infinite linear chain [9] with the same scenario for the SB. Multiple bifurcations to the symmetry breaking solutions were demonstrated by Wang et al. [10] for the nonlinear Schrödinger equation with a square four-well potential. Remarkably, the above system can also support a stable state with a nodal point, i.e., quantum vortex [11]. In the framework of the nonlinear Schrödinger equation one can achieve bifurcation to the states with broken symmetry varying the chemical potential which is equivalent to the variation of the population of the nonlinear sites or, analogously, of the constant in the nonlinear term of the Hamiltonian. In practice, however, one would resort to the optical counterparts of the quantum nonlinear systems where the variation of the amplitude of the injected wave affects the strength of Kerr nonlinearity (see Refs. [12–23] for optical examples of symmetry breaking).

A wave injected into the nonlinear quantum or optical systems accomplishes manyfold functions. Similar to the linear systems scattered wave carries information about inner structure of the closed system. However this wave can also switch on/off corresponding inner states. Application of wave pulses can cause transitions between stable solutions, realizing all-optical switching [24]. Other words, these pulses shake the nonlinear system to give rise to bifurcations [25]. The phase of wave function in the open nonlinear quantum system can also become a key player. The quantum system can be opened by many ways. Examples of the open symmetrical quantum dimer with attached one-dimensional wires are shown in Fig. 17.1.

The open quantum dimer shown in Fig. 17.1a is equivalent to the system in which two nonlinear cavities are aligned along the waveguide considered by Maes et al. [18, 19]. That system is symmetric relative to the inversion of the transport axis if equal power is injected on both sides of the coupled cavities. Maes et al. has shown that nevertheless the reflected output power might be different on both sides of the defects due to nonlinear effects, i.e., the symmetry of the system can be broken under effect of input power. The same result was demonstrated by Brazhnyi and Malomed in a linear discrete chain (Schrödinger lattice) with two nonlinear sites shown in Fig. 17.1b [9].

The next important question on the open nonlinear quantum systems is whether the solutions in the linear waveguides could be stationary monochromatic plane

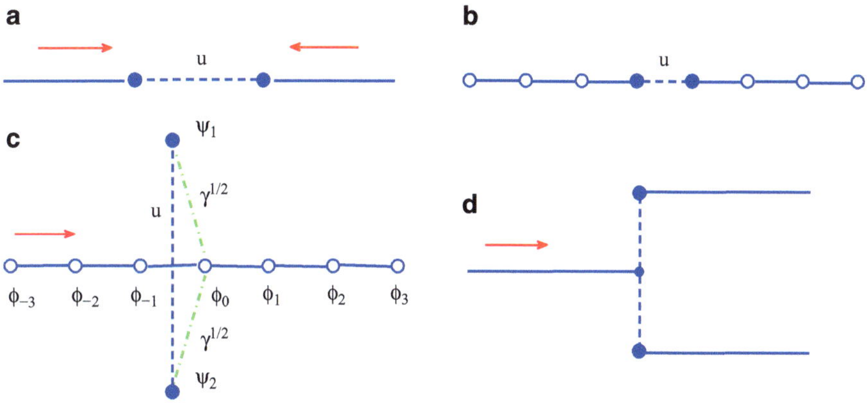

Fig. 17.1 Different designs of open quantum nonlinear dimer shown by *bold closed circles*. Linear wires attached to the dimer are shown by *solid lines*. Waves injected into nonlinear quantum dimer are shown by *red arrows*

waves, reflected $\psi(n,t) = R\exp(-ikn - iE(k)t)$ and transmitted $\psi(n,t) = T\exp(ikn - iE(k)t)$, where R and T are the reflection and transmission amplitudes. If the answer is positive then we can apply the Feshbach projection technique [26–31] and implement the formalism of non-Hermitian effective Hamiltonian (now nonlinear) which acts on the nonlinear sites only thus truncating the Hilbert space to the scattering region [22, 31]. When the radiation shifts of the energy levels are neglected that formalism reduces to the well known coupled mode theory (CMT) equations [32–35]. In the present paper we will show that there are domains in the parameter space where there are no stable stationary solutions. We will demonstrate numerically that a plane wave incident onto a nonlinear object gives rise to emission of multiple satellite waves with energies (frequencies) different from the energy (frequency) of the probing wave.

17.2 Symmetry Breaking for Transmission Through Nonlinear Dimer

We consider the nonlinear dimer coupled with linear tight-binding chain shown in Fig. 17.1c. The chain supports continuum of plane waves

$$\phi_j(t) = \frac{1}{\sqrt{2\pi|\sin k|}}\exp(\pm ikj - iE(k)t) \tag{17.2}$$

with the propagation bands given by

$$E(k) = -2\cos k, -\pi \le k \le \pi. \tag{17.3}$$

The nonlinear dimer is coupled with the linear chain via the coupling constant $\sqrt{\gamma}$ shown in Fig. 17.1c by dash lines. In the case of the nonlinear dimer the Schrödinger equation takes the following form

$$\begin{aligned} i\dot{\phi}_j &= -\phi_{j+1} - \phi_{j-1} + \sqrt{\gamma}\delta_{j,0}(\psi_1 + \psi_2), \\ i\dot{\psi}_1 &= \lambda|\psi_1|^2\psi_1 - u\psi_2 + \sqrt{\gamma}\phi_0, \\ i\dot{\psi}_2 &= \lambda|\psi_2|^2\psi_2 - u\psi_1 + \sqrt{\gamma}\phi_0. \end{aligned} \tag{17.4}$$

Let us, first, search for the solution of the Schrödinger equation (17.4) in the form of a stationary wave

$$\phi_j(t) = \phi_j e^{-iEt} \tag{17.5}$$

where the discrete space variable j and the time t are separated. The absence of nonlinearity in the waveguides drastically simplifies analysis of Eq. (17.4). Assuming that a wave is incident from the left we can write the solutions in the left $\phi_j^{(L)}$ and right $\phi_j^{(R)}$ waveguides as:

$$\phi_j^{(L)} = A_0 \exp(ikj) + r\exp(-ikj), \quad \phi_j^{(R)} = t\exp(ikj), \tag{17.6}$$

where parameter A_0 is introduced to tune the intensity of the probing wave. Notice that Eq. (17.6) implicitly defines reflection and transmission amplitudes r and t. One can now match the solutions Eq. (17.6) to the equations for the nonlinear sites to obtain a set of nonlinear equations for on-site and reflection/transmission amplitudes. Computationally, however, it is more convenient to use the approach of the non-hermitian Hamiltonian [29, 31] in which the number of unknown variables equals to the number of nonlinear sites. Following Ref. [22] we write the nonlinear analogue of the Lippmann-Schwinger equation

$$(E - H_{eff})|\psi\rangle = \sqrt{\gamma}\phi_0 \tag{17.7}$$

where the effective Hamiltonian is given in Appendix for the tight-binding wire. If the eigen-energies of the nonlinear dimer are spaced in the centrum of propagation band of wire, i.e., near zero we can approximate the effective Hamiltonian as follows

$$H_{eff} = \begin{pmatrix} -i\gamma + \lambda|\psi_1|^2 & -u - i\gamma \\ -u - i\gamma & -i\gamma + \lambda|\psi_2|^2 \end{pmatrix}. \tag{17.8}$$

which corresponds to phenomenological approach [26, 27] in the scattering theory. One can easily see that in the limit $\gamma \to 0$ the dimer is decoupled from the waveguides and Eq. (17.7) limits to the standard nonlinear Schrödiner equation of the closed nonlinear dimer (17.1).

On the other hand, let us write the effective Hamiltonian (17.8) in the representation of the eigen-states of the linear dimer

$$|s\rangle = \frac{1}{\sqrt{2}}\begin{pmatrix}1\\1\end{pmatrix}, \quad |a\rangle = \frac{1}{\sqrt{2}}\begin{pmatrix}1\\-1\end{pmatrix} \tag{17.9}$$

where the indexes s, a note the symmetric and antisymmetric states with the corresponding eigen-energies $E_{s,a} = \mp u$. In this representation Eq. (17.7) will take the following form

$$\begin{aligned}[E - E_s - \lambda(|\psi_1|^2 + |\psi_2|^2) + i\gamma]A_s - \lambda(|\psi_1|^2 - |\psi_2|^2)A_a &= \sqrt{\gamma}\phi_0,\\ -\lambda(|\psi_1|^2 - |\psi_2|^2)A_s + [E - E_a - \lambda(|\psi_1|^2 + |\psi_2|^2)]A_a &= 0.\end{aligned} \tag{17.10}$$

This equation is the coupled mode theory (CMT) equation for the stationary transmission [32–35]. After Eq. (17.7) is solved one easily obtains transmission/reflection amplitudes from Eq. (17.4)

$$r = 2\pi i V^{+}\frac{1}{E+i0-H_{eff}}V = i\hat{\Gamma}\frac{1}{E+i0-H_{eff}}\hat{\Gamma}^{+} = \sqrt{\gamma}A_s, \tag{17.11}$$

$$t = \phi_0 - R \tag{17.12}$$

where $\hat{\Gamma} = \sqrt{\gamma}(1 \;\; 1)$.

Thus, we reduced the problem of scattering of wave through the nonlinear quantum dimer to the computational problem of self-consistent solution of CMT equation (17.7). The results of computation are presented in Fig. 17.2 (amplitudes of symmetric and antisymmetric modes) and Fig. 17.3 (transmission). Equation (17.10) has three solutions. The first symmetry preserving solution inherits the linear case, i.e. limits to the solution of the linear Lippmann-Schwiner equation for $\phi_0 \rightarrow 0$ or $\lambda \rightarrow 0$ and shown by dash lines. Wave injected into the dimer has zero coupling with the antisymmetric mode and therefore can not excite it as shown in Fig. 17.2b. Respectively, the transmission through the dimer would have resonance dip at the eigen-energy of the dimer at $E_s = -u$. However because of the nonlinear shift $\lambda|\psi_1|^2$ the dip is spaced at the energy $E > E_s$ as seen from Fig. 17.2c.

If to assume a symmetry breaking with $|\psi_1| \neq |\psi_2|$ the antisymmetric mode A_a couples with the injected wave also as seen from Eq. (17.10). Therefore, there should be the second symmetry breaking (SB) solution in which symmetric and antisymmetric modes both are excited as shown in Fig. 17.2a, b by solid red lines. Respectively we observe the second resonance dip at the energy right to the antisymmetric eigen energy $E_a = u$ in Fig. 17.2c. Stability of the SB solution shown by open circles is restricted.

The symmetry preserving and symmetry breaking solutions exist provided that the determinant of matrix $|E - H_{eff}| \neq 0$. There could be the singular case when the $|E - H_{eff}| = 0$ when an inverse of the matrix does not exist. In the linear case that

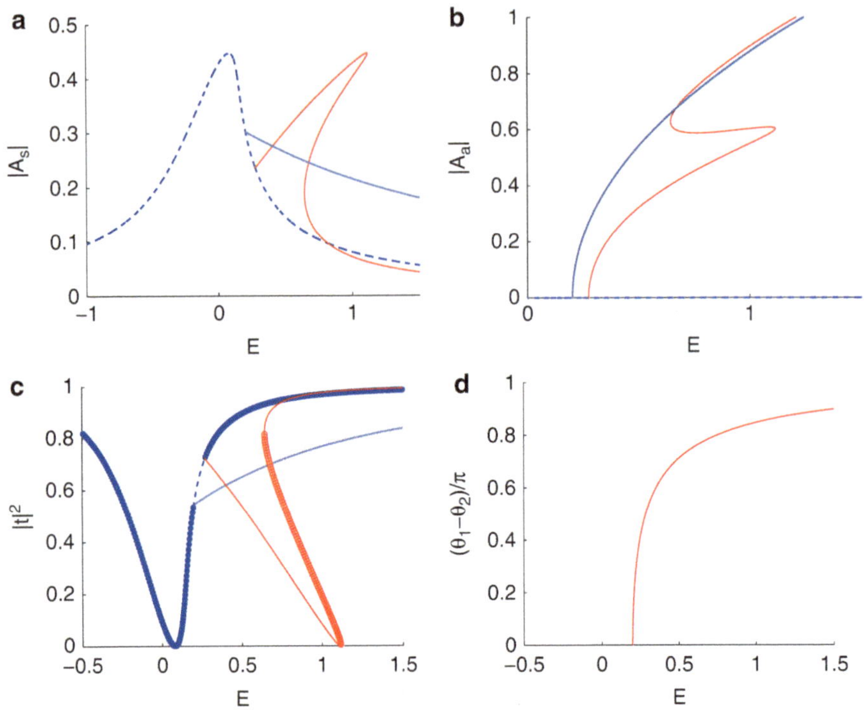

Fig. 17.2 Energy behavior of (**a**) the symmetric amplitude $|A_s|$, (**b**) the antisymmetric amplitude $|A_a|$, (**c**) transmission, and phase difference $(\theta_1 - \theta_2)/\pi$ of the model shown in Fig. 17.1c for, $\lambda = 0.5, \gamma = 0.04, u = 0.1, \phi_0 = 0.2$. Stability of the solutions is marked by *closed* and *open circles*

can occur only for discrete points in the space of the model parameters and results in rather exotic bound states in continuum (BSC) [36–39]. Nonlinear systems open new page in the BSC phenomenon. In framework of a generic two-level nonlinear Fano-Anderson model it was shown that the BSC arises by self-consistent way [40] without necessity to tune physical parameters as it was in the linear case. This phenomenon is generic and recently was applied to photonic crystals with defects made of Kerr media [22, 41] and nonlinear crystals [42].

For the open nonlinear dimer we have the following equation for $|E - H_{eff}| = 0$:

$$|\psi_1| = |\psi_2| = I, E = E_a + 2\lambda I. \tag{17.13}$$

After substitution of this equation into Eq. (17.10) we obtain

$$A_s = \frac{\sqrt{\gamma}\phi_0}{E_a - E_s + i\gamma} = -\frac{\sqrt{\gamma}\phi_0}{2u - i\gamma} = \frac{\psi_1 + \psi_2}{\sqrt{2}} = \sqrt{\frac{I}{2}}(e^{i\theta_1} + e^{i\theta_2}),$$

$$\cos^2(\theta_1 - \theta_2)/2 = \frac{\lambda\gamma\phi_0}{4u^2 + \gamma^2}. \tag{17.14}$$

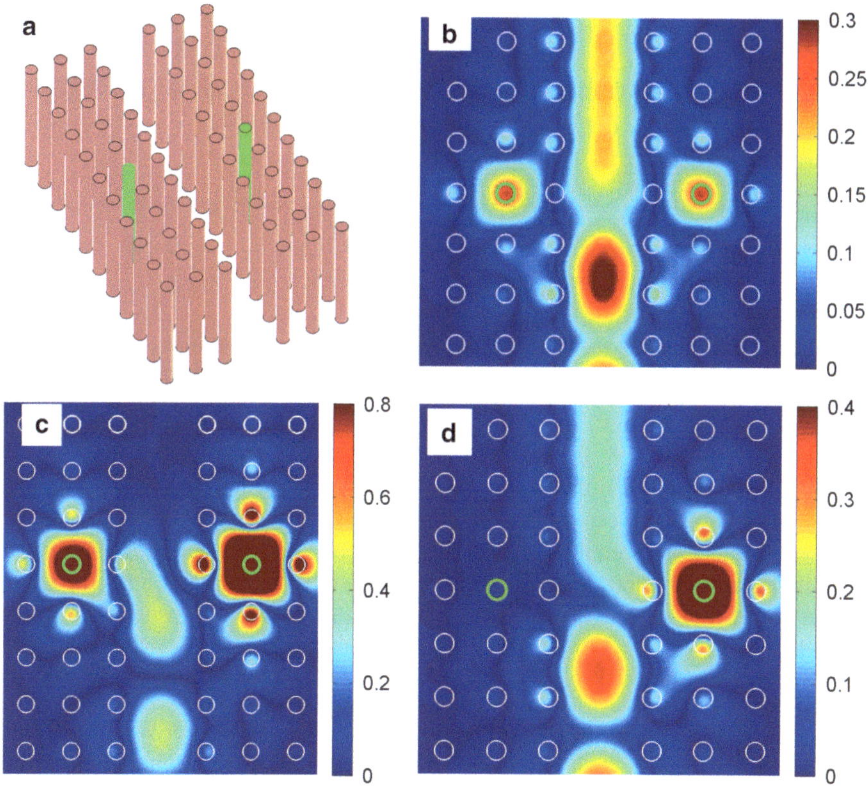

Fig. 17.3 (**a**) View of the waveguide coupled with two off-channel defect rods made from Kerr media. The absolute value of EM field solution for (**b**) the symmetry preserving solution $\omega a/2\pi c = 0.355$, (**c**) the symmetry breaking solution $\omega a/2\pi c = 0.355$ and (**d**) $\omega a/2\pi c = 0.358$. The EM wave incidents at the left of the waveguide. the lattice constant $a = 0.5\,\mu$m, the cylindrical dielectric rods have radius $0.18a$ and dielectric constant $\epsilon = 11.56$ (GaAs at the wave length $1.5\,\mu$m) in air

We name such a solution as the phase symmetry breaking solution. Figure 17.2 shows the energy behavior of the amplitudes and phase difference by gray solid line. As seen from Fig. 17.2c this solution is unstable.

The Schrödinger equation exactly corresponds to the Maxwell equations for TM modes in two-dimensional photonic crystals where the z-component of electric field plays role of the quantum mechanical wave function [43]. In particular the considered open quantum dimer is equivalent to the directional photonic crystal waveguide with two off-channel defect cavities [22]. Figure 17.3 demonstrates the symmetry preserving and symmetry breaking solutions of the nonlinear Maxwell equations in the two-dimensional photonic crystal system with optical cavities made from a Kerr media. In particular, Fig. 17.3d demonstrates the symmetry breaking solution where one of the cavities remains completely dark.

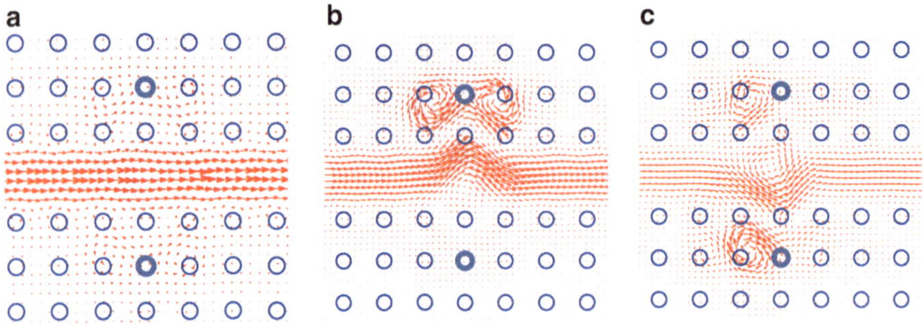

Fig. 17.4 Current flows for the symmetry preserving solution which inherits linear case (**a**), the symmetry breaking solution (**b**), and (**c**) the phase symmetry breaking one at $a\omega/2\pi c = 0.35$. *Bold open circles* mark the nonlinear defects

Because of the phase difference $\theta_1 \neq \theta_2$ for the phase symmetry breaking branch a current can flow between the nonlinear sites similar to tunneling current between two superconducting samples. Multiplying equation (17.4) by complex conjugated amplitudes and subtracting the complex conjugated terms one can obtain the value of the power flow current flowing between the chain at the "0"-th site and cavities as follows

$$j_{0\to 1,2} = \sqrt{\gamma} Im(\phi_0 \psi_{1,2}^*). \tag{17.15}$$

Similar manipulations with the cavity amplitudes give the current between the cavities

$$j_{1\to 2} = uIm(\psi_1 \psi_2^*) = u\delta\epsilon I \sin(2\theta). \tag{17.16}$$

The current from the "-1"-th site to the "0"-th site of the chain coincide with current from the "0"-th site to the "1"-th one. Therefore the currents (17.15) and (17.16) coincide too in accordance with the Kirchhoff rule. Thus, the input power induces vortical current between the waveguide and cavities via the coupling $\sqrt{\gamma}$ and between the cavities via the coupling u. The Josephson analogue of current can be demonstrated in photonic crystal which reveals the stability domain for the phase symmetry breaking solution [22]. This solution does not visualize broken symmetry in modulus of wave function but clearly shows breaking of symmetry in the Poynting currents in Fig. 17.4.

17.3 Switching of Quantum Flows in T-Shape Structure

Figure 17.3d demonstrates the resonance state where only one of nonlinear sites is excited while opposite site is completely dark. Note there is equivalent resonance state with symmetry inversion of the dimer. Then this dark site will blockade the quantum transmission through the site. Therefore in the T-junction shown in Fig. 17.1d one of the outputs will be switched off. Moreover applying pulses of injected wave we can switch these flows from one output to another. These ideas were realized in our Ref. [23].

For the present system we approximate the tight-binding wires with finite propagation band with continual wires with infinite propagation bands. That allows us to use the CMT formulated in Refs. [33, 35]:

$$\begin{aligned} i\dot{\psi}_1 &= (\lambda|\psi_1|^2 - i\gamma)\psi_1 + i\sqrt{\gamma}\sigma_{2-}e^{i\phi} \\ i\dot{\psi}_2 &= (\lambda|\psi_2|^2 - i\gamma)\psi_2 + i\sqrt{\gamma}\sigma_{3-}e^{i\phi} \end{aligned} \tag{17.17}$$

The phase $\phi = kL$ where k is the wave vector and L is a distance between junction and the dimer sites as shown in Fig. 17.5a. These CMT equations are to be complemented by the equations for light amplitudes at each site

$$\begin{aligned} S_{2-} &= \sigma_{2-}e^{i\phi} - \sqrt{\gamma}\psi_1, \\ S_{3-} &= \sigma_{3-}e^{i\phi} - \sqrt{\gamma}\psi_2, \\ \sigma_{2+}e^{-i\phi} &= -\sqrt{\gamma}\psi_1, \\ \sigma_{3+}e^{-i\phi} &= -\sqrt{\gamma}\psi_2. \end{aligned} \tag{17.18}$$

The T-connection connects ingoing and outgoing amplitudes by the S-matrix as follows [23]

$$\begin{pmatrix} S_{1-} \\ \sigma_{2-} \\ \sigma_{3-} \end{pmatrix} = \frac{1}{3}\begin{pmatrix} -1 & 2/3 & 2 \\ 2 & -1 & e \\ 2 & 2 & -1 \end{pmatrix}\begin{pmatrix} S_{1+} \\ \sigma_{2+} \\ \sigma_{3+} \end{pmatrix}. \tag{17.19}$$

Equations (17.17)–(17.19) form full system of equations for nine amplitudes: $A_1, A_2, \sigma_{2+}, \sigma_{2-}, \sigma_{3+}, \sigma_{3-}, S_{1-}, S_{2-}, S_{3-}$. In Fig. 17.5b we present the stationary solution of these nonlinear equations as the transmission probabilities from the input wire into the output wires $|t_{12}|^2 = |S_{2-}|^2/|\phi_0|^2, |t_{13}|^2 = |S_{3-}|^2/|\phi_0|$. There are at least two solutions. The first solution shown by grey solid line preserves the symmetry and therefore is not interesting for T-switching. The second SB solution shown by red and blue lines in Fig. 17.5b provides almost 100 % blocking of quantum flow to the second output waveguide, and is stable. In Fig. 17.6 we show realization of these effects of the symmetry breaking in the T-shaped photonic crystal waveguide [23].

That result is extremely important for the switching of the output flows from one output into another one. In order to switch the system from one asymmetric state to the other we following Refs. [13, 18] apply pulses of the input power injected into the waveguide 1. The direct numerical solution of the temporal CMT equation (17.17) with $S_{1+}(t) = \phi_0(t)e^{iEt}$ is shown in Fig. 17.7 which demonstrates the switching effect. The stepwise time behavior of amplitude $\phi_0(t)$ is shown by gray line. One can see that after the first pulse of the input amplitude the oscillations of the cavity

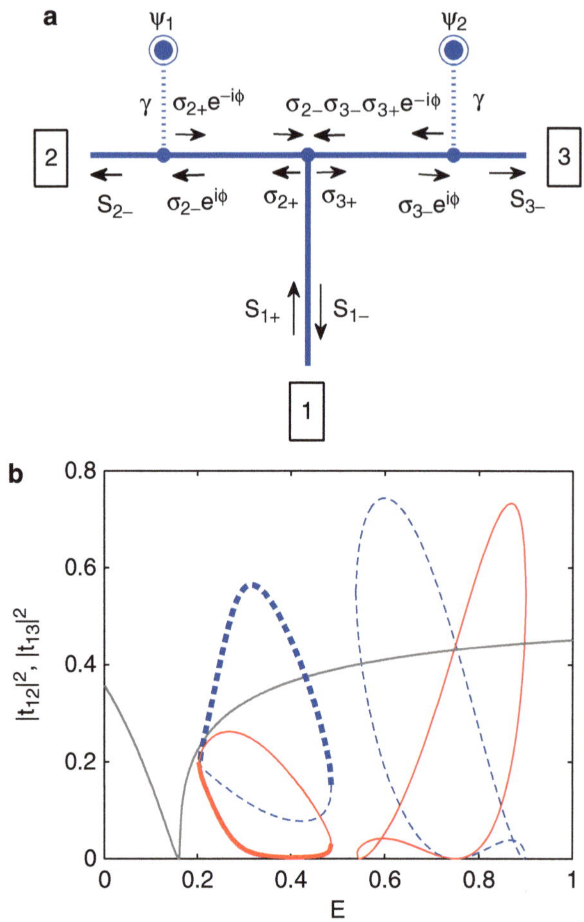

Fig. 17.5 (**a**) Continual version of the open quantum dimer shown in Fig. 17.1c. The input wave labelled as $S_{1+} = \phi_0 e^{iEt}$ is applied through waveguide 1. (**b**) The transmission probabilities into the output waveguides 2 and 3 vs energy. Stability of solutions is marked by *thicker lines*

amplitude relax into the stable stationary solutions with broken symmetry. Moreover after each next pulse the state of the system transmits from one asymmetric state to the other as was observed by Maes et al. [18].

17.4 Four-Site Nonlinear Plaquette

In the previous section we demonstrated how the phenomenon of SB occurs in the nonlinear dimer due to resonant excitation of both symmetric and antisymmetric modes A_s, A_a. In this section we consider the transmission through four-site plaquette shown in Fig. 17.8. The closed system supports two degenerate modes

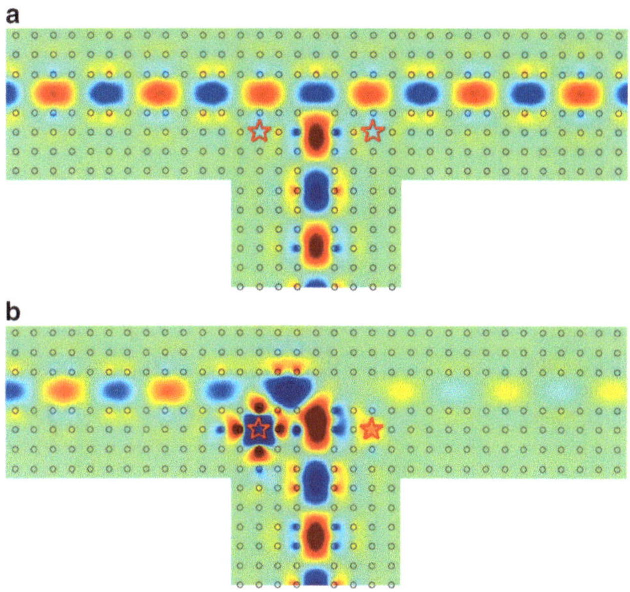

Fig. 17.6 The EM field solution for (**a**) the symmetry preserving solution $\omega a/2\pi c = 0.3442$ and (**b**) the symmetry breaking solution the for $\omega a/2\pi c = 0.3442$ in the T-shaped photonic crystal made of the same material as given in Fig. 17.3. Defect rods fabricated from Kerr media which play role of nonlinear sites of the nonlinear dimer are shown by *stars*

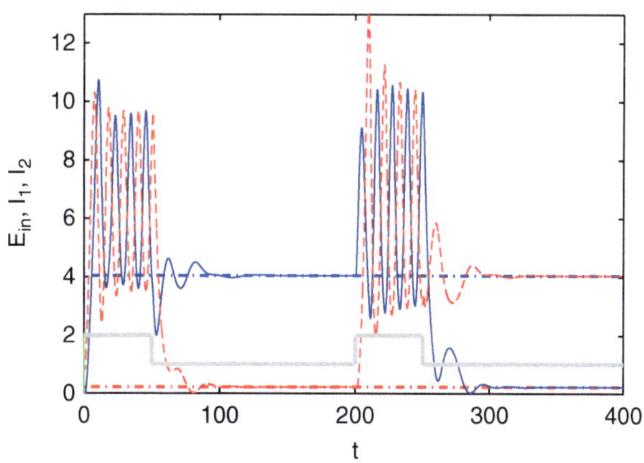

Fig. 17.7 The time dependence of the amplitudes $|\psi_1|, |\psi_2|$ in the cavities (*solid* and *dashed* respectively) which follow the impulses of the input amplitude ϕ_0 (*gray*)

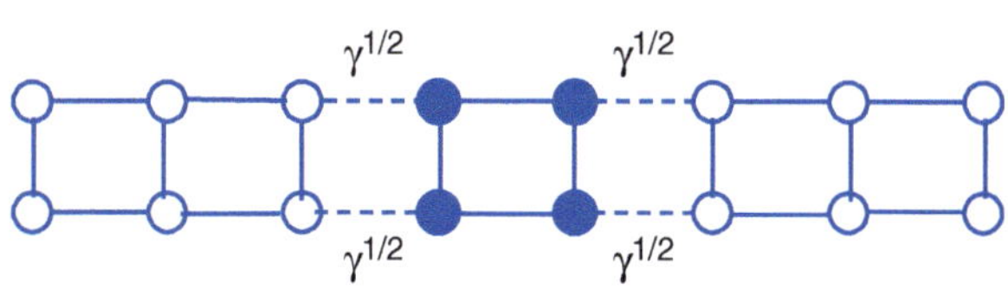

Fig. 17.8 Two-channel waveguides in form of binary tight-binding chain are attached to the four-site nonlinear plaquette via the coupling constant $\sqrt{\gamma}$

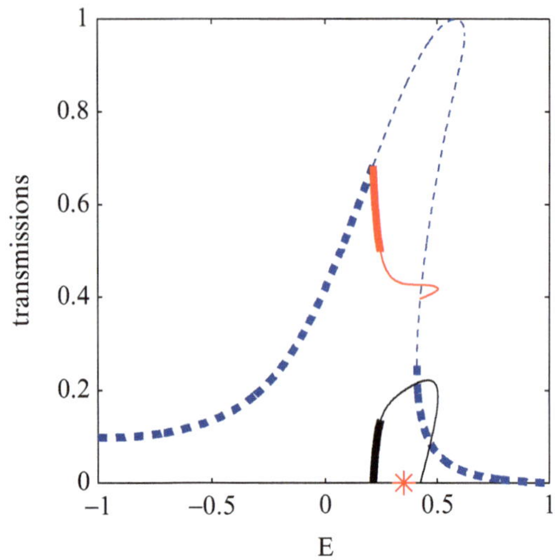

Fig. 17.9 Response of nonlinear plaquette to symmetric probing wave $u = 0.4, \gamma = 1, \lambda = 0.09$, transmission probability $|t_{11}|^2$ for symmetry preserving solutions – *dashed blue line*, $|t_{11}|^2$ SB solutions – *solid red line*, $|T_{12}|^2$ SB solutions – *solid black*. Transmissions vs. energy, $\phi_0 = 1$. One can see windows were neither symmetry preserving nor symmetry breaking stable solutions exist where stability is marked by *thicker lines*

of different symmetries $\langle A_s| = \frac{1}{2}(1, 1, -1, -1)$ and $\langle A_a| = \frac{1}{2}(1, -1, 1, -1)$. The corresponding degenerate eigenvalue is $E_{s,a} = 0$. In the case of the four-site plaquette the effective Hamiltonian takes the following form [29, 31]

$$H_{eff} = \begin{pmatrix} \hat{h}(1,2) & \hat{v} \\ \hat{v} & \hat{h}(3,4) \end{pmatrix},$$

$$\hat{h}(i,j) = \begin{pmatrix} -\gamma(e^{ik_1} + e^{ik_2})/2 + \lambda|\psi_i|^2 & -u - \gamma(e^{ik_1} - e^{ik_2})/2 \\ -u - \gamma(e^{ik_1} - e^{ik_2})/2 & -\gamma(e^{ik_1} + e^{ik_2})/2 + \lambda|\psi_j|^2 \end{pmatrix},$$

$$\hat{v} = \begin{pmatrix} -u & 0 \\ 0 & -u \end{pmatrix}, \tag{17.20}$$

Eq. (17.7) should now be solved for the state vector of the plaquette $\langle\psi| = (\psi_1, \psi_2, \psi_3, \psi_4)$ with the source term $\langle in| = (1, 0, \pm 1, 0)$. The results of numerical solution are presented in Fig. 17.9.

The result is similar to the case of nonlinear dimer including that there is now a domain the parameter space where *all* stationary solutions are unstable as shown in Figs. 17.2 and 17.7. Respectively, the solution of the transmission problem in such system can be described neither by transmission and reflection amplitudes equations (17.6) nor by the Feshbach projection method, i.e. by the effective non-Hermitian Hamiltonian equation (17.7). The problem of plane wave scattering from the nonlinear plaquette can only be solved through numerical simulation of the time-dependent equation.

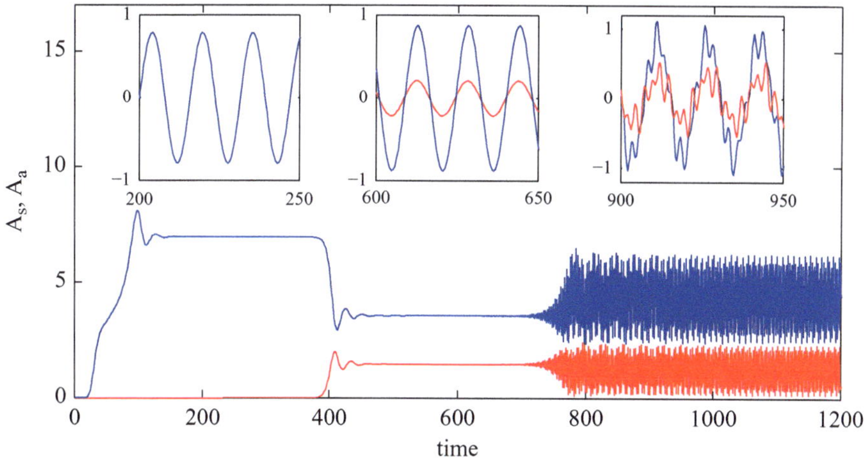

Fig. 17.10 Time evolution of populations of symmetric B_s (*blue line*), and antisymmetric B_a (*red line*) resonant modes, $\epsilon = 0.4, \gamma = 1, \lambda = 0.09E = 0.4$. The *insets* show the real parts of amplitudes u_3 (*blue line*), and v_3 (*red line*) vs. time t in the corresponding regimes. One can clearly see that in course of time the system evolves to a non-stationary solution

As numerical tests we perform the wave-front simulations with the initial state

$$|\psi_{in}(n)\rangle = f(n)\psi_1^{(+)}(n,0), \tag{17.21}$$

where

$$f(n) = \begin{cases} 1 & \text{if } n \leq -100\,; \\ e^{-(n+100)^2/250} & \text{if } n > -100, \end{cases} \tag{17.22}$$

is an auxiliary function that provides a smooth increase of incident plane wave amplitude. Figure 17.10 shows the time evolution of populations A_s, A_a of symmetric and antisymmetric resonant modes. At $t = 0$ a symmetric wave front (17.21) is sent from the left waveguide towards the plaquette with its energy and amplitude within the domain of unstable stationary solutions $E = 0.4$ (shown by red star in Fig. 17.7a). First when $t \in [150, 350]$ only symmetric state of the plaquette is excited; $A_a = 0$. However, the symmetry preserving solution is unstable. That causes transition to the symmetry breaking solution $A_a > 0$ at $t \approx 400$. As a result the plaquette emits stationary plane waves of both symmetries with the same energy as the probing wave when $t \in [450, 650]$. Since this solution is also unstable the system transits to another regime at $t \approx 700$. It is clearly seen in Fig. 17.10 that in this regime the solution is also symmetry breaking, however, what is more interesting, it is *non*-stationary. The Fourier power spectrum $F(E)$ Fig. 17.11 of the amplitude ψ_3 for $t \geq 800$ clearly shows the presence of three peaks, the central peak with energy $E = 0.4$ and two satellites with energies $E_1 = -1.71$, $E_2 = 2.51$.

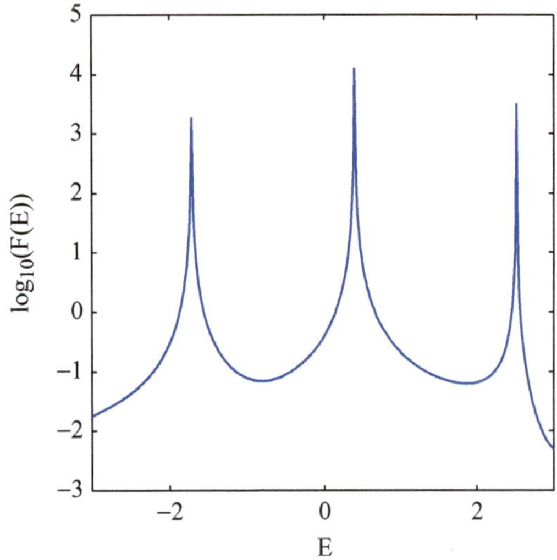

Fig. 17.11 Logarithmic plot of Fourier power spectrum of output amplitude u_3 in the non-stationary regime Fig. 17.9

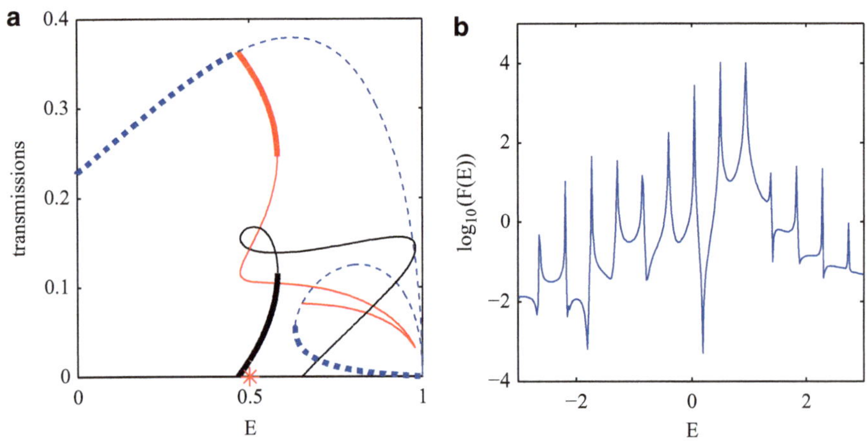

Fig. 17.12 (**a**) Response of nonlinear plaquette to symmetric probing wave $\epsilon = 0.4, \gamma = 1, \lambda = 0.4, A_0 = 1$. Transmission probability $|T_{11}|^2$ for symmetry preserving solutions – *dashed blue line*, $|T_{11}|^2$ SB solutions – *solid red line*, $|T_{12}|^2$ SB solutions – *solid black*. (**b**) Corresponding Fourier power spectrum of amplitude u_3 at $E = 0.5$ (*red star* in (**a**))

With the growth of the nonlinearity constant λ which is equivalent to growth of the amplitude of the injected wave the dynamical properties of the nonlinear plaquette change drastically. Figure 17.12a shows the transmission probabilities vs. the incident energy. One can see that compared against Fig. 17.9 the transmission peak is now shifted towards the edge of the propagation band. When the energy of the incident wave belongs to the instability window of the symmetry preserving solution $E = 0.5$ (red star in Fig. 17.11a) the system rapidly evolves to non-stationary symmetry breaking regime. The time-Fourier power spectrum with several equidistant satellite peaks is shown in Fig. 17.12b. It should be noticed that although there now is a pair of stable symmetry breaking solutions at $E = 0.5$ the system nevertheless does not access them but immediately transits to the non-stationary regime. That phenomenon of satellite peak generation obviously differs from the second harmonic generation where the waves of twice the energy (frequency) would be emitted [44].

17.5 Summary and Discussions

In this paper we considered the simplest nonlinear open systems whose closed analogues allow for symmetry breaking, namely, dimer and four-site square plaquette [2–8]. The term "open" means that linear waveguides are now attached to the nonlinear objects. The waveguides are chosen in the form of tight-binding double chains. As shown in Fig. 17.1 this architecture preserves the mirror symmetry with respect to the center-line of the waveguides. Then *if* there are stationary solutions (17.5) the standard procedure of matching reflected/transmitted waves (17.6) can be applied to obtain the transmision/reflection coefficients. It is more convenient, however, to use the Feshbach projection technique to project the total Hilbert space onto the space of the inner states that describe the scattering region only [26,28,29,31]. The resulting equation could be seen as a nonlinear equivalent of the Lippmann-Schwinger equation (17.7) where H_{eff} is the nonlinear non hermitian effective Hamiltonian whose matrix elements depend in turn on the amplitude of the injected wave. The corresponding equations are written down for both dimer by Eq. (17.8) and four-site square nonlinear plaquette equation (17.20). The effective Hamiltonian differs from the nonlinear Hamiltonian considered in Refs. [2–7, 10] due to the presence of dissipative terms $\epsilon^2 \exp(ik_p)$ where ϵ is the hopping matrix element that controls the coupling between the closed system and the waveguides.

In case of transmission of a symmetric plane wave through the nonlinear dimer we found two families of solutions. In the symmetry preserving family the incident symmetric wave is reflected and transmitted into the same symmetric channel. The second family, however, violates the symmetry of the probing wave. It means that when a symmetric wave is injected into the system the SB gives rise to emission of the antisymmetric plane waves and vice versa. Therefore the nonlinear dimer is capable for the mode conversion, although, with maximum efficiency around 50 %. We found that the direct solution of the time-dependent Schrödinger equation with a wave front incident to the nonlinear dimer gives the same results for the transmission probabilities as found from the approach of the non-Hermitian Hamiltonian. It should be pointed out that the key feature that makes possible to access the SB solutions is the presence of domains in the parameter space where all symmetry preserving solutions are unstable. It means that in course of time the symmetry preserving solution will eventually collapse due to the presence of noise. The second important aspect about the open nonlinear dimer is that symmetry is broken by both intensity and phase of the scattering function.

Similar consideration was made for the open nonlinear four-site plaquette (see Fig. 17.1b). Its closed counterpart has the symmetry group D_4 that provides many opportunities for the symmetry breaking [10]. However the presence of the waveguides in the design shown in Fig. 17.1b substantially reduces this symmetry to the symmetry of the open dimer. Therefore one can expect a similar scenario for SB. However, in the case of plaquette four nonlinear degrees of freedom participate in the transmission which dramatically changes the dynamical picture. The standard theory of stability [45, 46] based on small perturbation technique reveals that there are domains in the parameter space where *none* of the stationary solutions (neither symmetry preserving nor symmetry breaking) are stable. It means that the scattering problem could not be reduced to stationary equations. Direct solution of the time-dependent Schrödinger equation revealed the emission of satellite waves at the energies different from the energy of incident wave provided that this energy is chosen within the domain where the symmetry preserving solution is unstable. The number of satellite wave and their energies depend mostly on the intensity of injected wave (or equivalently on the nonlinearity constant). This effect is different from the second harmonic generation with satellite energies not equal twice the injected wave energy. Emergence of additional equidistant peaks in the Fourrier power spectrum of four-cite nonlinear system was reported almost 30 years ago in the seminal paper by Eilbeck et al. [2]. We believe that nowadays with the ongoing development of experimental techniques, in particular in handling photonic crystal waveguides, that phenomenon opens a new opportunity for harmonics generation. Another interesting possibility for constructing nonlinear quantum double-chain set-ups could be Bose-Hubbard ladders in optical lattices [47].

Appendix

The solution of the infinite tight-binding one-dimensional wire is

$$\langle j|E\rangle = \psi_E(j) = \frac{1}{\sqrt{2\pi|\sin k|}}\exp(ikj), \tag{17.23}$$

where $E = -2\cos k$, and $\langle E|E'\rangle = \delta(E - E')$. In the site representation we have for example the matrix element at the the first off-channel site

$$\begin{aligned}
&\langle\psi_1|V^\dagger\frac{1}{E + i0 - H_w}V|\psi_1\rangle \\
&= \gamma\langle\psi_1|V^\dagger|j = 0\rangle\int_{-2}^{2} dE'\psi_{E'}(j = 0)\frac{1}{E + i0 - E'}\psi^*_{E'}(j = 0) \\
&= \frac{i\gamma}{2|\sin k(E)|} + \frac{\gamma}{4\pi}P\int_{-2}^{2} dx\frac{1}{\sqrt{1 - x^2}(x - E/2)} = \frac{i\gamma}{2|\sin k(E)|}.
\end{aligned} \tag{17.24}$$

The same expression we obtain for other matrix elements. Here P have a meaning of principal integral. Therefore we have

$$H_{eff} = \begin{pmatrix} -\frac{i\gamma}{2|\sin k(E)|} + \lambda|\psi_1|^2 & -u - \frac{i\gamma}{2|\sin k(E)|} \\ -u - \frac{i\gamma}{2|\sin k(E)|} & -\frac{i\gamma}{2|\sin k(E)|} + \lambda|\psi_2|^2 \end{pmatrix}. \tag{17.25}$$

References

1. Akhmediev NN (1982) Novel class of nonlinear surface waves: asymmetric modes in a symmetric layered structure. Sov Phys JETP 56:299
2. Eilbeck JC, Lomdahl PS, Scott AC (1985) The discrete self-trapping equation. Physica 16D:318
3. Tsironis GP, Kenkre VM (1988) Initial condition effects in the evolution of nonlinear dimer. Phys Lett A 127:209
4. Kenkre VM, Wu H-L (1989) Interplay of quantum phases and non-linearity in the non-adiabatic dimer. Phys Lett A 135:120
5. Kenkre VM, Kuś M (1994) Bifurcations in the quantum nonlinear dimer. Phys Rev B 49:5956
6. Bernstein LJ (1991) Nonlinear self-trapping in a quantum dimer. Physica D 53:240
7. Tsironis GP, Deering WD, Molina MI (1993) Applications of self-trapping in optically coupled devices. Physica D 68:135
8. Molina MI (1999) Self-trapping on a generalized nonlinear tetrahedron. Mod Phys Lett 13:225
9. Brazhnyi VA, Malomed BA (2011) Spontaneous symmetry breaking in Schrodinger lattices with two nonlinear sites. Phys Rev A 83:053844
10. Wang C, Theocharis G, Kevrekidis PG, Whitaker N, Law KJH, Frantzeskakis DJ, Malomed BA (2009) Two-dimensional paradigm for symmetry breaking: the nonlinear Schrodinger equation with a four-well potential. Phys Rev E 80:046611
11. Law KJH, Qiao L, Kevrekidis PG, Kevrekidis IG (2008) Stability of quantized vortices in a Bose-Einstein condensate confined in an optical lattice. Phys Rev A 77:053612
12. Yabuzaki T, Okamoto T, Kitano M, Ogawa T (1984) Optical bistability with symmetry breaking. Phys Rev A 29:1964
13. Otsuka K, Ikeda K (1987) Hierarchical multistability and cooperative flip-flop operation in a bistable optical system with distributed nonlinear elements. Opt Lett 12:599
14. Haelterman M, Mandel P (1990) Pitchfork bifurcation using a two-beam nonlinear Fabry-Perot interferometer: an analytical study. Opt Lett 15:1412
15. Babushkin IV, Logvin YuA, Loǐko NA (1998) Symmetry breaking in optical dynamics of two bistable thin films. Quantum Electron 28:104
16. Kevrekidis PG, Chen Zh, Malomed BA, Frantzeskakis DJ, Weinstein MI (2005) Spontaneous symmetry breaking in photonic lattices: theory and experiment. Phys Lett A 340:275
17. Shwartz S, Weil R, Segev M, Lakin E, Zolotoyabko E, Menon VM, Forrest SR, El-Hanany U (2006) Light-induced symmetry breaking and related giant enhancement of nonlinear properties in CdZnTe: V crystals. Opt Express 14:9385
18. Maes B, Soljačić M, Joannopoulos JD, Bienstman P, Baets R, Gorza S-P, Haelterman M (2006) Switching through symmetry breaking in coupled nonlinear micro-cavities. Opt Express 14:10678
19. Maes B, Bienstman P, Baets R (2008) Symmetry breaking with coupled Fano resonances. Opt Express 16:3069
20. Aydin K, Pryce IM, Atwater HA (2010) Symmetry breaking and strong coupling in planar optical metamaterials. Opt Express 18:13407
21. Li R, Lv F, Lu Li, Xu Z (2011) Symmetry breaking and manipulation of nonlinear optical modes in an asymmetric double-channel waveguide. Phys Rev A 84:033850

22. Bulgakov EN, Pichugin KN, Sadreev AF (2011) Symmetry breaking for transmission in a photonic waveguide coupled with two off-channel nonlinear defects. Phys Rev B 83:045109; Light induced Josephson like current between two coupled nonlinear cavities coupled with a symmetrically positioned photonic crystal waveguide. J Phys Cond Mat 23:065304
23. Bulgakov EN, Sadreev AF (2011) Symmetry breaking in a T-shaped photonic waveguide coupled with two identical nonlinear cavities. Phys Rev B 84:155304
24. Maier AA (1982) Optical transistors and bistable devices utilizing nonlinear transmission of light in systems with undirectional coupled waves. Sov J Quantum Electron 12:1490
25. Gibbs HM (1985) Optical bistability: controlling light with light. Academic, New York
26. Feshbach H (1958) Unified theory of nuclear reactions. Ann Phys (NY) 5:357; Feshbach H (1962) Unified theory of nuclear reactions, II. Ann Phys (NY) 19:287
27. Mahaux C, Weidenmuller HA (1969) Shell model approach in nuclear reactions. North-Holland, Amsterdam
28. Rotter I (1991) A continuum shell model for the open quantum mechanical nuclear system. Rep Prog Phys 54:635
29. Datta S (1995) Electronic transport in mesoscopic systems. Cambridge University Press, Cambridge
30. Dittes F-M (2000) The decay of quantum systems with a small number of open channels. Phys Rep 339:215
31. Sadreev AF, Rotter I (2003) S-matrix theory for transmission through billiards in tight-binding approach. J Phys A Math Gen 36:11413
32. Haus HA (1984) Waves and fields in optoelectronics. Prentice-Hall, New York
33. Manolatou C, Khan MJ, Fan S, Villeneuve PR, Haus HA, Joannopoulos JD (1999) Coupling of modes analysis of resonant channel add/Drop filters. IEEE J Quantum Electron 35:1322
34. Fan S, Suh W, Joannopoulos JD (2003) Temporal coupled-mode theory for the Fano resonance in optical resonators. J Opt Soc Am A 20:569
35. Suh W, Wang Z, Fan S (2004) Temporal coupled-mode theory and the presence of non-orthogonal modes in lossless multimode cavities. IEEE J Quantum Electron 40:1511
36. von Neumann J, Wigner E (1929) Ubermerkwuridge diskrete Eigenwerte. Phys Z 30:465
37. Bulgakov EN, Sadreev AF (2008) Bound states in the continuum in photonic waveguides inspired by defects. Phys Rev B 78:075105
38. Moiseyev N (2009) Suppression of Feshbach resonance widths in two-dimensional waveguides and quantum dots: a lower bound for the number of bound states in the continuum. Phys Rev Lett 102:167404
39. Bulgakov EN, Sadreev AF (2011) Formation of bound states in the continuum for a quantum dot with variable width. Phys Rev B 83:235321
40. Bulgakov EN, Sadreev AF (2009) Resonance induced by a bound state in the continuum in a two-level nonlinear Fano-Anderson model. Phys Rev B 80:115308
41. Bulgakov EN, Sadreev AF (2010) Bound states in photonic Fabry-Perot resonator with nonlinear off-channel defects. Phys Rev B 81:115128
42. Molina MI, Miroshnichenko AE, Kivshar YuS (2012) Surface bound states in the continuum. Phys Rev Lett 108:070401
43. Joannopoulos J, Johnson SG, Winn JN, Meade RD (2008) Photonic crystals: molding the flow of light. Princeton University Press, Princeton
44. New GHC, Ward JF (1967) Optical third-harmonic generation in gases. Phys Rev Lett 19:556
45. Litchinitser NM, McKinstrie CJ, de Sterke CM, Agrawal GP (2001) Spatiotemporal instabilities in nonlinear bulk media with Bragg gratings. J Opt Soc Am B 18:45
46. Cowan AR, Young JF (2003) Optical bistability involving photonic crystal microcavities and Fano line shapes. Phys Rev E 68:046606
47. Dhar A, Maji M, Mishra T, Pai RV, Mukerjee S, Paramekanti A (2012) Bose-Hubbard model in a strong effective magnetic field: emergence of a chiral Mott insulator ground state. Phys Rev A 85:041602(R)

Chapter 18
Charge Separation and Transport in Third Generation Hybrid Polymer-Fullerene Solar Cells

B.L. Oksengendler, Oksana B. Ismailova, M.B. Marasulov, N.N. Turaeva, Davron Matrasulov, and J.R. Yusupov

Abstract Exciton dissociation in polymer-fullerene hybrid organic solar cells is studied within the quantum mechanical and statistical approaches. The different mechanisms for splitting of exciton into electron and hole is discussed.

18.1 Introduction

Third generation solar cells based on organic photovoltaic materials are being considered as a serious alternative for silicon based ones. Usually such solar cells are fabricated using polymer structures as donors and fullerene embedded into polymers as acceptor. In some cases quantum dots can be embedded into the polymer matrix. The maximum conversion efficiency of such solar cells are expected to reach 80 % [10] due to so-called multiple exciton generation effect [12]. The mechanism such effect is still subject for discussions and extensive studies. Topics of the past studies

B.L. Oksengendler • M.B. Marasulov
Institute of Polymer Chemistry and Physics, Tashkent, Uzbekistan

O.B. Ismailova (✉)
Institute Ion-Plasma and Laser Technology, Uzbekistan Academy of Sciences, Tashkent, Uzbekistan
e-mail: oksana-ib@yandex.com

N.N. Turaeva
Biological Department, Webster University, Louis, MO, USA

D. Matrasulov • J.R. Yusupov
Turin Polytechnic University in Tashkent, Tashkent, Uzbekistan

D. Matrasulov and H.E. Stanley (eds.), *Nonlinear Phenomena in Complex Systems: From Nano to Macro Scale*, NATO Science for Peace and Security Series C: Environmental Security, DOI 10.1007/978-94-017-8704-8_18,

of third generation solar cells both polymer-fullerene and polymer-quantum dot based ones can be classified as follows:

1. Mechanism of the multiple exciton generation by quantum dots induced by single photon absorption;
2. Transmission of generated excitons through the quantum-dot-polymer or polymer-fullerene interface and possible splitting of exciton into electron and hole during this separation;
3. The exciton transport and charge separation along the polymer chain;
4. Transport of electrons and holes through in the polymer-quantum dot or polymer-fullerene networks;
5. The charge carrier collection at the metal-polymer interface;
6. Degradation of solar cell.

Among the above issues (2), (3), and (6) are still remaining as less studied, despite the extensive studies during last few years [2–4, 10–12, 16].

In this paper we develop microscopic mechanism for exciton splitting and charge transport in organic solar cells on the basis of polymers and nanoparticles (e.g. quantum dot, fullerene).

18.2 Charge Separation via Dopant Charge Exchange

Assuming that electron is moving along linear polymer chain and the mass of the hole is large enough, splitting of the exciton can be considered as similar to charge exchange between the hydrogen atom and a potential well induced by effective positive charge (Fig. 18.1).

In [13, 14] the process of exciton dissociation on fullerene via the reaction

$$ex + F^0 \rightarrow h + F^-, \tag{18.1}$$

was considered as a resonance charge exchange studied in [17]. However, in general case potential wells induced by electron and hole on fullerene trap have different depths. Therefore one needs to strict treatment of charge separation as a charge exchange process. This can be done within the Landau-Zener theory [9].

If exciton moves along linear polymer chain and approaches to fullerene molecule, the molecular terms corresponding to states "electron-on hole+fullerene" and "electron-hole on fullerene" (Fig. 18.1) can be written as

$$\begin{cases} U_1(R) = U_0^{(1)} + \Delta E_{ex} \\ U_2(R) = U_0^{(2)} - e^2/\varepsilon R \end{cases} \tag{18.2}$$

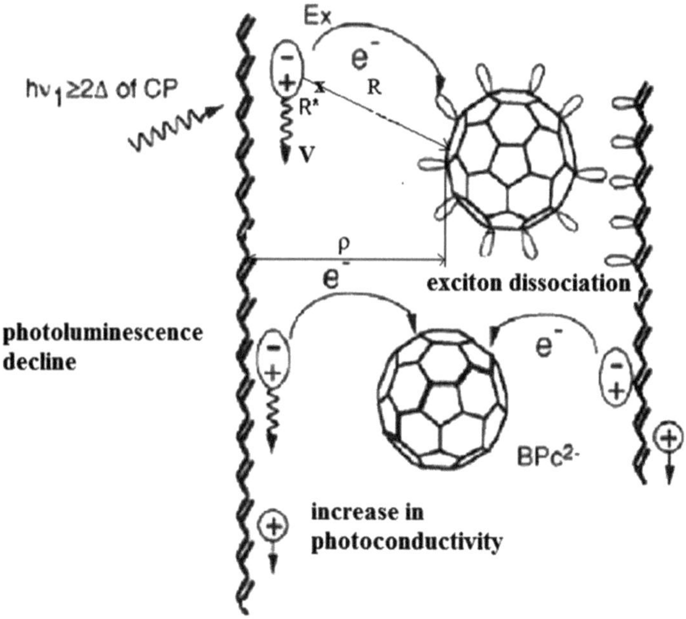

Fig. 18.1 Schematic representation of exciton dissociation via resonance rechargeable

Fig. 18.2 Schematic representation of the crossing terms $V_1(R)(\text{exciton} + F^0)$ and $V_2(R)(\text{hole} + F^-)$

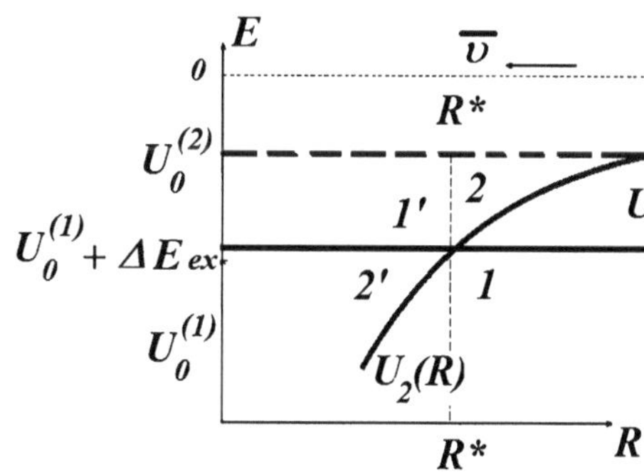

where $U_0^{(1)}$ and $U_0^{(2)}$ are the terms of electron on polymer (highest occupied orbital, HOMO) and trapped into fullerene correspondingly; ΔE_{ex} is the exciton perturbation energy, ε is dielectric constant of polymer matrix.

As it can be seen from Fig. 18.2, the terms $U_1(R)$ and $U_2(R)$ are crossing at the point:

$$R^* = \frac{e^2/\varepsilon}{U_0^{(2)} - U_0^{(1)} - \Delta E_{ex}} \tag{18.3}$$

In its motion along the polymer chain with the velocity υ, exciton passes through the point R^* (Fig. 18.2) that leads to radiationless transition from term 1 to 2 (Landau-Zener transition) and subsequently the reaction given by Eq. (18.1) occurs. Furthermore, the "free hole" continues to move along the chain and reaches the point $\tilde{R}^*$. If the electron hole system remains at the term 2 this will correspond to the splitting of exciton into electron (which will remain at the fullerene) and hole that will move along the chain. Then the probability for such decay can be written as [9]:

$$\mathrm{P}(\tilde{\upsilon}) = \omega_{1\to 2'}(R^*)\left[1 - \omega_{2'\to 1}(\tilde{R}^*)\right] \tag{18.4}$$

$$\omega_{1\to 2'}(R^*) = \omega_{2'\to 1}(\tilde{R}^*) = 2\left[1 - e^{-\frac{2\pi|\hat{V}|^2}{\hbar\tilde{\upsilon}|F_1 - F_2|}}\right] e^{-\frac{2\pi|\hat{V}|^2}{\hbar\tilde{\upsilon}|F_1 - F_2|}}, \tag{18.5}$$

where $F_1(R^*)$ and $F_2\left(\tilde{R}^*\right)$ are values for forces of terms 1 and 2 at the crossing point R^* and $\tilde{R}^*$, $\left|\hat{V}\right|^2$ is the matrix element describing transition from term 1 to term 2 in second order of perturbation theory; $\tilde{\upsilon} = \upsilon\sqrt{1 - \rho^2/R^{*2}}$ is the velocity component directed along fullerene (Fig. 18.2). Using Eqs. (18.4) and (18.5), for the of exciton splitting cross-section we have

$$\sigma_{ex}^{dec}(\upsilon) = 2\pi \int_0^{R^*} \mathrm{P}\left[\tilde{\upsilon}(\rho)\right] \rho d\rho \tag{18.6}$$

if the perturbation is week enough, i.e., $\left|\hat{V}\right|^2 \ll \hbar\upsilon\left|F_1 - F_2\right|$, by taking into account Eqs. (18.4) and (18.5) from Eq. (18.6) we obtain

$$\sigma_{ex}^{dec} \approx \frac{8\pi^2\left|\hat{V}\right|^2\left(e^2/\varepsilon\right)^3}{\hbar\left[U_0^{(2)} - U_0^{(1)} - \Delta E_{ex}\right]^4}\frac{1}{\upsilon} \tag{18.7}$$

As it can be seen from this result, exciton splitting cross section depends on (υ) and (ΔE_{ex}). This fact imposes certain restrictions for using of polymer matrices with high mobility of excitons. Having known the cross section, σ_{ex}^{dec} for exciton splitting one can easily estimate the relaxation time for emitted excitons [17]:

$$\tau_{ex}^{dec} = 1/\sigma_{ex}^{dec} N_F \cdot \upsilon, \tag{18.8}$$

here N_F is the concentration of fullerene molecules in the polymer matrix (network).

18.3 Statistical Model for Exciton Dissociation in Regular Polymers

One of the possible mechanisms for splitting of exciton into electron and hole is in transition via the polymer network vertex. If one draws a sphere of $\rho\tilde{}\hbar/\sqrt{2m_0\Delta E_{ex}}$ around the polymer vertex where lying at the crossing N bonds of the polymer chain, it follows from the uncertainty principle that exciton can be in dissociated state inside such sphere. To some extent such a state of analog of well-known compound states from nuclear physics [1]. The life time of such state can be estimated as [1,9] $\tau_{kc} \approx \hbar/\Delta E_{ex}$. After this time exciton can pass through the vertex as a whole system or splits into electron and hole which will move along different bonds of a polymer chain (Fig. 18.3). The probability for splitting of exciton at the vertex within the statistical physics based approach can be written as [8]

$$F_{stat} = Z_{sep}/Z_{sum} = (2C_N^2 e^{-\Delta E_{ex}/kT})/[N + 2C_N^2 e^{-\Delta E_{ex}/kT}], \qquad (18.9)$$

where are Z_{sep} and Z_{sum} are the partition functions of splitted and whole systems, respectively.

For $\Delta E_{ex}/kT \ll 1$ Eq. (18.9) can be written as

$$F_{stat} = \frac{N-1}{N}. \qquad (18.10)$$

It is clear from this expression that in large N limit the probability approaches 1. For $N = 3$ we have $F_{stat} = 2/3$.

For $\Delta E_{ex}/kT \gg 1$ from Eq. (18.9) we get

$$F_{stat} = 2(N-1)\exp(-\Delta E_{ex}/kT) \qquad (18.11)$$

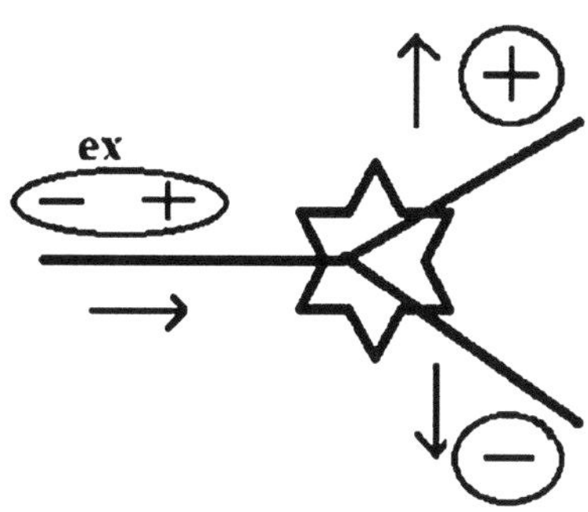

Fig. 18.3 The exciton separation at the polymer chain vertex connecting three bonds

Having known the probability for exciton splitting, F_{stat} from Eqs. (18.9) to (18.11) one can get estimate for the exciton dissociation time as

$$\tau_{ex}^{cs} = 1/\sigma_{ex}^{cs} \cdot N_0 \upsilon \cdot F_{st} = 1/N_0 \upsilon \pi \left(\frac{\hbar}{\sqrt{2m\Delta E_{ex}}} \right)^2 \cdot F_{stat}, \qquad (18.12)$$

where N_0 is the concentration of excitons in polymer matrix.

For $N_0 = 10^{21} sm^{-3}$, $\upsilon = 10^5 sm/s$, $\Delta E_{ex} \approx 5e$, $VN = 3$ we have $\tau_{ex}^{cs} = 1.5 \times 10^{-11} s$.

18.4 Quantum Mechanical Model: Perturbation Theory

Constructing of quantum-mechanical model for exciton dissociation in its transmission through the polymer network vertex is of importance for understanding microscopic mechanisms of the process. In this work we develop perturbation theory based approach for such study. The main question we are interested to explore is: What kind of explicit form has the perturbation potential acting to the exciton at the polymer chain vertex?

We assume that the potential acting to exciton at the vertex of the polymer chain has delta-function form:

$$U = U_0 a^3 \delta(\vec{r}). \qquad (18.13)$$

where U_0, a^3 and δ are measured in eV, $\dot{A}^3$ and cm^{-3}, respectively. For the vertex having N-bonds this potential can be written as $V = N \times U$.

Furthermore, we consider exciton moving along a bond of the polymer chain and approaching a vertex of the chain. Fixing the origin of the coordinate system at the center of mass of exciton one can consider interaction of the exciton with external potential and a scattering of a vertex on exciton. Introducing two characteristic times, reorganization time of the exciton, (τ_{in}) and the time during which interaction with a vertex occurs, (τ_{ext}), we can consider two limiting situations for which we can use different types of perturbation theory. Namely, Landau and Livshic [9] for $\tau_{in} \gg \tau_{ext}$, one can use sudden perturbation approximation, while for the case $\tau_{in} \ll \tau_{ext}$, one can apply adiabatic perturbation theory.

To find in under which conditions each regime is possible we need to estimate the time during which exciton reaches the vertex of the polymer chain, i.e. the area where external perturbation acts. If the motion of exciton from bond to bond is jump-like motion, then each jump is characterized by diffusion coefficient which can be written as [16],

$$D \cong \frac{1}{6} a^2 / \tau_j, \qquad (18.14)$$

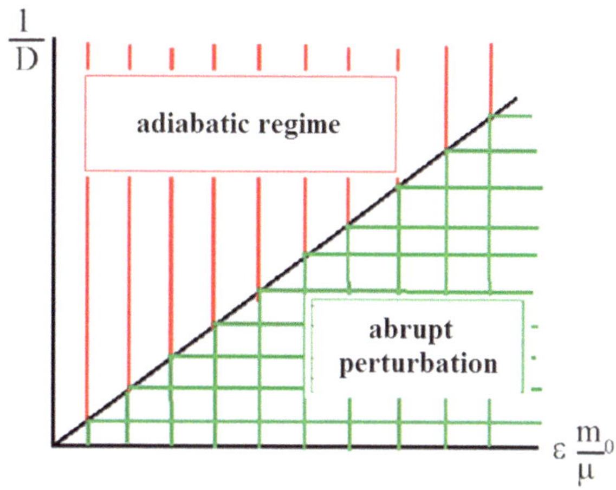

Fig. 18.4 "Phase diagram" based on Eq. (18.16)

where $1/\tau_j$ is the jump frequency. On the other hand, within the Bohr atom framework, the internal reorganization time can be estimated as

$$\tau_{in} = \frac{2\pi\sqrt{r^3\mu}}{e}, \tag{18.15}$$

where r is the effective exciton radius, μ is the reduced mass of exciton $\frac{1}{\mu} = \frac{1}{m_e} + \frac{1}{m_n}$, $\tau_{in} = 1.8 \times 10^{-16}\varepsilon\left(\frac{m_0}{\mu}\right)c$, dielectric permittivity ε of the polymer, m_0 is the electron mass, $\tau_{ext} = 1.8 \times 10^{-16}\frac{1}{D}c$. Thus the boundary between two regimes can be defined as

$$\varepsilon\left(\frac{m_0}{\mu}\right) = \frac{1}{D} \tag{18.16}$$

Equation (18.16): The "phase diagram" describing these regimes can be represented as in the Fig. 18.4.

Consider the case when the following condition is obeyed:

$$\frac{1}{D} < \varepsilon\frac{m_0}{\mu}. \tag{18.17}$$

Potential acting to exciton at the vertex connecting N-bonds of the polymer chain can be written as

$$U_{vertex} = N \times U_0 a^3 \delta\left(\vec{r}\right)$$

Then the probability for transition of exciton from the ground state, $\varphi_{ground} = \sqrt{\frac{\pi}{\tilde{a}^3}}e^{-r/n\tilde{a}}$ to an excited state, $\varphi_{ex} = \sqrt{\frac{\pi}{\tilde{a}^3 n^3}}e^{-r/n\tilde{a}}$ can be written as

$$W_{gr\to ex} = \frac{\left|\langle\varphi_{ex}\left|U_{vertex}\right|\varphi_{ground}\rangle\right|^2}{(\hbar\omega)^2} = \frac{|M|^2}{|\hbar\omega|^2}, \tag{18.18}$$

where $\hbar\omega \approx E_{ex}^{(n)} - E_{ex}^{(gr)}$.

We assume that the exciton shaking processes consist of two stages: Excitation of the electron in exciton into the higher orbit and its transition into conductance band induced by thermal motion. Subsequently, the hole appears in lower band. Such transition is possible under the condition $\left|E_{ex}^{(n)}\right| \approx kT$, and we have $\hbar\omega = -\frac{\Delta E_{ex}}{n^2} + \Delta E_{ex} = \Delta E_{ex}\left(1 - \frac{1}{n^2}\right)$. Using the wave functions of ground and excited states, the matrix element in the expression for the transition probability can be written as

$$M = \langle\varphi_{ex}\left|U_{vertex}\right|\varphi_{gr}\rangle = \frac{\pi}{\tilde{a}^3}U_0 N\frac{1}{n^{3/2}} \tag{18.19}$$

Then for probability we have

$$\tilde{W}_{gr\to ex} = \pi^2\left(\frac{NU_0}{\Delta Eex}\right)^2\left(\frac{kT}{\Delta E_{ex}}\right)^{3/2}\left(\frac{1}{1-\frac{kT}{\Delta E_{ex}}}\right)^2, \tag{18.20}$$

The total probability for dissociation of exciton into electron and hole being distributed over the bonds of the polymer chain can be written as

$$W = \left(\tilde{W}_{gr\to ex}\right)\frac{N-1}{N} \tag{18.21}$$

In case of the adiabatic perturbation occurring under the condition

$$\frac{1}{D} > \varepsilon\frac{m_0}{\mu}, \tag{18.22}$$

one can use Landau-Zener theory [9] dissociation theory. For this purpose we need to plot the energy terms of the systems “exciton-vertex” and “unbounded electron-hole-vertex” (Fig. 18.5). To find the explicit functional forms of these terms we assume that they can be considered as the sum of the ground state energy of exciton and Van der Waals interaction between the exciton and chain vertex. The latter will be modeled as the dielectric ball. Assuming that the Van der Waals interaction is weak enough we can plot the terms presented in (Fig. 18.5).

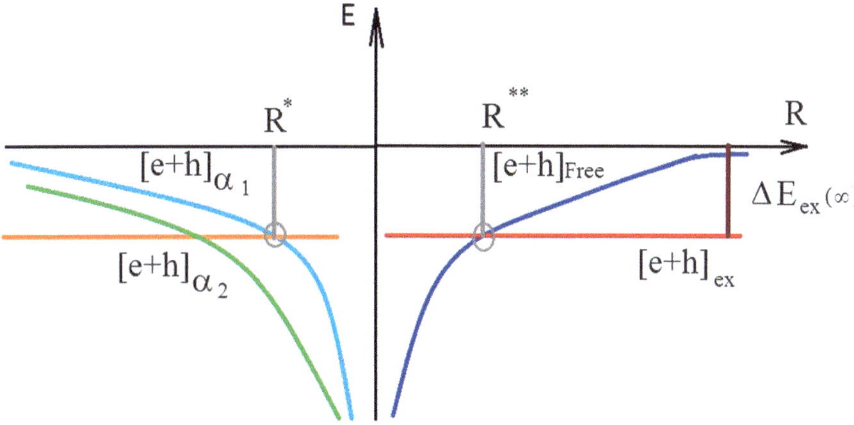

Fig. 18.5 Schematic representation of exciton dissociation and association via scattering at the vertex

Fig. 18.6 Polarization of the vertex B, that is considered modeled as a dielectric ball

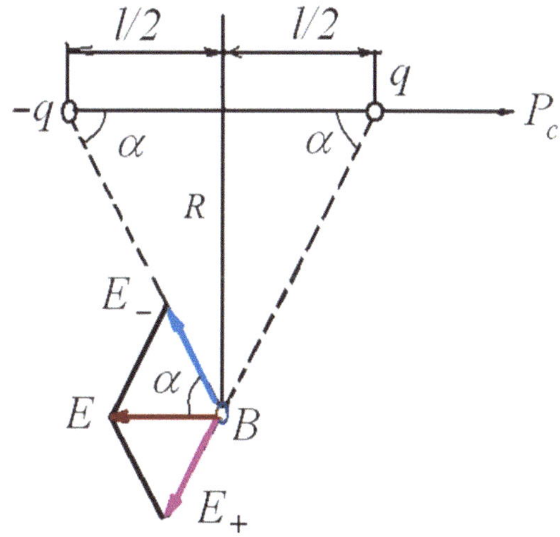

The potential of interaction between the vertex and the free carriers can be considered as a result of the polarization of electron and hole in dielectric medium (Fig. 18.6).

Then the effective electrostatic field caused by electron-hole pair and acting on a vertex can be written as [18]:

$$|\mathbf{E}_{+} + \mathbf{E}_{-}| = \frac{1}{\varepsilon}\frac{2q\cos\alpha}{R^2 + l^2/4} = \frac{1}{2\varepsilon_0}\frac{2ql}{(R^2 + l^2/4)^{3/2}}. \tag{18.23}$$

It is clear that in case of separation of electron and hole over the different bonds of the chain, one can write $\frac{l}{2} = Rctg\alpha$. For this case the energy term can be written as

$$E_{ex}^{(e+h)}(R) = -\frac{1}{2}\alpha_{pol}\left|E_{+} + E_{-}\right|^{2} - \frac{e^{2}}{R\times 2\varepsilon\times ctg\alpha}$$
$$= -\frac{1}{2}\frac{\alpha_{pol}}{(\varepsilon)^{2}}\frac{4e^{2}\cos^{2}\alpha}{R^{4}\left(1+ctg^{2}\alpha\right)^{2}} - \frac{e^{2}}{R\times 2\varepsilon\times ctg\alpha}, \quad (18.24)$$

where a_{pol} is the vertex polarization angle α depends on the number of bonds connected at the vertex:

$$\alpha = 90^{\circ}\left(1-\frac{2}{N}\right) \quad (18.25)$$

The term illustrated in Fig. 18.5 crosses with that of delta-like potential at the R^{*}:

$$R^{*} \approx \left(\frac{e\cos\alpha}{1+ctg^{2}\alpha}\right)^{1/2}\left(\frac{2\alpha_{pol}}{\Delta E_{ex(\infty)}}\right)^{1/4}. \quad (18.26)$$

The probability for inirradiative Landau-Zener transition describing of exciton dissociation at the point R^{*} can be written as [9]:

$$1 - W_{2} = 1 - \frac{4\pi V^{2}}{\hbar v\left|F_{2}\right|}, \quad (18.27)$$

where V^{2} is the square of the adiabatic coupling operator matrix element, v is velocity of exciton

$$F_{2} = \left(\frac{\Delta Eex(\infty)}{2}\right)^{5/4}\frac{8}{\alpha_{pol}^{1/4}(\varepsilon)^{2}}\left(\frac{1+ctg^{2}\alpha}{e\cos^{2}\alpha}\right)^{1/2} \quad (18.28)$$

second term force at the point R^{*}. Furthermore, for the case when exciton amoves along the chain the crossing point of terms can be found from the following equation:

$$U_{o}a^{3}\delta(R) - \Delta E_{ex}(\infty) = -\frac{1}{2}\alpha_{pol}\frac{1}{2}\frac{2ql_{0}}{\left(R^{2}+\frac{l_{0}^{2}}{4}\right)^{3/2}}, \quad (18.29)$$

where l_{0} is the dipole moment of the electron-hole pair in two electron bands.

The total exciton dissociation probability can be written as a product of two probabilities as

$$W = W_{1}\times(1 - W_{2}), \quad (18.30)$$

where $W_1 = \frac{4\pi^2|V|^2}{\hbar v|F(R^{**})|}$ is the probability for transfer from one term to another one at the point R^{**} and $(1 - W_2)$ is the probability for staying of free electrons and holes at the point R^*.

18.5 Quantum Mechanical Model: Quantum Graphs Based Approach

An alternative to the above approaches for description of exciton splitting in polymer network is so-called quantum graph based description. In physics quantum graphs were introduced as a "toy" model for studies of quantum chaos by Kottos and Smilansky [7]. However, the idea for studying of a system confined to a graph dates back to Pauling [15] who suggested to use such systems for modeling free electron motion in organic molecules. During last two decades quantum graphs found numerous applications in modeling different discrete structures and networks in nanoscale and mesoscopic physics (e.g., see reviews [5–7] and references therein). However, mathematical properties of quantum graphs were extensively studied in eighties of the last century. Later quantum graphs became subject for extensive research in different topics of mesoscopic and nanoscale physics and quantum chaos theory (see, e.g. review papers [5–7] and references therein).

Graphs are the systems consisting of bonds which are connected at the vertices. The bonds are connected according to a rule which is called topology of a graph. Topology of a graph is given in terms of adjacency matrix [5, 7]:

$$C_{ij} = C_{ji} = \begin{cases} 1, & \text{if } i \text{ and } j \text{ are connected} \\ 0, & \text{otherwise} \end{cases} \qquad i, j = 1, 2, \ldots, V.$$

Quantum dynamics of a particle on a graph is described by one-dimensional Schrödinger equation [5, 7] (in the units $\hbar = 2m = 1$):

$$i\frac{\partial \Psi_b(x,t)}{\partial t} = -\frac{\partial^2 \Psi_b(x,t)}{\partial x^2} + V(x,t)\Psi_b(x,t) \qquad (18.31)$$

where b denotes a bond connecting ith and jth vertices, and for each bond b, the component Ψ_b of the total wavefunction Ψ_b is a solution of the Eq. (18.31).

In this work we consider simplest topology, so-called star graph, i.e. three or more bonds connected at the single vertex. In this case the boundary conditions can be written as

$$\begin{cases} \phi_1|_{y=0} = \phi_2|_{y=0} = \ldots = \phi_N|_{y=0}, \\ \phi_1|_{y=l_1} = \phi_2|_{y=l_2} = \ldots = \phi_N|_{y=l_N} = 0, \\ \sum_{j=1}^{N} \frac{\partial}{\partial y}\phi_j|_{y=0} = 0. \end{cases} \tag{18.32}$$

We note that to some extent, quantum graphs can be convenient and effective tool to describe charge separation and transport in polymer networks which are the underlying structure of organic solar cells. In particular, exciton can be described as wave packet and splitting of exciton into electron and hole can be easily treated within such approach. The Schrodinger equation can be solved with the initial condition given in the form of Gaussian wave packet:

$$\psi_1(x_1, t = 0) = \frac{1}{\sqrt{\sqrt{2\pi}\sigma}} e^{-\frac{(x-\mu)^2}{2\sigma^2} + ip_0 x}. \tag{18.33}$$

Figure 18.7 presents wave packet dynamics (which effectively can describe exciton dynamics) on quantum star graph with three bonds. The profile of the packet is plotted for different time moments. Sensitivity of the dynamics with respect to the changes of the initial velocity of the packet and packet width can be seen from the Fig. 18.8. Using more realistic initial wave function, i.e. wave function of exciton allows to make the results more realistic. Also, including into the Schrodinger equation the potential for interaction of exciton with the vertex should improve the quality of the obtained results. Thus quantum graphs can be rather effective tool for description of the exciton dynamics and charge separation in polymer networks. Treating the process for other (than star graph) more complicated topologies is of importance for practical application of the method to poly-conjugated polymers.

18.6 Conclusion

A microscopic mechanism of the exciton dissociation in the polymer network is treated within the quantum mechanical approaches. Considering exciton as an analog of hydrogen atom allows one to consider exciton separation as ionization process occurring in the "collision" of exciton with the vertex of the polymer chain. In this case, depending on the "collision energy" sudden – and adiabatic perturbation theories can be used to describe splitting of exciton at the chain vertex. Alternative approach, which is based on modeling of polymer network by quantum graphs provides possibility to study the dependence of the dissociation process on the network/chain topology.

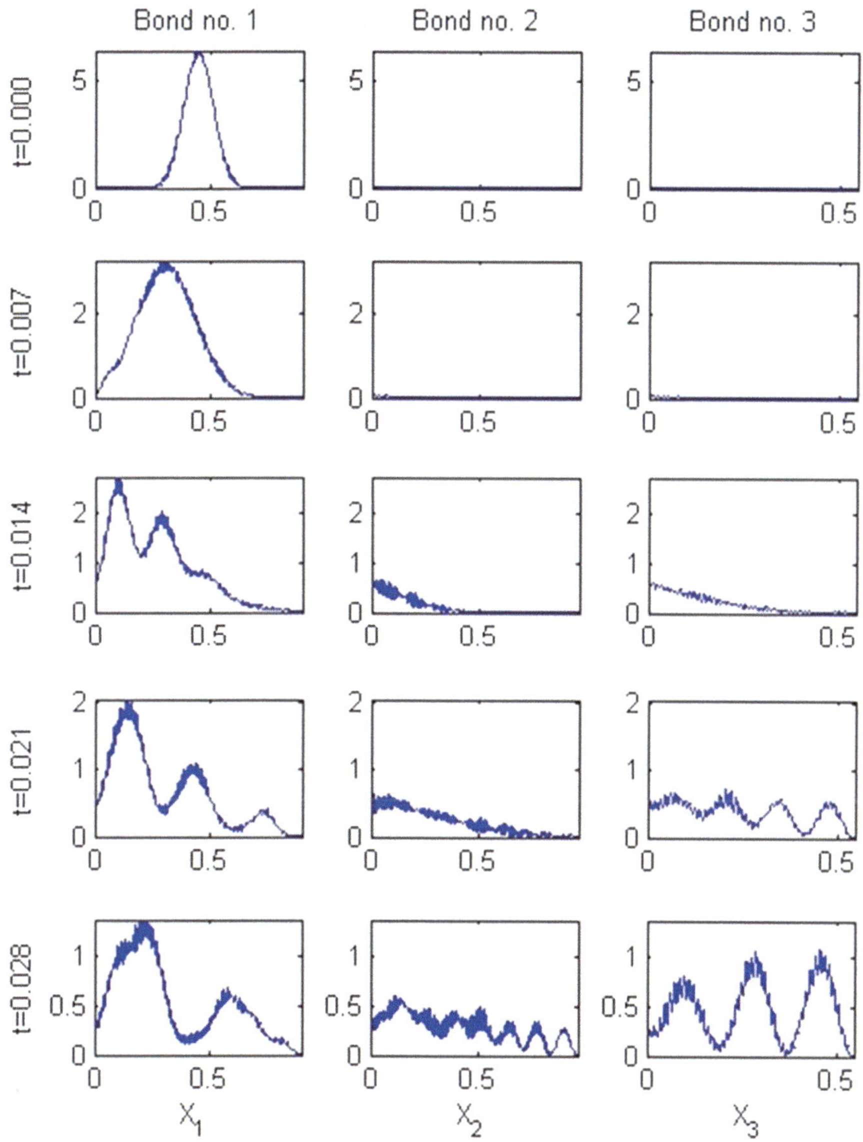

Fig. 18.7 Gaussian wavepacket evolution in quantum star graph. The bonds of the graph have different lengths

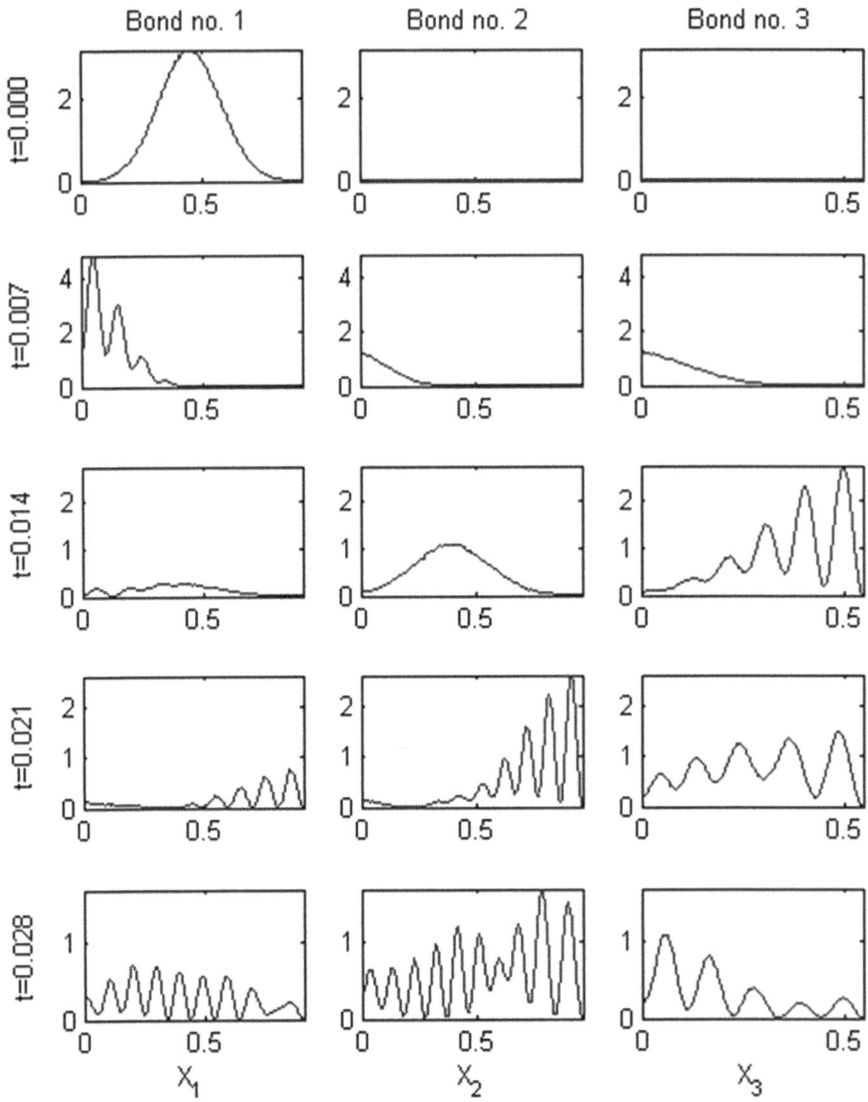

Fig. 18.8 Gaussian wavepacket evolution in quantum star graph. The same graph in Fig. 18.7 with the different initial velocity of the packet and packet width

References

1. Blatt JM, Weisskopf VF (1952) Theoretical nuclear physics. Wiley, New York
2. Brabec C, Scherf U, Dyakonov V (2008) Organic photovoltaics. Wiley-VCH, Woinheim
3. Brendas J-L, Norton J, Corniel J (2009) Molecular understanding of organic solar cell: the challenges. Acc Chem Res 42(11):1691–1699
4. Deibel C, Djyakonov V (2010) Polymer-fullerene bulk heterojunction solar cells. Rep Progr Phys 73:09401
5. Gnutzmann S, Smilansky U (2006) Adv Phys 55:527
6. Gnutzmann S, Keating JP, Piotet F (2010) Ann Phys 325:2595
7. Kottos T, Smilansky U (1999) Ann Phys 76:274
8. Kubo R (1965) Statistical mechanics: an advanced course with problems and solutions. North-Holland, Amsterdam
9. Landau LD, Livshic EM (1974) Quantum mechanics. Nauka, Moskv, p 576
10. Lewis N, Crabtel, Nozik A et al (2005) Basic research need for solar energy utilization. Report of the basic energy sciences workshop on solar energy utilization, April 2005, Second Printing, October 2005, US Department of Energy
11. Nicholson P, Castro F (2010) Organic photovoltaics: principals and techniques for nanometers scale characterizations. Nanotechnology 21:492001
12. Nozik A (2008) Chem Phys Lett 457:3–11
13. Oksengendler B, Turaeva N, Marasulov M et al (2012) Appl Sol Energy 48(3):22
14. Oksengendler BL, Marasulov MB, Nurgaliev IN (2013) Doklady AN RU 2:22
15. Pauling L (1936) J Chem Phys 4:673
16. Pope M, Swenberg C (1999) Electronic processes in organic crystals and polymers, 2nd edn. Oxford University Press, New York
17. Smirnov BM (2001) Physics of ionized gases. Wiley, New York
18. Yavorsky B, Detlaf A (1982) A modern handbook of physics. Mir Publisher, Moscow, p 712

Chapter 19
Complex Antenna Optimization

Haojiong Liu, Ibrahim Tekin, Oksana Manzhura, and Edip Niver

Abstract Complex electromagnetic problems arise due to various applications in science and technology that are becoming a necessity in our daily activities. As systems become more complex their design and implementation require novel topologies and sophisticated optimization tools to meet the challenges. One of the key components in commercial and military communication systems is an antenna which serves as an energy conversion device from an electrical source into radiated electromagnetic fields. However, system specifications require optimized design for antennas in terms but not limited to higher gain, broader bandwidth, smaller size, proper radiation pattern and polarization and lower cost. Addressing these complex issues requires an optimized solution based on sophisticated numerical electromagnetic solvers. Here, Fano-Chu (1950) limits which were proposed for electrically small antennas have been extended to rather larger structures. The basic premise for the concept is that volume antenna (more conducting elements within specified volume) yields higher gain. Starting from this assumption an antenna structure based on Moxon antenna has been developed to produce circularly polarized antenna for satellite communications (SATCOM), RFID tag reader and Global Positioning System (GPS). Cross-Moxon elements were redesigned to optimize gain and lower cross-polarization as well as maintain low profile and

H. Liu • O. Manzhura • E. Niver (✉)
New Jersey Institute of Technology, Newark, NJ, USA
e-mail: hl96@njit.edu; oksana.manzhura@njit.edu; edip.niver@njit.edu

I. Tekin
Electronics Engineering, Sabanci University, Istanbul, Turkey
e-mail: tekin@sabanciuniv.edu

D. Matrasulov and H.E. Stanley (eds.), *Nonlinear Phenomena in Complex Systems: From Nano to Macro Scale*, NATO Science for Peace and Security Series C: Environmental Security, DOI 10.1007/978-94-017-8704-8_19,

adequate bandwidth. Designed antennas were fabricated and measured leading to significant size reduction, improved higher gain, reduced cross-polarization and lower cost compared to commercially available state-of-the-art antennas.

19.1 Introduction

Moxon antenna [1] is basically a two-element Yagi-Uda antenna [2], with a bent dipole element to reduce its height and is commonly preferred antenna for HAM operators due to its size, forward gain and wide band impedance match. A systematic sequence of topologies starting from a single vertical element to two cross vertical elements of the Moxon arms fed through a hybrid coupler to achieve Circular Polarization is implemented. Then widened strip arm elements were investigated to understand the effects on widening the bandwidth. The logic in this evolution was to obtain maximized gain based on Fano-Chu limits [3], which suggests that more metalization in the radiating configuration that fills the volume would yield higher gain for electrically small antenna [4–6]. Extending the width of the strip into tapered shape and splitting of the equivalent dipole elements with additional bends at the extended tips of these tapered bow tie [7] arms lead to a wider bandwidth and improved cross-polarization ratio.

Novel Circularly Polarized (CP) UHF SATCOM antenna [8–10] is developed based on Moxon antenna (bent dipole element over a ground plane), and further extended for RFID and GPS applications. For RFID mobile applications [11], tag reader antenna is required to have high performance including a broadband operation, circular polarization as well as a large angular coverage from horizon to zenith. For systems at these frequencies, wavelength could be on the order of third to quarter of a meter and conventional antennas may be too big for commercial use. For GPS applications, antennas [12] are required to have very precise narrow band performance at specific frequency bands (L1 and L2 bands). Novel tag reader RFID antenna and another dual band GPS antenna based on extended Moxon antenna were proposed.

In overall, the measured antenna produced lower physical height, higher gain, wider bandwidth, cross-polarization and lower back lobe radiation compared to commercial counterparts, such as an eggbeater currently used in SATCOM practice as well as similar antennas in RFID and GPS applications.

19.2 Prototype Antenna Topologies and Performance

Moxon antenna is known for its compact size and its directive properties due to the presence of the ground plane. A sketch of one of the bent dipole antennas is shown in Fig. 19.1. The length of the one arm of the dipole is L + W, and the arm is bent

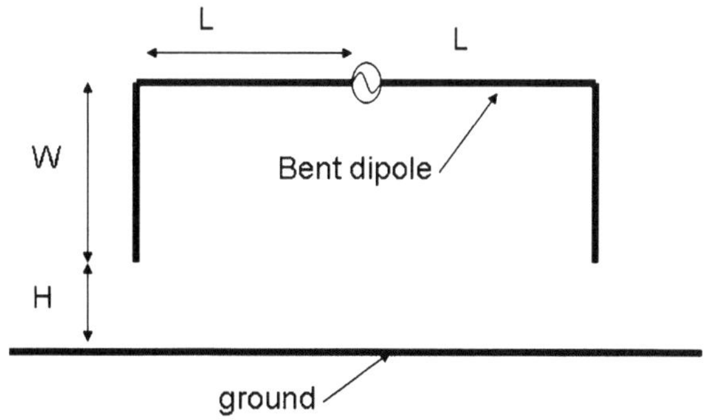

Fig. 19.1 Moxon like bent dipole antenna over a ground plane

toward a ground plane from L distance away from the center of the dipole. The end point of the bent dipole antenna is H away from the ground plane as in Fig. 19.1.

The bent dipole is fed from the center of the antenna with a differential input. The RHCP can be obtained simply by placing two dipole bent antennas perpendicular to each other, one in x-z plane, the other in y-z plane and feeding through a hybrid quadrature coupler.

The electromagnetic simulations are performed using HFSS (for antenna) and ADS (matching circuit) software. The ground plane is finite and its dimensions are 4L × 4L. The power handling capability has not been taken into consideration for simulations. The material for the antenna conductor is chosen as copper. The two dipole antennas are fed by a 90° phase shift from the two lumped ports. After optimization antennas are prototyped and measurements are taken.

19.3 General Design and Optimization of the New Split Bow Tie Moxon Based Antenna

The proposed antenna consists of two bent Moxon type split bowtie antennas. The evolution of the Moxon antenna through widening and tapering the arms, and further splitting arms kept at an optimum angle with terminated bends resulted in the antenna shown in Fig. 19.2. The two bent antennas, located perpendicular to each other as shown in Fig. 19.2, are fed at the center via differential input through a hybrid coupler to produce Right Hand Circular Polarization (RHCP). An expanded view of a split-arm (element) is shown in Fig. 19.3 marked with numbers to identify the optimization parameters used in numerical simulations. The detailed optimization procedure is outlined in Table 19.1 is subject to extensive numerical simulations.

These Moxon based prototype antennas have been applied to UHF SATCOM, RFID tag reader and GPS applications. Simulation results as well as experimental measurements of these prototype antennas are outlined below.

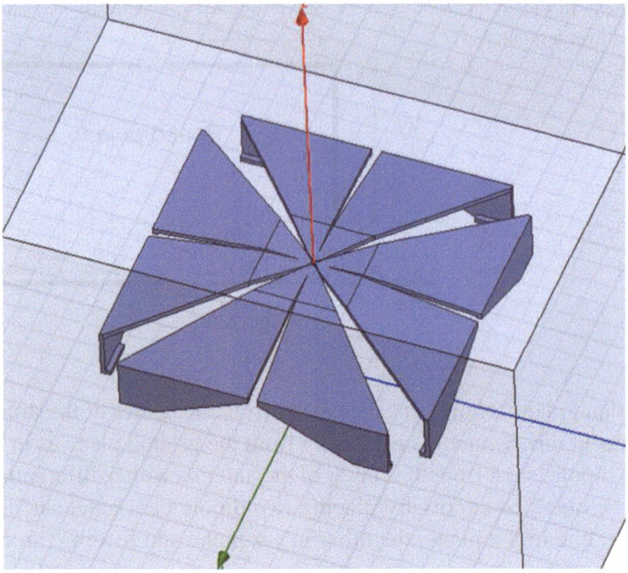

Fig. 19.2 Proposed Moxon based antenna in 3D plane over a ground plane

19.3.1 Moxon Based UHF SATCOM Antenna

UHF antenna is designed to operate in 200–400 MHz frequency range. Assembled antenna over a ground plane is shown on Fig. 19.4.

This antenna was initially designed without design detail No. 10, which, when used, allows to tune antenna with a lower overall height No. 12.

Characteristic dimensions of a single triangular shaped antenna conductor is shown on Fig. 19.5. Most of the dimension values are as follows:

OL	Overall radial length	138 mm
W	Length of vertical arms	45 mm
H	Distance from the ground plane	80 mm
α	Angle of element bend towards ground plane (angle to horizontal)	22°
OH	Overall height of the antenna	200 mm
OD	Overall diameter of the antenna	280 mm

Input Impedance of this antenna model is simulated and shown on Fig. 19.6. It shows that $S11_{min} = -8\text{dB}@275$ and 390 MHz, and 3 dB bandwidth of the input impedance is over $(f_2 - f_1)$ @3 dB $=$ 170 MHz. Measured Return loss of this antenna is compared with measured return loss of the commercial eggbeater antenna and presented in Fig. 19.7. UHF antenna is fed through a quadrature coupler

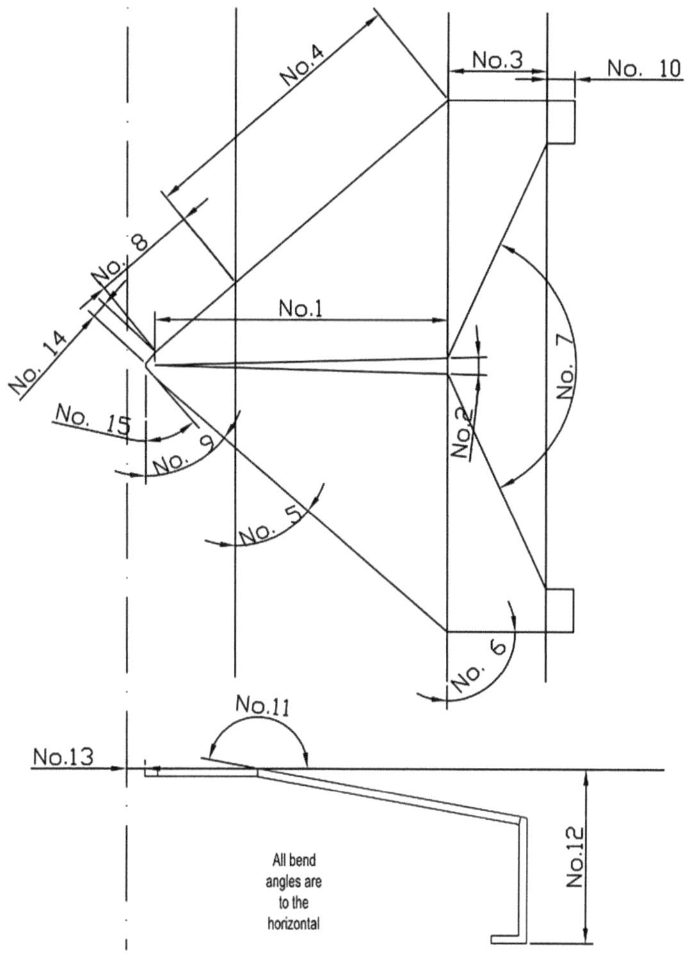

Fig. 19.3 Single split arm element of a Moxon based antenna

MINICIRCUITS ZX10Q-2-3+ in the frequency range of 220–470 MHz. Measured antenna return loss is better then 18 dB for the entire measured range. Radiation patterns of the UHF split bow tie antenna were simulated for various frequencies of the frequency range. The results of RHCP and LHCP Gain simulations are presented in Table 19.2. Typical radiation pattern of this antenna, simulated at 340 MHz, is presented on Fig. 19.8.

During design and simulations of the split bow tie UHF antenna surface currents were calculated and are presented in Fig. 19.9. Currents are shown at 280 MHz, the lower S11 resonance point. Currents are presented on the plot in the 1–25 A/m range, but reach as high as 60 A/m in the most active (shown as red) areas.

Table 19.1 Optimization of geometrical parameters in numerical simulations

Number	Parameter description	RFID antenna behavior	GPS antenna behavior
1	Length of the wedge cutout	Moving wedge tip closer to the Z axis, effectively makes the first section of the wedge $\Longleftrightarrow$, shifts central frequency $\Downarrow$ and BW $\Rightarrow\Leftarrow$	
2	Spread angle of the wedge cutout	$\Rightarrow\Leftarrow$ the angle, i.e. sharpening the wedge cutout, $\Longleftrightarrow$ BW and shifts central frequency (or resonance) $\Uparrow$	
3	Vertical length	$\Longleftrightarrow$ the length, low resonance point $\Downarrow$ in frequency but $\Uparrow$ in S11, high resonance point $\Downarrow$ in frequency and $\Downarrow$ in S11. Total bandwidth decreases	
		$\Rightarrow\Leftarrow$the length, low resonance point $\Uparrow$ in frequency but $\Downarrow$ in S11, high resonance point $\Uparrow$ in frequency and $\Uparrow$ in S11. Total bandwidth increases	
4	Length of the first bend	$\Longleftrightarrow$ the length, low resonance point $\Downarrow$ in frequency but $\Uparrow$ in S11, high resonance point $\Downarrow$ in frequency and $\Uparrow$ S11. Total bandwidth $\Uparrow$	$\Longleftrightarrow$ the length, low resonance point $\Downarrow$ in frequency but $\Uparrow$ in S11, high resonance point $\Downarrow$ in frequency and $\Uparrow$ S11. Total bandwidth $\Downarrow$
		$\Rightarrow\Leftarrow$ the length, low resonance point $\Uparrow$ in frequency but $\Downarrow$ in S11, high resonance point $\Uparrow$ in frequency and $\Downarrow$ S11. Total bandwidth $\Downarrow$	$\Rightarrow\Leftarrow$ the length, low resonance point $\Uparrow$ in frequency but $\Downarrow$ in S11, high resonance point $\Uparrow$ in frequency and $\Downarrow$ S11. Total bandwidth $\Uparrow$
5	Outer angle of the first bend	$\Uparrow$ the outer angle, low resonance point $\Downarrow$ in frequency but $\Uparrow$ in S11, high resonance point $\Downarrow$ in frequency and $\Downarrow$ in S11. Total BW $\Downarrow$	$\Uparrow$ the outer angle, low resonance point $\Uparrow$ in frequency but $\Downarrow$ in S11, high resonance point $\Downarrow$ in frequency and $\Downarrow$ in S11. Total BW $\Downarrow$
		$\Downarrow$ the outer angle, low resonance point $\Uparrow$ in frequency but $\Downarrow$ in S11, high resonance point $\Uparrow$ in frequency and $\Uparrow$ in S11. Total BW $\Uparrow$	$\Downarrow$ the outer angle, low resonance point $\Downarrow$ in frequency but $\Uparrow$ in S11, high resonance point $\Uparrow$ in frequency and $\Uparrow$ in S11. Total BW $\Uparrow$
6	Outer angle of the vertical section (90°)	$\Rightarrow\Leftarrow$ outer angle of the vertical section, i.e. sharpening the angle, improves Reflection Impedance around lower resonance frequency, while looses some match around higher resonance frequency. No significant loss of bandwidth is observed with sharper outer angle	
7	Inner angle of the vertical section	$\Longleftrightarrow$ the inner angle, low resonance point $\Downarrow$ in frequency but $\Downarrow$ in S11, high resonance point $\Downarrow$ in frequency and $\Downarrow$ in S11. Total BW stays unchanged	$\Longleftrightarrow$ the inner angle, low resonance point $\Uparrow$ in frequency but $\Downarrow$ in S11, high resonance point $\Downarrow$ in frequency and $\Uparrow$ in S11. Total BW $\Downarrow$
		$\Rightarrow\Leftarrow$ the inner angle, low resonance point $\Uparrow$ in frequency but $\Uparrow$ in S11, high resonance point $\Uparrow$ in frequency and $\Uparrow$ in S11. Total BW stays unchanged	$\Rightarrow\Leftarrow$ the inner angle, low resonance point $\Downarrow$ in frequency but $\Uparrow$ in S11, high resonance point $\Uparrow$ in frequency and $\Downarrow$ in S11. Total BW $\Uparrow$

(continued)

Table 19.1 (continued)

Number	Parameter description	RFID antenna behavior	GPS antenna behavior
8	Horizontal length (without tip)	⟺ the length, low resonance point ⇓ in frequency but ⇑ in S11, high resonance point ⇓ in frequency and ⇓ S11. Total BW ⇓	
		⇒⇐ the length, low resonance point ⇑ in frequency but ⇓ S11, high resonance point ⇑ in frequency and ⇑ in S11. Total BW ⇑	
9	Outer angle of the horizontal section	⟺ the outer angle, low resonance point ⇑ in frequency but ⇓ S11, high resonance point ⇓ in frequency and ⇑ in S11. Total BW ⇓	
		⇒⇐ the outer angle, low resonance point ⇓ in frequency but ⇑ in S11, high resonance point ⇑ in frequency and ⇓ in S11. Total BW ⇑	

Only 9 out of 14 parameters are presented here
⟺-longer(larger), ⇒⇐-shorter(smaller), ⇑-higher (increase), ⇓-lower (decrease)

Fig. 19.4 Assembled UHF antenna over a ground plane

19.3.2 Moxon Based RFID Tag Reader Antenna

RFID tag reader antenna is designed to operate in 850–1,050 MHz range. Characteristic dimensions of a single triangular shaped antenna conductor are shown on Fig. 19.10. Assembled antenna over a ground plane is shown on Fig. 19.11.

The Return Loss of the RFID antenna is simulated and shown on Fig. 19.12 and compared to measured performance of the prototype shown on Fig. 19.13. RFID antenna has simulated S11 3 dB range of 710–1,200 MHz. Measured S11 is better then 10 dB in 800–1,180 MHz.

Antenna gain is simulated to be approximately 7 dB and front to rear ratio is 15 dB. When assembled antenna is measured over a small (compared to the size of the antenna) ground plane and compared with a commercially available RFID of known gain, the measured gain of the Moxon type RFID antenna can be judged to be around 15–17 dB (Fig. 19.14).

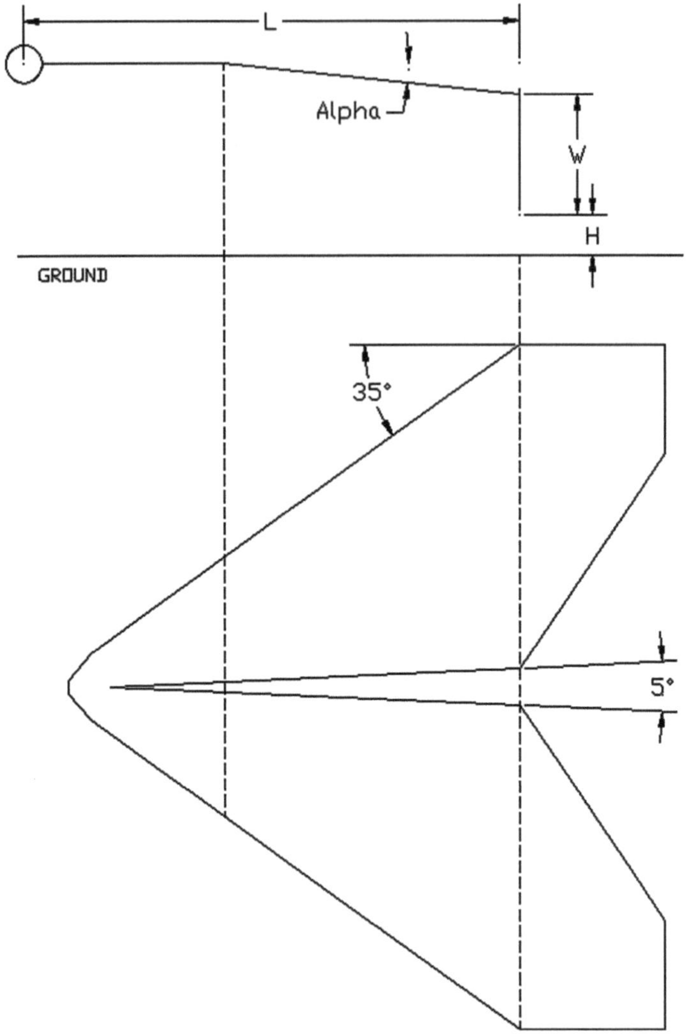

Fig. 19.5 Single element of the Moxon UHF antenna over a ground plane

19.3.3 Moxon Based GPS Antenna

Moxon based type antenna for GPS applications is designed to work in two GPS bands, 1,227.60 ± 10.23 and 1,575.42 ± 10.23 MHz. Characteristic dimensions of a single split antenna arm are shown in Fig. 19.15.

Assembled antenna over a ground plane is shown on Fig. 19.16.

The Return Loss of the GPS antenna is simulated and shown on Fig. 19.17 and compared to measured performance of the prototype shown on Fig. 19.18. GPS antenna has simulated S11 3 dB range of 1,000–1,720 MHz.

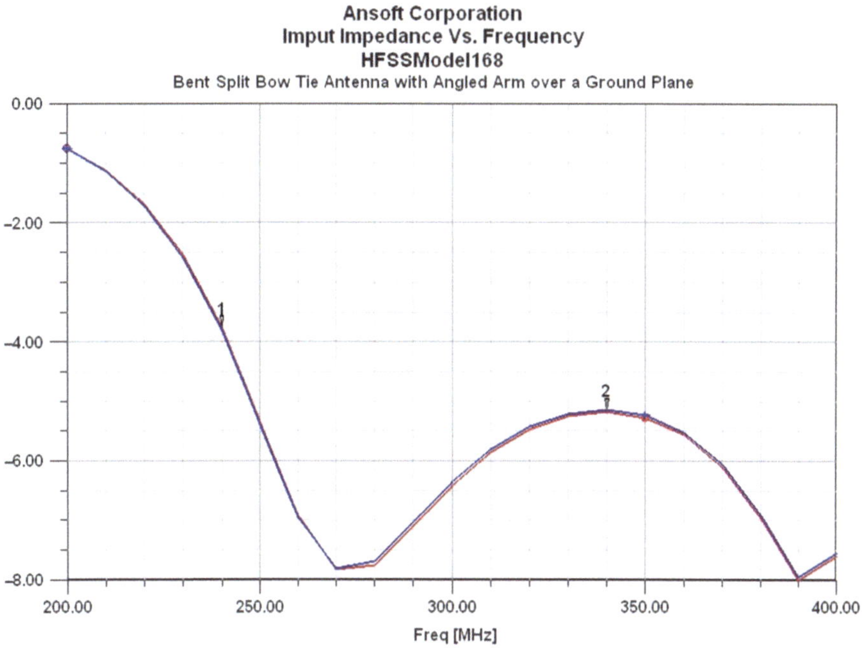

Fig. 19.6 Input impedance S11 and S22 of the Moxon UHF antenna over a ground plane

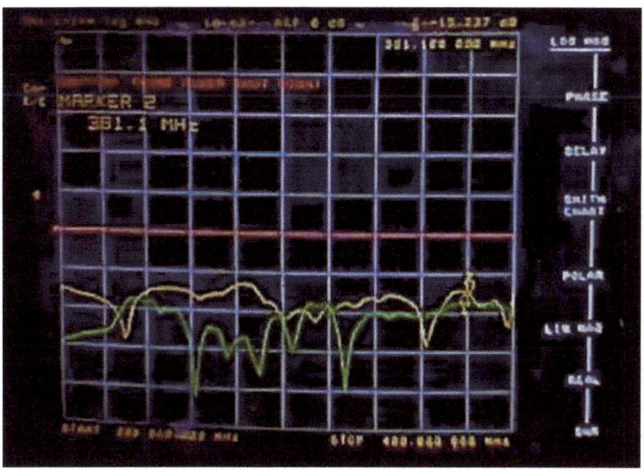

Fig. 19.7 Measured S11 of the bent bow tie Maxon antenna at the input terminal of the hybrid coupler. The higher curve belongs to the commercial eggbeater antenna

Measured S11 is better then 10 dB in 1,000–1,600 MHz and features deep resonances around both bands of interest, where RL is better then 30 dB. Antenna gain is simulated to be approximately 6.6 dB at 1,227 MHz and 8.25 dB at 1,575 MHz, while front to rear ratio is better then 14 dB.

Table 19.2 Maximum RHCP and LHCP gains of Moxon based split bow tie antenna

Frequency (MHz)	Max gain (dB-RHCP)	Max gain (dB-LHCP)
240	11.4	− 1.6
280	7.4	− 7.6
340	8.6	−11.9
380	8.2	−15
400	8	−14

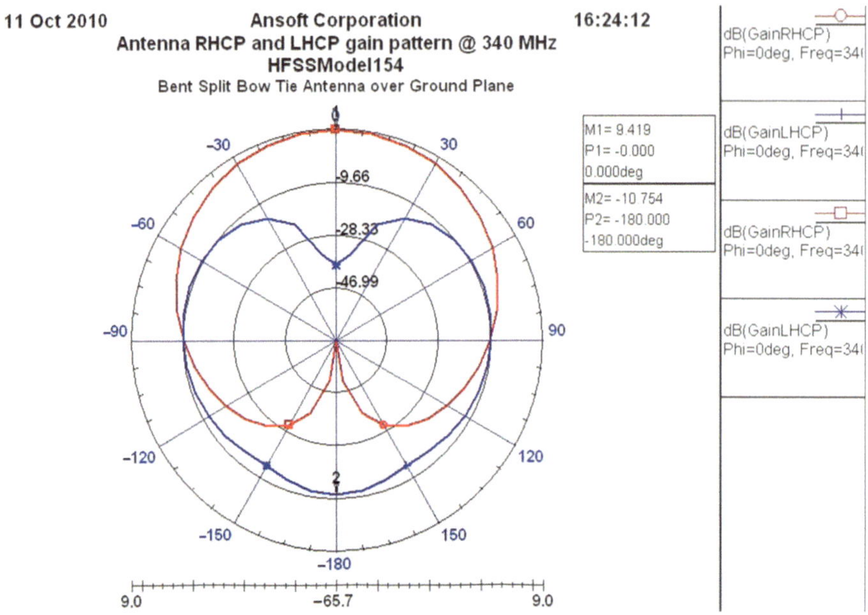

Fig. 19.8 RHCP and LHCP Gain of the UHF Moxon based split bow tie antenna at 340 MHz

Measured relative power delivered by the antenna is shown on Fig. 19.19 for both ranges. Higher band shows somewhat lower power then the lower band due to mismatch and alignment issues during the experiment.

19.4 Comparisons of Split Bow Tie Moxon Type Antenna Sizes and Performance to Commercially Available Antennas for All Three Applications

In case of UHF SATCOM application the proposed antenna design yields size of at least a third size by volume as compared to the standard UHF eggbeater antenna. Proposed antenna also outperforms a standard eggbeater antenna in frequency

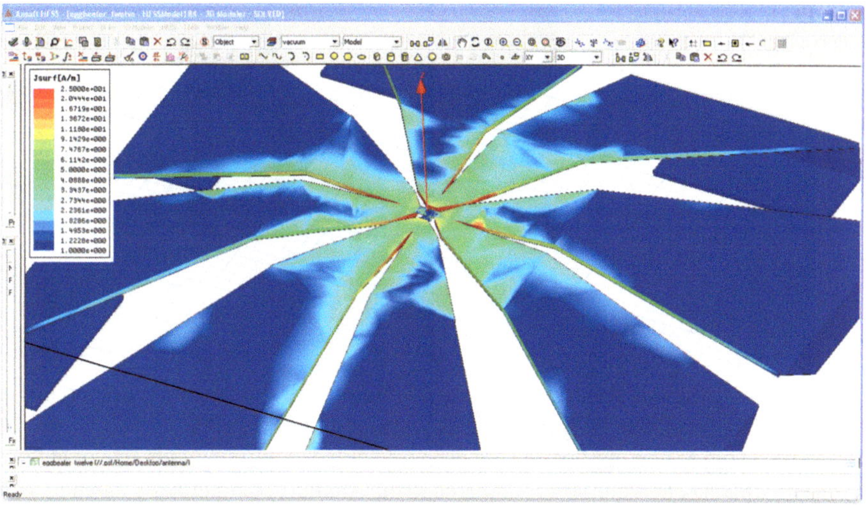

Fig. 19.9 Surface current distribution on the split bow tie UHF antenna at 280 MHz

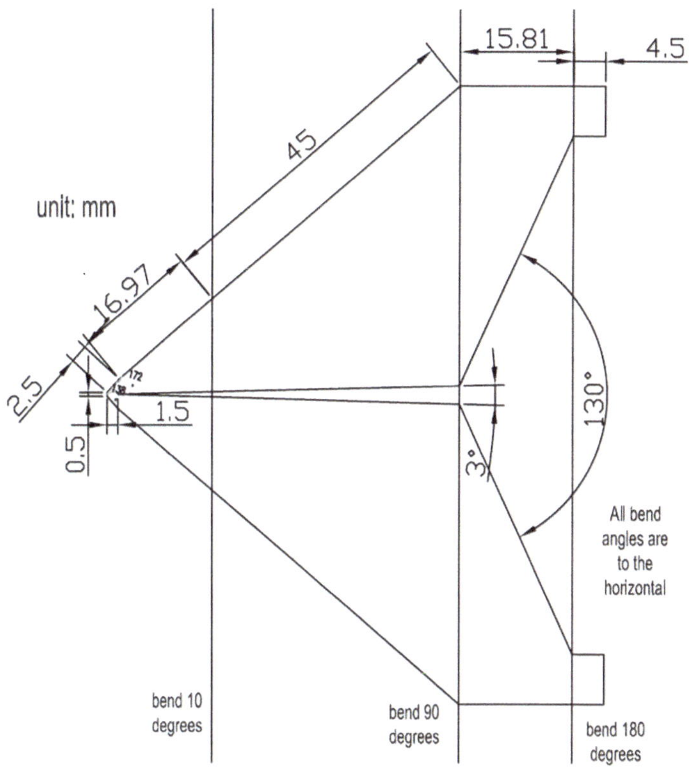

Fig. 19.10 Dimensions of a single split arm of the RFID tag reader antenna arm

Fig. 19.11 Assembled RFID tag reader antenna over a ground plane

bandwidth , gain and front to back ratio. Summary of the antenna comparison to commercially available UHF antennas is presented in Table 19.3.

In RFID tag reader application the proposed antenna design yields size of at least a one quarter size by volume as compared to the smallest commercially antenna with similar parameters. Summary of the antenna comparison to commercially available RFID tag reader antennas is presented in Table 19.4.

In case of GPS application the proposed antenna design yields size of at three quarter size by volume as compared to the smallest commercially available antenna with similar parameters, while providing better frequency range and gains in both bands of interest. Summary of the antenna comparison to commercially available RFID tag reader antennas is presented in Table 19.5.

19.5 Conclusion

Proposed circularly polarized (CP) antenna configurations based on Moxon type antenna (bent dipole element over a ground plane) are presented for broadband UHF SATCOM, RFID tag reader and GPS applications.

Extensive numerical simulations based on optimization of various parameters on the antenna structure were carried out to achieve higher gain, wide band impedance match, high cross-polarization and low profile. Two vertical elements of the Moxon arms, widened strip arm elements, structures with bends at 90° for achieving broadband operation are simulated, and measured. For the antennas that are prototyped, return loss S11 measurements were performed, and gains are simulated using HFSS.

For the band of 225–400 MHz, antennas have reasonable CP gain and can be used for UHF SATCOM band. It was observed that height of the proposed antennas was reduced by at least 50 % compared to conventional eggbeater antenna making them suitable to comply with the aerodynamic structure of a radome that can be placed on a body of a helicopter or a fixed wing aircraft.

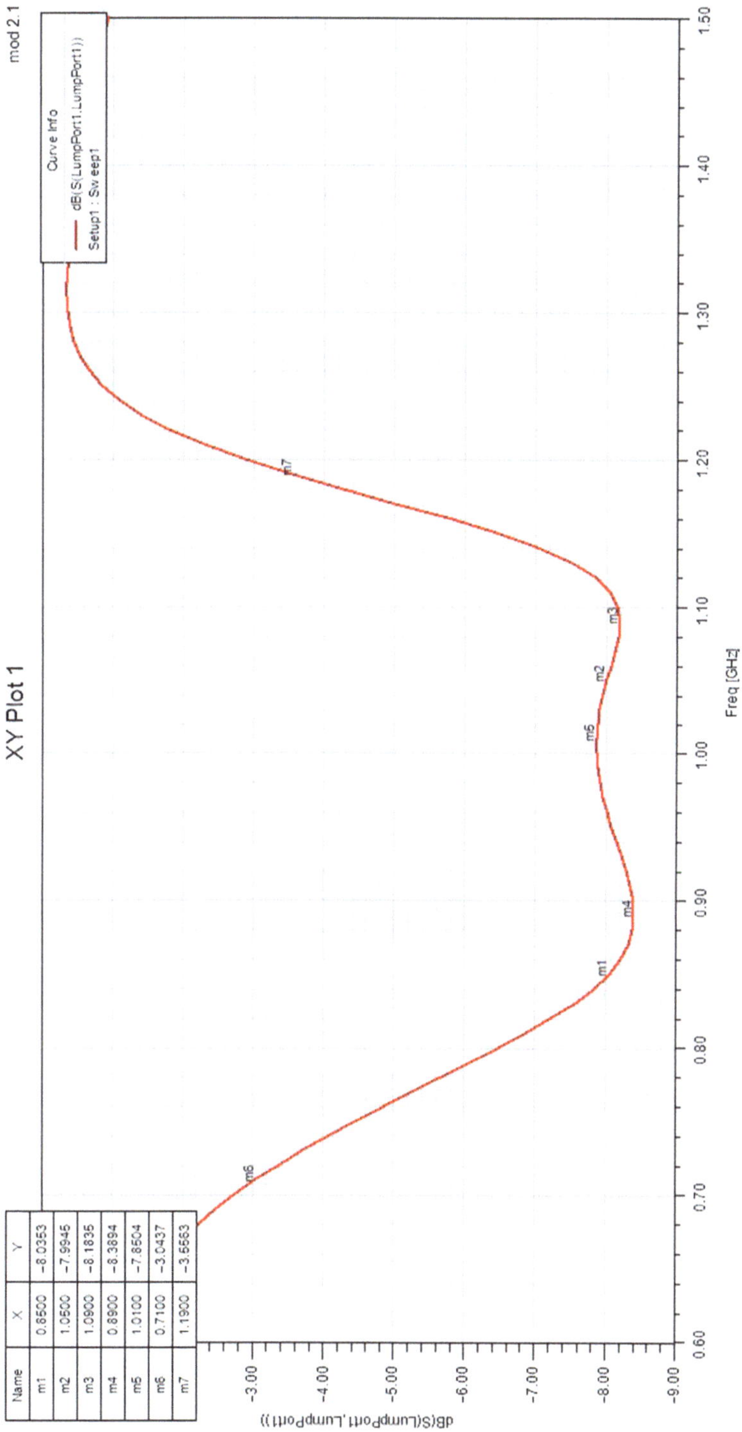

Fig. 19.12 Simulated return loss of the RFID antenna

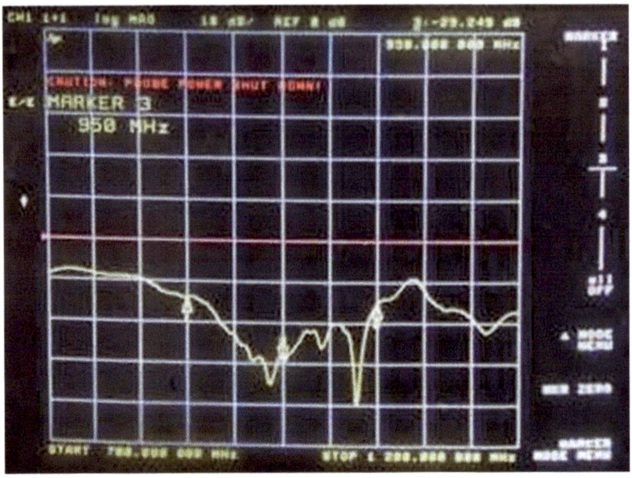

Fig. 19.13 Measured return loss of the RFID antenna

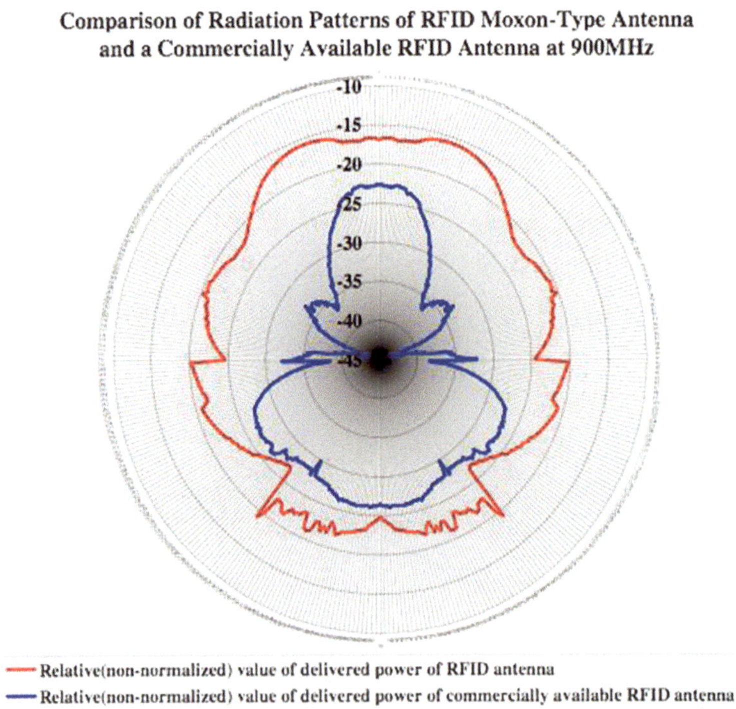

Fig. 19.14 RFID measured antenna pattern comparison

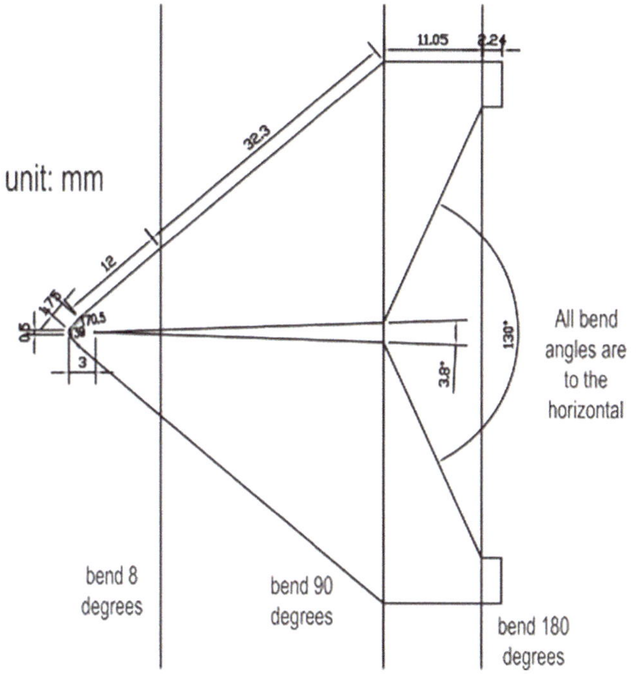

Fig. 19.15 Dimensions of a single triangular shaped GPS antenna arm

Fig. 19.16 Assembled GPS antenna over a ground plane

Furthermore, prototype antennas were compared with commercial counterparts and were observed that RFID tag reader antenna was almost four times smaller in physical dimensions for a higher gain of 17 dB. In case of GPS antenna the overall gain was observed to increase for the comparable physical dimensions.

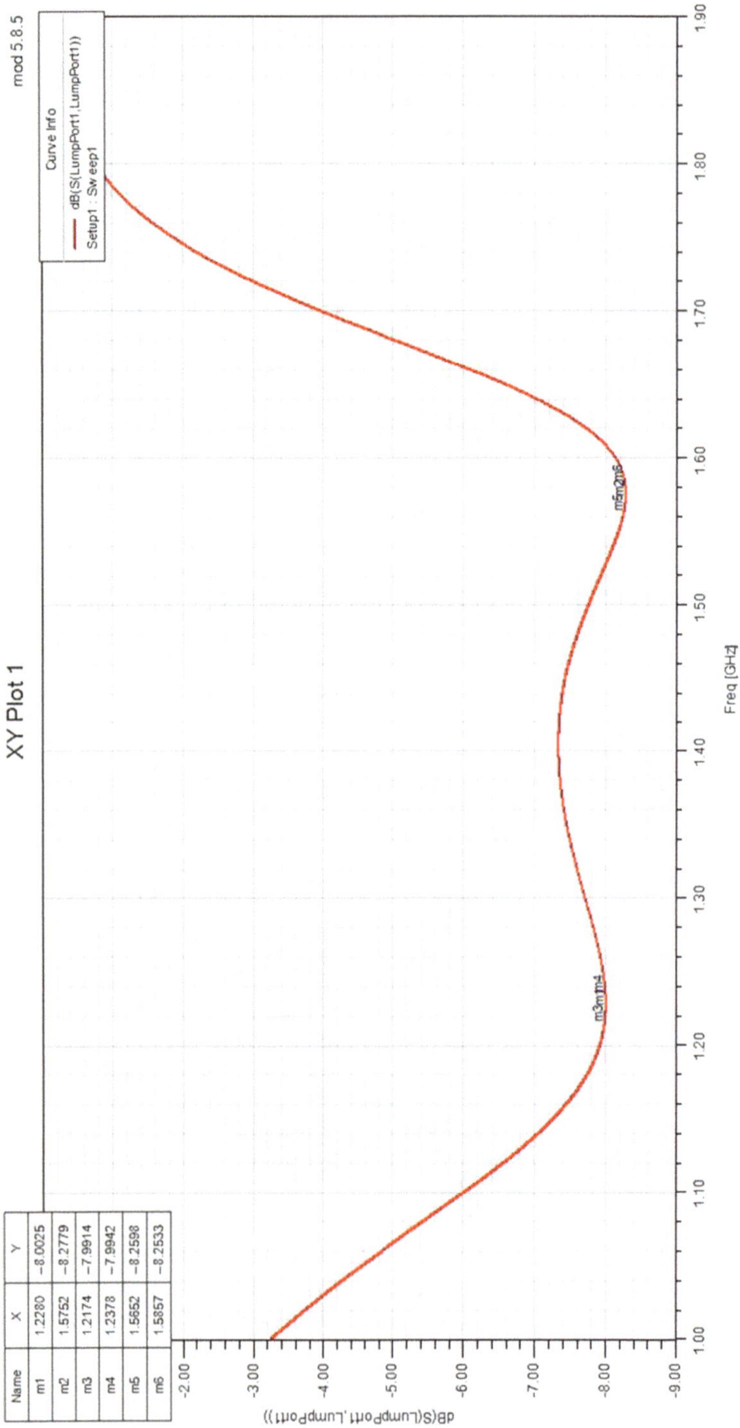

Fig. 19.17 Simulated return loss of the GPS antenna over a ground plane

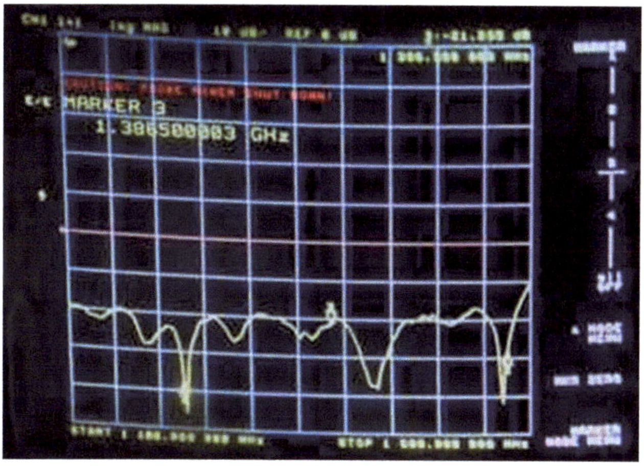

Fig. 19.18 Measured return loss of the GPS antenna over a ground plane

Fig. 19.19 Measured radiation patterns of the GPS antenna at L1 and L2 bands

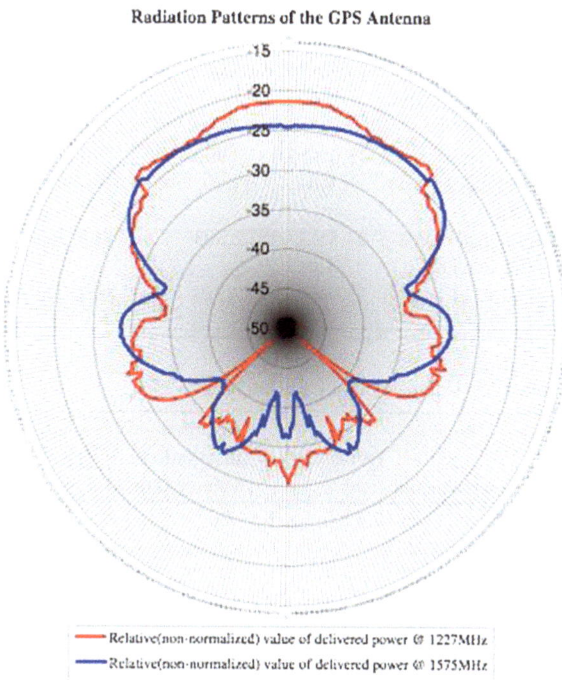

Table 19.3 Comparison of UHF SATCOM antennas

Manufacturer/model	Dimensions (cm)	Required installation volume (cm^3)	Frequency range (MHz)	Gain (dB)	Front-to-back ration (dB)
Proposed bent bow tie Moxon type UHF antenna	⌀28 × 16.5	10,169	225–400	8	22
Myers engineering international mini-UHF SATCOM eggbeater antenna	20×20×27	10,800	243–318	6	
Standard UHF eggbeater antenna	⌀34 × 30	27,237	240–360	5.9	11

Table 19.4 Comparison of RFID tag reader antennas

Manufacturer/model	Dimensions (cm)	Required installation volume (cm^3)	Frequency range (MHz)	Gain (dB)	Front-to-back ration (dB)	Beam width°
Proposed Moxon type bent bow tie RFID antenna	10.0×10.0× 2.2	220	850–1,050	6.75	15.35	80° @3.5
Laird tech S8656-X special application antenna	19.2×19.2× 2.4	884	865–870	6		80° @3.0
AvaLAN wireless 6 dBi indoor antenna	15.0×15.0× 4.0	900	890–960	605	12	
Poynting patch A 0025 antenna	24.5×23.5× 4.0	2,303	860–960	7		
IA33A INTELLITAG	25.9×25.9× 3.8	2,549	902–928	7	18	65° @3.0

Table 19.5 Comparison of GPS antennas

Manufacturer/model	Dimensions (cm)	Required installation volume (cm^3)	Frequency range (MHz)		Gain (dB)	
			L1	L2	L1	L2
Proposed bent bow tie Moxon type GPS antenna	7.3 × 7.3 × 1.5	80	1,217.4–1,237.8 (20.4)	1,528.1–1,607.8 (79.7)	60.6	8.3
GPS source L1/L2 DARG antenna	6.6 × 6.6 × 2.4	104	1,212.6–1,242.6 (30)	1,560.5–1,590.5 (30)	4	7
GPS source ruggedized L1/L2 GPS passive antenna	6.6 × 6.6 × 2.4	104	1,217.5–1,237.8 (20.3)	1,565–1,586 (21)	5	5
ALLICOM SB240 Marine GPS antenna	⌀12.0 × 20.65	113		1,575.42±10	4	

References

1. Moxon L (2002) HF antennas for all locations, 2nd edn. Radio Society of Great Britain, Bedford
2. Elliott RS (1981) Antenna theory and design. Prentice-Hall, Englewood Cliffs
3. Chu LJ (1948) Physical limitations on omni-directional antennas. J Appl Phys 19:1163–1175
4. Schwering F (1976) Workshop on electrically small antennas: background and purpose. In: Proceedings of the ECOM-ARO workshop on electrically small antennas, Fort Monmouth, 6–7 May 1976
5. Hansen RC, Collin RE (2011) Small antenna handbook. Wiley, Hoboken
6. Jofre L, Martinez-Vazquez M, Serrano R, Roqueta G (2012) Handbook on small antennas. EurAAP, Bruxelles
7. Nagatoshi M, Tanaka S, Horiuchi S, Morishita H (2010) Downsized bow-tie antenna with folded elements. IEICE Trans Electron E93-C(7):1098–1104
8. Jahoda JR, Shergold SD (1992) Broadband antenna. US Patent 5,111,213, 5 May 1992
9. Regala F (2006) Portable co-located LOS and SATCOM antenna. US Patent 7,019,708, 28 Mar 2006
10. Tekin I, Manzhura O, Niver E (2011) Broadband circularly polarized antennas for UHF SATCOM. In: IEEE general assembly and scientific symposium, 2011 XXXth URSI, Istanbul, 13–20 Aug 2011
11. Marrocco G (2008) The art of UHF RFID antenna design: impedance-matching and size-reduction techniques. IEEE Antennas Propag Mag 50(1):66–79
12. Parkinson BW, Spilker JJ Jr (1996) Global positioning system: theory and applications, vols 1 and 2. AIAA, Washington, DC

Chapter 20
Complex Nonlinear Riccati Equations as a Unifying Link in Fundamental Physics

Dieter Schuch

Abstract Even if the discovery of the Higgs boson should be confirmed, the theoretical physics building is far from being complete. Essential everyday experience like dissipation and irreversibility are not naturally included in the formalism of classical and quantum mechanics. Quantum mechanics, as a supposedly linear theory, is then by definition not able to include the majority of processes in nature that obey nonlinear evolution equations. Also thermodynamics as a phenomenological theory and cosmology (including gravity) reside on yet different floors in this building. In this paper, it will be shown how a reformulation of quantum mechanics in terms of complex Riccati equations might allow for a unification of these aforementioned aspects in terms of the same formalism.

20.1 Introduction

With the formalisms of Lagrange and Hamilton at the beginning of the nineteenth century, classical mechanics was not only an instrument for practical purposes, but an almost elegant piece of art derived from first principles and based on energy as a conserved quantity. However, in this picture there was no direction of time (only canonical transformations) and no dissipation of mechanical energy (the Hamiltonian function is constant). The universe appeared to be a gigantic clockwork, everything seemed to be determined and could – in principle – be calculated. The picture of a cold, inhuman universe did not really fit in with the romantic Zeitgeist of that time and its slogans like "back to nature".

D. Schuch (✉)
Institut für Theoretische Physik, J.W. Goethe-Universität Frankfurt am Main, Max-von-Laue-Str. 1, D-60438 Frankfurt am Main, Germany
e-mail: Schuch@em.uni-frankfurt.de

D. Matrasulov and H.E. Stanley (eds.), *Nonlinear Phenomena in Complex Systems: From Nano to Macro Scale*, NATO Science for Peace and Security Series C: Environmental Security, DOI 10.1007/978-94-017-8704-8_20,

So it was not too surprising that, as a kind of reaction, phenomenological theories based on the observation of nature were developed and gained interest. Particularly thermodynamics, based on three empirical laws, turned out to be extremely successful and the industrial revolution in the nineteenth century would not have been possible without it. Also electrodynamics, based on Maxwell's equations, added to the success of a description of nature in terms of continuous quantities and flows instead of particles or even atomistic models. Scientists like Mach dominated; scientists like Boltzmann were depressed.

But at the beginning of the twentieth century it got even worse. The same system could, in some experiments, behave like a discrete particle, in others like a continuous wave. This wave-particle duality was only resolved by quantum mechanics, specifically in Schrödinger's formulation of wave mechanics. However, since Schrödinger's theory is somehow based on classical (Hamiltonian) mechanics, we face the same problems as in classical mechanics: there is no direction of time (only unitary transformations) and no dissipation of energy (the (hermitian) Hamiltonian operator is a constant of motion). Nevertheless, quantum mechanics is probably the most successful and influential (also for the economy) physical theory so far.

Towards the end of the twentieth century, a rather different theory, nonlinear (NL) dynamics, became very popular. Not only because it was able to produce aesthetically attractive pictures of so-called fractals [1], but also because it was able to describe complex phenomena that can be observed in nature, like the weather, different growth processes and much more (including evolutionary processes with a direction of time and dissipation of energy). A typical pattern in fractals, as well as in natural growth processes, is a spiral, i.e. an object that combines an angular motion with a change of radius leading to something like a logarithmic spiral (see the shell of a nautilus or the horn of a ram).

Now at the beginning of the twenty-first century, physics is in a somewhat schizophrenic state: quantum mechanics (and other alleged fundamental theories) is reversible, conservative and linear (thus providing a desired superposition principle), whereas the observable (macroscopic) nature (e.g., as described by NL dynamics) is essentially irreversible, dissipative and nonlinear. (Not to mention that also thermodynamics and cosmology (including gravity) are not naturally compatible with quantum mechanics.)

Is it therefore possible to find a unifying formulation for all these different aspects of physics? In order to answer this question let us have a closer look at which properties are supposedly essential for quantum mechanics and decide which one(s) is(are) disputable. (1) According to Planck, the *action* (not necessarily the energy!) is *quantized* (in portions of Planck's constant h or $\hbar = \frac{h}{2\pi}$). (2) C.N. Yang stated in his talk on the occasion of Schrödinger's 100th birthday [2] that quantum mechanics is essentially based on *complex quantities* (in the mathematical sense). (3) The formal structure of quantum mechanics is that of a *linear* theory, thus having a superposition principle which goes well with the wave aspect of material systems.

In a modified formulation, phenomena like irreversibility and dissipation should also be covered so it shall not be assumed that reversibility and conservation of energy are actually essential for quantum mechanics.

Which of the above-mentioned properties can be sacrificed to allow it to link with other fields like NL dynamics? I claim it is the "credo" of linearity since linking it with NL dynamics is otherwise impossible by definition. Apart from this, there are certain NL differential equations for which a kind of superposition principle still exists (related to the linearizability). An equation of that kind is the quadratically NL Riccati equation.

In the following, therefore, it will be shown how one can find complex Riccati equations in time-dependent (TD) quantum mechanics (Sect. 20.2) and time-independent (TI) quantum mechanics (Sect. 20.3). Section 20.4 explains how aspects like irreversibility and dissipation can be incorporated into this formulation of quantum mechanics. The formal similarity with other fields like NL dynamics, thermodynamics, Bose–Einstein condensates and even cosmology (to mention a few examples) will be depicted in Sect. 20.5 and some concluding remarks given in Sect. 20.6.

20.2 Complex Riccati Equations in TD Quantum Mechanics

In the following, TD Schrödinger equations (SE) with at most quadratic Hamiltonian (particularly, the harmonic oscillator (HO) with constant or TD frequency ω, and the free motion, $V = 0$ in the limit $\omega \to 0$) in one dimension shall be considered,

$$i\hbar \frac{\partial}{\partial t} \Psi(x,t) = \left\{ -\frac{\hbar^2}{2m} \frac{\partial^2}{\partial x^2} + \frac{m}{2} \omega^2 x^2 \right\} \Psi(x,t) \; . \tag{20.1}$$

In these cases, Gaussian wave packet (WP) solutions can be obtained that can be written in the form

$$\Psi_{WP}(x,t) = N(t) \; \exp\left\{ i \left[y(t) \tilde{x}^2 + \frac{1}{\hbar} < p > \tilde{x} + K(t) \right] \right\} \tag{20.2}$$

with $\tilde{x} = x - < x > = x - \eta$ where the mean value of position is given by $< x > = \int_{-\infty}^{+\infty} dx \Psi^* x \Psi = \eta(t)$, and $< p > = m\dot{\eta}$ represents the classical momentum. The (possibly TD) normalization factor $N(t)$ and the purely TD function $K(t)$ in the exponent are not relevant to the following discussion. The TD coefficient of the quadratic term in the exponent is assumed to be complex, $y(t) = y_R + i y_I$, where the imaginary part is related to the position uncertainty $< \tilde{x}^2 > = < x^2 > - < x >^2$ via $y_I = \frac{1}{4 < \tilde{x}^2 >}$. The maximum of the WP is located at $x = < x > (t) = \eta$

and, thus, follows the classical trajectory determined by the Newtonian equation of motion that can be obtained by inserting (20.2) into (20.1),

$$\ddot{\eta} + \omega^2 \eta = 0 \,, \tag{20.3}$$

where overdots denote derivatives with respect to time.

The time-evolution of the WP width can be determined in the same way and is governed by the complex NL Riccati equation

$$\dot{\mathscr{C}} + \mathscr{C}^2 + \omega^2 = 0 \tag{20.4}$$

where the complex variable $\mathscr{C}(t)$ is up to a constant factor identical to $y(t)$, i.e., $\mathscr{C} = \frac{2\hbar}{m} y$. Equation (20.4) actually contains all the information about the WP dynamics (including that of Eq. (20.3)) that can also be obtained from the TDSE (20.1), thus providing the desired NL description of the corresponding quantum system. However, due to the nonlinearity, also further information that is not obvious in the linear form can be extracted. This shall be demonstrated subsequently.

There are different ways of treating the (inhomogeneous) Riccati equation. It can be solved directly by transforming it into a homogeneous NL (complex) Bernoulli equation if a particular solution $\tilde{\mathscr{C}}$ of the Riccati equation is known. The general solution of Eq. (20.4) is then given by $\mathscr{C} = \tilde{\mathscr{C}} + \mathscr{V}(t)$ where $\mathscr{V}(t)$ fulfils the Bernoulli equation

$$\dot{\mathscr{V}} + 2\tilde{\mathscr{C}}\mathscr{V} + \mathscr{V}^2 = 0 \,. \tag{20.5}$$

This equation can be linearized via $\mathscr{V} = \frac{1}{\kappa}$ to yield

$$\dot{\kappa} - 2\tilde{\mathscr{C}}\kappa = 1 \tag{20.6}$$

which can be solved straightforwardly. The solution depends sensitively on the initial value κ_0 and contains an integral depending on the particular solution $\tilde{\mathscr{C}}$, $\mathscr{I}(t) = \int^t dt' \, {}^{-\int^{t'} dt'' 2\tilde{\mathscr{C}}(t'')}$. The general solution of the Riccati equation (20.4) can then be written as

$$\mathscr{C} = \tilde{\mathscr{C}} + \frac{d}{dt} \ln\left[\kappa_0 + \mathscr{I}(t)\right] \,, \tag{20.7}$$

or, in the case of constant $\tilde{\mathscr{C}}$, simply in the form

$$\mathscr{C} = \tilde{\mathscr{C}} + \frac{e^{-2\tilde{\mathscr{C}}t}}{\frac{1}{2\tilde{\mathscr{C}}}\left(1 - e^{-2\tilde{\mathscr{C}}t}\right) + \kappa_0} \,, \tag{20.8}$$

defining a one-parameter family of solutions depending on the (complex) initial value κ_0 as parameter. Note that for imaginary $\tilde{\mathscr{C}}$, the exponential functions in (20.8) turn into trigonometric functions and provide, e.g. for the HO with constant ω, Gaussian WPs with oscillating width.

Comparison with supersymmetric (SUSY) quantum mechanics [3, 4] shows that this solution is formally identical to the most general superpotential $W(x)$, fulfilling a *real* Riccati equation and leading to a one-parameter family of isospectral potentials that have the same supersymmetric partner potential (see, e.g., [5–7]). A major difference between the SUSY situation and the one in our TD case (apart from replacing the spatial variable by a temporal one) is the fact that the variables of the NL equations (20.4) and (20.5) are *complex*, whereas $W(x)$ is usually real. Also, the parameter $\kappa_0 = \mathscr{V}_0^{-1}$ in our case is generally complex and determines the initial conditions.

The *complex* NL Riccati equation (20.4) can be rewritten as a *real* NL equation if a new (real) variable $\alpha(t)$ is introduced via $\left(\frac{2\hbar}{m} y_I\right) = \mathscr{C}_I = \frac{1}{\alpha^2}$ where $\alpha(t)$ is directly proportional to the WP width. Inserting this into the imaginary part of Eq. (20.4) allows one to determine the real part of the variable as $\left(\frac{2\hbar}{m} y_R\right) = \mathscr{C}_R = \frac{\dot{\alpha}}{\alpha}$, which, when inserted into the real part of (20.4) together with the above definition of $\mathscr{C}_I$, finally turns the complex Riccati equation into the real NL so-called Ermakov equation[1] for $\alpha(t)$,

$$\ddot{\alpha} + \omega^2 \alpha = \frac{1}{\alpha^3} \,. \tag{20.9}$$

It had been shown by Ermakov [11] in 1880, i.e., 45 years before quantum mechanics was formulated by Schrödinger and Heisenberg, that from the pair of Eqs. (20.3) and (20.9), coupled via ω^2, by eliminating ω^2 from the equations, a dynamical invariant, the Ermakov-invariant

$$\begin{aligned} I_L &= \frac{1}{2}\left[(\dot{\eta}\,\alpha - \dot{\alpha}\,\eta)^2 + \left(\frac{\eta}{\alpha}\right)^2\right] = \frac{1}{2}\alpha^2\left[\left(\dot{\eta} - \frac{\dot{\alpha}}{\alpha}\,\eta\right)^2 + \left(\frac{\eta}{\alpha^2}\right)^2\right] \\ &= \frac{1}{2}\alpha^2\left[(\dot{\eta} - \mathscr{C}_R\eta)^2 + (\mathscr{C}_I\,\eta)^2\right] = \frac{1}{2}\alpha^2\left[(\dot{\eta} - \mathscr{C}\eta)\,(\dot{\eta} - \mathscr{C}^*\eta)\right] = \text{const.} \end{aligned} \tag{20.10}$$

can be obtained (this invariant was rediscovered by several authors, also in a quantum mechanical context; see, e.g., [12–14]).

This invariant has (at least) two remarkable properties: (i) it is also a constant of motion for $\omega = \omega(t)$, in the case where the corresponding Hamiltonian does not have this property; (ii) apart from a missing constant factor m, i.e., mass of the system, it has the dimension of an *action*, not of an energy. The missing factor m is due to the fact that Ermakov used the *mathematical* Eq. (20.3), whereas in a *physical* context, Newton's equation of motion, i.e., Eq. (20.3) multiplied by m, is relevant.

---

[1] This equation had been studied already in 1874 by Adolph Steen [8]. However, Steen's work was ignored by mathematicians and physicists for more than a century, because it was published in Danish in a journal not usually containing many articles on mathematics. An English translation of the original paper [9] and generalizations can be found in [10].

Since the WP solutions of the TDSE (particularly in the context of quantum optics) can also be considered as coherent states (CS), it shall now be shown how the complex Riccati variable can be used to define generalized creation/annihilation operators. These can be used to construct CS with TD width that are no minimum uncertainty WPs but fulfil the Schrödinger–Robertson uncertainty relation [15, 16].

The standard creation/annihilation operators can be obtained by factorizing the Hamiltonian operator H_{op} of the HO [17, 18] or an operator related to it via

$$\hat{H}_{op} = \frac{H_{op}}{\hbar\omega_0} = \left(a^+ a + \frac{1}{2}\right) \tag{20.11}$$

where $a^+ a$ is the so-called number operator and the creation and annihilation operators are defined by

$$a^+ = -i\sqrt{\frac{m}{2\hbar\omega_0}}\left(\frac{p_{op}}{m} + i\,\omega_0\,x\right) = \frac{1}{2\hbar\omega_0}\left(-\frac{\hbar}{\sqrt{m}}\frac{\partial}{\partial x} + \sqrt{m}\omega_0\,x\right) \tag{20.12}$$

$$a = i\sqrt{\frac{m}{2\hbar\omega_0}}\left(\frac{p_{op}}{m} - i\,\omega_0\,x\right) = \frac{1}{2\hbar\omega_0}\left(\frac{\hbar}{\sqrt{m}}\frac{\partial}{\partial x} + \sqrt{m}\omega_0\,x\right) \tag{20.13}$$

where $p_{op} = \frac{\hbar}{i}\frac{\partial}{\partial x}$ and a is the adjoint operator of a^+.

The number that is the eigenvalue of $a^+ a$ is the number of quanta of the action $\hbar$ since $\frac{H_{op}}{\omega_0}$ has the dimension of an *action*! With the help of a, the ground state wave function can be obtained and from this, by successive application of a^+, the excited states can be created. Via superposition of all these states, Schrödinger obtained a stable Gaussian WP (with constant width) [19]. Generalizations of Schrödinger's approach were achieved for the description of coherent light beams emitted by lasers in terms of what is now called coherent state (CS).

One of at least three different definitions of CSs is that these are eigenstates of the annihilation operator a with (complex) eigenvalue z, $a\,|z>= z\,|z>$. Comparing the CS $|z>$ for the HO with the minimum uncertainty WP solution in the form of Eq. (20.2), it shows that $\omega_0 = \mathscr{C}_I = \frac{1}{\alpha_0^2}$. So, in definitions (20.12) and (20.13), $i\;\omega_0$ can be replaced by $i\;\mathscr{C}_I$. Therefore, for the more general case of WPs or CSs with TD width, ($\mathscr{C}_R \neq 0$), $i\;\omega_0$ must be replaced by the full complex quantity $\mathscr{C}$ in a and by $\mathscr{C}^*$ in the adjoint operator a^+. If one then substitutes $\frac{1}{\sqrt{\omega_0}} = \alpha_0$ in front of the brackets with $\alpha(t)$, the generalized creation and annihilation operators take the form

$$a^+(t) = -i\sqrt{\frac{m}{2\hbar}}\,\alpha(t)\,\left(\frac{p_{op}}{m} - \mathscr{C}^*\,x\right) \tag{20.14}$$

$$a(t) = i\sqrt{\frac{m}{2\hbar}}\,\alpha(t)\,\left(\frac{p_{op}}{m} - \mathscr{C}\,x\right). \tag{20.15}$$

These operators can even be turned into constants of motion if an additional phase factor is taken into account. But in the case of CSs, as discussed here, this factor can be absorbed into the phase of the CS and will therefore be omitted in the following (for further details see [20]).

Employing the above definition of the CS, but now with our generalized annihilation operator, i.e., $a(t)\ |z>= z\ |z>$, the CS (in position representation) can be obtained in complete agreement with our WP definition in Eq. (20.2).

The complex eigenvalue z of $a(t)$ can be determined as

$$z = \sqrt{\frac{m}{2\hbar}}\ \alpha\ [\mathscr{C}_I \eta + i\ (\dot{\eta} - \mathscr{C}_R \eta)\ ] = i\ \sqrt{\frac{m}{2\hbar}}\ \alpha\ (\dot{\eta} - \mathscr{C}\eta)\,, \tag{20.16}$$

which looks familiar when compared with the Ermakov invariant (20.10). Indeed, the absolute square of z is, up to a constant factor, identical to I_L,

$$I_L = \frac{\hbar}{m}\left(z_I^2 + z_R^2\right) = \frac{\hbar}{m} z z^* = \frac{\hbar}{m}|z|^2\,. \tag{20.17}$$

An operator, corresponding to I_L can then be written in analogy to $\hat{H}_{op} = \frac{H_{op}}{\hbar\omega_0}$ as

$$\frac{m}{\hbar}\ I_{L,op} = \left[a^+(t)\ a(t) + \frac{1}{2}\right]\,. \tag{20.18}$$

Factorization of this operator was also used [21] to find generalized creation and annihilation operators for the HO with TD frequency but these operators were expressed in terms of α and $\dot{\alpha}$ instead of $\mathscr{C}$.

Another property of the Riccati equation, which is particularly interesting in a quantum mechanical context, is the existence of a superposition principle for this NL differential equation [22–24]. This is related to the fact that the Riccati equation can always be linearized. In our case, this can be achieved using the ansatz

$$\mathscr{C} = \frac{\dot{\lambda}}{\lambda}\,, \tag{20.19}$$

with complex $\lambda(t)$, leading to

$$\ddot{\lambda} + \omega^2 \lambda = 0 \tag{20.20}$$

which has the form of the Newton-type equation (20.3) of the corresponding problem, but now for a complex variable.

First, a kind of geometric interpretation of the motion of λ in the complex plane shall be given. Expressed in Cartesian coordinates, λ can be written as $\lambda = u + i\,v$, or in polar coordinates as $\lambda = \alpha e^{i\varphi}$. Inserting the polar form into Eq. (20.19) leads to

$$\mathscr{C} = \frac{\dot{\alpha}}{\alpha} + i\ \dot{\varphi} \tag{20.21}$$

where the real part is already identical to $\mathscr{C}_R$, as defined above.

The quantity α defined above in $\mathscr{C}_I$ as being proportional to the position uncertainty is identical to the absolute value of λ if it can be shown that

$$\dot{\varphi} = \frac{1}{\alpha^2} \, . \tag{20.22}$$

This, however, can be proven by simply inserting real and imaginary parts of (20.21) into the imaginary part of the Riccati equation (20.4). Comparing relation (20.22), that can also be written in the form

$$\dot{v}u - \dot{u}v = \alpha^2 \dot{\varphi} = 1 \, , \tag{20.23}$$

with the motion of a particle under the influence of a central force in two-dimensional physical space, it shows that this relation corresponds to the "conservation of angular momentum", but here for the motion in the *complex* plane!

Relation (20.23) also shows that real and imaginary parts, or phase and amplitude, respectively, of the complex quantity are not independent of each other but uniquely coupled. This coupling is due to the quadratic nonlinearity in the Riccati equation. We will find an analogous situation also in the TI case, discussed in the next section.

Finally, using the aforementioned results it can be shown that from the solution of the Riccati equation (20.4), also the solution of the Newtonian equation (20.3), and thus the complete information about the dynamics of the WP or the quantum system can be gained. The square root of the inverse of $\mathscr{C}_I$ supplies $\alpha(t)$ integration of $\mathscr{C}_I$ provides φ (t), so $\lambda = \alpha e^{i\varphi}$ can (up to a constant) be determined and it can be shown that the imaginary part of $\lambda = u + iv$ is up to a constant factor identical with the classical trajectory $\eta(t)$, i.e., $v = \frac{m}{\alpha_0 p_0} \eta(t)$ (for details see, e.g., [25]).

20.3 Complex Riccati Equations in TI Quantum Mechanics

In the TD case the real and imaginary parts, or phase φ and amplitude α, respectively, of the complex variable $\lambda(t) = \alpha e^{i\varphi}$ which fulfils the linear equation (20.20) (that is obtained from the Riccati equation (20.4) via Eq. (20.19)) are not independent of each other but coupled via the conservation law (20.23). A similar situation exists when considering the TISE but now in the space-dependent case.

This can be shown using Madelung's hydrodynamic formulation of quantum mechanics [26] where the wave function is written in polar form as

$$\Psi(\mathbf{r}, t) = \varrho^{1/2}(\mathbf{r}, t) \, \exp\left(\frac{i}{\hbar} S(\mathbf{r}, t) \right) \tag{20.24}$$

with the square root of the probability density $\varrho = \Psi^* \Psi$ as amplitude and $\frac{1}{\hbar} S$ as phase ($\mathbf{r}$ is the position vector in three dimensions).

Inserting this form into the TDSE (20.1) (now in three dimensions, and replacing $\frac{\partial}{\partial x}$ by the nabla operator ∇), leads to a modified Hamilton–Jacobi equation for the phase,

$$\frac{\partial}{\partial t}S + \frac{1}{2m}(\nabla S)^2 + V - \frac{\hbar^2}{2m}\frac{\Delta \varrho^{1/2}}{\varrho^{1/2}} = 0\,, \tag{20.25}$$

and a continuity equation for the amplitude,

$$\frac{\partial}{\partial t}\varrho + \frac{1}{m}\nabla(\varrho\,\nabla S) = 0\,. \tag{20.26}$$

Already here, the coupling of phase and amplitude can be seen clearly since the Hamilton–Jacobi equation for the phase S contains a term (misleadingly called "quantum potential", V_{qu}) depending on ϱ, and the continuity equation for the density ϱ contains ∇S. It can be shown that also in the TI case this coupling is not arbitrary but related to a conservation law.

In 1994, G. Reinisch [27] presented a nonlinear formulation of TI quantum mechanics. Since in this case $\frac{\partial}{\partial t}\varrho = 0$ and $\frac{\partial}{\partial t}S = -E$ are valid, the continuity equation (20.26) (we now use the notation $\varrho^{1/2} = |\Psi| = a$) turns into

$$\nabla(a^2 \nabla S) = 0 \tag{20.27}$$

and the modified Hamilton–Jacobi equation into

$$-\frac{\hbar^2}{2m}\Delta a + (V - E)\,a = -\frac{1}{2m}(\nabla S)^2 a\,. \tag{20.28}$$

Equation (20.27) is definitely fulfilled for $\nabla S = 0$, turning (20.28) into the usual TISE for the real wave function $a = |\Psi|$ with position-independent phase S. (N.B.: the kinetic energy term divided by a is just identical to V_{qu}!)

However, Eq. (20.27) can also be fulfilled for $\nabla S \neq 0$ if only the conservation law

$$\nabla S = \frac{C}{a^2} \tag{20.29}$$

is fulfilled with constant (or, at least, position-independent) C.

This relation now shows explicitly the coupling between phase and amplitude of the wave function and is equivalent to Eq. (20.22) in the TD case. Inserting (20.29) into the rhs of Eq. (20.28) changes this into the Ermakov equation

$$\Delta a + \frac{2m}{\hbar^2}(E - V)\,a = \left(\frac{1}{\hbar}\nabla S\right)^2 a = \left(\frac{C}{\hbar}\right)^2 \frac{1}{a^3}\,, \tag{20.30}$$

equivalent to Eq. (20.9) in the TD case.

So far the energy E occurring in Eq. (20.30) is still a free parameter that can take any value. However, solving this equation numerically for arbitrary values of E leads, in general, to solutions a that diverge for increasing x. Only if the energy E is appropriately tuned to any eigenvalue E_n of the TISE (see Eq. (20.32), below) does this divergence disappear and normalizable solutions can be found. So the quantization condition that is usually obtained from the requirement of the truncation of an infinite series in order to avoid divergence of the wave function is, in this case, obtained from the requirement of non-diverging solutions of the nonlinear Ermakov equation (20.30) by variation of the parameter E. This has been verified numerically in the case of the one-dimensional HO and the Coulomb problem and there is the conjecture that this property is "universal" in the sense that it does not depend on the potential V (see [27, 28]).

The corresponding complex Riccati equation is now given by

$$\nabla\left(\frac{\nabla\Psi}{\Psi}\right)+\left(\frac{\nabla\Psi}{\Psi}\right)^2+\frac{2m}{\hbar^2}(E-V)=0 \tag{20.31}$$

with the complex variable $\left(\frac{\nabla\Psi}{\Psi}\right)=\frac{\nabla a}{a}+i\frac{1}{\hbar}\nabla S$ which corresponds to $\left(\frac{2\hbar}{m}y\right)=C=\frac{\dot\lambda}{\lambda}=\frac{\dot\alpha}{\alpha}+i\,\dot\varphi$ in the TD problem.

It is possible to show straightforwardly that Eq. (20.31) can be linearized to yield the usual TISE

$$-\frac{\hbar^2}{2m}\Delta\Psi+V\,\Psi=E\,\Psi\,, \tag{20.32}$$

but in this form the information on the coupling of phase and amplitude, expressed by Eq. (20.29) and originating from the quadratic NL term in Eq. (20.31), gets lost.

20.4 Dissipative Systems

The conventional way of treating open dissipative systems uses the system-plus reservoir approach, i.e., the system of interest is coupled to some (in the limit infinitely many) environmental degrees of freedom (often harmonic oscillators) and system plus reservoir together are considered a closed Hamiltonian system. Taking certain limits and applying averaging processes finally leads to an irreversible dissipative equation of motion for the system of interest. One of the most often-quoted approaches of that kind is the one of Caldeira and Leggett [29, 30]. A similar idea, but with the most minimalistic environment, namely only one additional position variable (plus the corresponding momentum) is behind the Bateman Hamiltonian [31] that represents a constant of motion and provides an irreversible equation of motion for the system,

$$\ddot{x} + \gamma\,\dot{x} + \frac{1}{m}\frac{\partial}{\partial x}V = 0\,, \tag{20.33}$$

i.e., a Newtonian equation with an additional linear velocity (or momentum) dependent friction force (with friction coefficient γ); actually the Langevin equation without stochastic force.

Since the environmental degrees of freedom are eliminated or ignored in the end anyway, several approaches exist where only the effect of the environment on the observable system is taken into account without considering the individual environmental degrees of freedom. This can lead to modifications of the classical Lagrange/Hamilton formalism where the corresponding modified (linear) Schrödinger equation is obtained via subsequent canonical quantization. The canonical variables of these approaches are related with the physical position and momentum variables via non-canonical transformations in the classical case, corresponding to non-unitary transformations in the quantum mechanical case (for further details see [32, 33]). The most frequently applied approach of that kind is the one of Caldirola [34] and Kanai [35] which is uniquely related to one using an exponentially expanding coordinate system [36], leading to a Hamiltonian that is still a constant of motion. These approaches can be linked directly to the aforementioned ones. Using standard methods to eliminate the environmental degrees of freedom, Yu and Sun [37, 38] have shown how the conventional approach of Caldeira–Leggett leads directly to the Hamiltonian operator of Caldirola–Kanai. It is also possible to eliminate the second set of variables of the Bateman approach by imposing time-dependent constraints [39] to get to the Hamiltonian of the expanding system. Furthermore, this approach and the one of Caldirola–Kanai are connected via an explicitly time-dependent canonical transformation. In our context it is interesting that for these two approaches also an exact Ermakov invariant exists. In the quantized version, Gaussian wave packet solutions can be obtained in the same cases as in the conventional reversible theory, but now the maximum follows a damped motion according to Eq. (20.33) and the time-dependence of the width is determined by a modified complex Riccati equation that can again be transformed into a (real) Ermakov-type equation.

Another type of effective approaches starts already on the quantum level by adding some friction terms $W(x, p_{op}, t; \Psi)$ to the Hamiltonian operator. This usually leads to NL Hamiltonians, $H_{NL} = H_L + W$, where quite different forms of nonlinearities are considered in the literature (some are NL only because some mean-value $< \ldots >$ occurs in W) ([40–47], Süssmann 1973, unpublished, [48–50]). Of these, an exact invariant was found [51,52] for only two approaches [48,49].

In the following, only those NLSEs possessing an Ermakov invariant shall be discussed explicitly since it can be shown that the canonical approaches are unambiguously related to these by a non-unitary transformation [32,33]. In particular, the equations of motion for the WP maximum and width can be uniquely transformed into each other [33]. The approach of Hasse [48] uses a combination of products of position and momentum operators and their mean values. The other one [58] is based on an irreversible Fokker–Planck-type equation for the probability density that is obtained from the usual continuity equation by adding a time-symmetry-breaking diffusion term. Following a method by Madelung [53] and Mrowka [54] this (real) so-called Smoluchowski equation can be separated into two complex equations: namely a modified SE for the wave function Ψ and its complex conjugate Ψ^*, provided the separation condition

$$- D \frac{\frac{\partial^2}{\partial x^2} \varrho}{\varrho} = \gamma \left(\ln \varrho - < \ln \varrho > \right) \tag{20.34}$$

with diffusion coefficient D is fulfilled (for details see, e.g., [49,55]).

This leads to the NLSE

$$i\hbar \frac{\partial}{\partial t} \Psi_{NL}(x,t) = \{ H_L + \gamma \frac{\hbar}{i} \left(\ln \Psi_{NL} - < \ln \Psi_{NL} > \right) \} \Psi_{NL}(x,t) \tag{20.35}$$

with complex logarithmic nonlinearity.

The additional NL term (W_{SCH}) can be written as real and imaginary contributions in the form

$$W_{\mathrm{SCH}} = W_R + i W_I = \frac{\gamma}{2} \frac{\hbar}{i} \left(\ln \frac{\Psi_{NL}}{\Psi_{NL}^*} - < \ln \frac{\Psi_{NL}}{\Psi_{NL}^*} > \right) + \frac{\gamma}{2} \frac{\hbar}{i} \left(\ln \varrho_{NL} - < \ln \varrho_{NL} > \right) \tag{20.36}$$

where the real part only depends on the phase of the wave function and provides the friction force in the averaged equation of motion. The imaginary part does not contribute to dissipation but introduces irreversibility into the evolution of the wave function. It corresponds to the diffusion term in the Smoluchowski equation, but still allows for normalizability due to the subtraction of the mean value of $\ln \varrho$. Comparison with the approaches mentioned earlier shows that the real part is just identical to Kostin's term [45] and the imaginary part corresponds to Beretta's term [42–44] introduced to describe non-equilibrium systems (without dissipation).

The imaginary part breaks the time-reversal symmetry on the level of the probability density, introduces a non-unitary time evolution and turns the Hamiltonian into a non-Hermitian one while still guaranteeing normalizable wave functions and real energy mean values since its mean value vanishes.

Also from the real part of W_{SCH} no additional term to the energy mean value occurs, so this is still given by the mean value of the operators of kinetic and potential energies. This real part is however not arbitrary but uniquely determined by the separation condition and provides the correct dissipative friction forces in the equation of motion for the mean values. Besides, the ratio of energy dissipation (for the classical contribution) is in agreement with the classical counterpart and arises because the mean values are calculated with Ψ_{NL} (the solution of Eq. (20.35)) instead of Ψ_L.

The real part, by itself, would provide dissipation but retain a unitary time-evolution of the wave function; whereas the imaginary part, on its own, would provide irreversibility via a non-unitary time-evolution but no dissipation. Consequently, only the combination of real and imaginary parts provides all the desired properties of the quantum system under consideration. The reason for this is the coupling of phase and amplitude of the wave function since W_R depends on the phase and W_I on the amplitude.

The relation between the two NL approaches is discussed in detail in [33] and can be traced back to a modification of the Riccati equation (20.31) by adding a linear term. The two NLSEs have the same WP solutions where, in both cases, the maximum $\eta(t)$ follows an equation of motion, like (20.33), with a linear friction force and the WP width is determined by the modified Riccati equation

$$\dot{\mathscr{C}}_{NL} + \gamma\,\mathscr{C}_{NL} + \mathscr{C}_{NL}^2 + \omega^2 = 0 \tag{20.37}$$

with an additional linear term depending on γ.

As in the conservative case, $\mathscr{C}_{NL,I} = \frac{\hbar}{2m<\tilde{x}^2>_{NL}} = \frac{1}{\alpha_{NL}^2}$ is valid but the real part of the complex Riccati variable now takes the modified form

$$\mathscr{C}_{NL,R} = \frac{\dot{\alpha}_{NL}}{\alpha_{NL}} - \frac{\gamma}{2}\,. \tag{20.38}$$

he corresponding Ermakov equation and invariant are now given by

$$\ddot{\alpha}_{NL} + \left(\omega^2 - \frac{\gamma^2}{4}\right)\alpha_{NL} = \frac{1}{\alpha_{NL}^3} \tag{20.39}$$

and

$$I_{NL} = \frac{1}{2}\alpha_{NL}^2\left[\left(\dot{\eta} - \left(\frac{\dot{\alpha}_{NL}}{\alpha_{NL}} - \frac{\gamma}{2}\right)\eta\right)^2 + \left(\frac{1}{\alpha_{NL}^2}\eta\right)^2\right]e^{\gamma t} = \text{const.} \tag{20.40}$$

From this it is obvious that, apart from the factor $e^{\gamma t}$, I_{NL} can be written in exactly the same form as in the conservative case if expressed in terms of η and $\mathscr{C}$ instead of α and $\dot{\alpha}$, i.e.,

$$I_{NL} = \frac{1}{2}\alpha_{NL}^2\left[(\dot{\eta} - \mathscr{C}_{NL,R}\,\eta)^2 + (\mathscr{C}_{NL,I}\,\eta)^2\right]e^{\gamma t}\,, \tag{20.41}$$

which again shows the more universal validity of relations when expressed in terms of the Riccati variable. Also, in this dissipative case, the invariant (without the exponential factor) can be factorized to yield generalized creation and annihilation operators where the CS obtained as eigenstates of the annihilation operator are identical to the WP solutions of the NLSEs (for details, see [20]).

20.5 Similarities with NL Dynamics and Other Fields of Physics

In NL dynamics, an important phenomenon is the Hopf bifurcation as it can be the first step on a route to turbulence and chaos [56]. A system displaying this property can be described by the NL evolution equation

$$\dot{r} + \Gamma r + r^3 = 0 \tag{20.42}$$

which has the solution

$$r^2(t) = \frac{\Gamma \, r_0^2 \, e^{-2\,\Gamma t}}{r_0^2 \, (1 - e^{-2\,\Gamma t}) + \Gamma} \,. \tag{20.43}$$

For $\Gamma \geq 0$, the trajectory approaches a fixed point (the origin); however, for $\Gamma < 0$, it spirals towards a limit cycle with radius $r_\infty = |\Gamma|^{1/2}$ [56]. The same type of differential equation is also discussed by Großmann with respect to self-similarity and scale-invariance (see [57]).

The relation to our Riccati equations (20.4) or (20.37) is easily seen by multiplying equation (20.42) by $4r$ and introducing a new variable $R = 2r^2$, leading to

$$\dot{R} + 2\,\Gamma\, R + R^2 = 0\,. \tag{20.44}$$

This is exactly the form of the Bernoulli equation (20.5) that can be obtained if a particular solution $\tilde{\mathscr{C}}$ of the Riccati equation is known. The coefficient 2Γ of the linear term in Eq. (20.44), corresponds to $2\,\tilde{\mathscr{C}}$ in Eq. (20.5) and, in the dissipative case, is simply replaced by $2\,\tilde{\mathscr{C}} + \gamma$. It has indeed been shown that, in the dissipative case, this bifurcation occurs and one obtains two different WP solutions with different behaviour in the spreading of the WP width, different uncertainties and different energies (for details, see [58]).

Similarities with statistical thermodynamics become obvious if solution (20.8) of the Riccati equation is rewritten as

$$\mathscr{C} = \tilde{\mathscr{C}} + \frac{2\,\tilde{\mathscr{C}}}{\kappa_0\, 2\,\tilde{\mathscr{C}}\, e^{2\tilde{\mathscr{C}}t} + \left(e^{2\tilde{\mathscr{C}}t} - 1\right)} \tag{20.45}$$

where the second term on the rhs is the solution of the corresponding homogeneous Bernoulli equation (see Eq. (20.5)). Choosing $\kappa_0 = 0$, this term just has the form of a Bose–Einstein distribution. Furthermore, with the substitutions $2\,\tilde{\mathscr{C}} = \hbar\,\omega$, $t = \frac{1}{kT}$, the (real) expression

$$\frac{\hbar}{2}\,\omega + \frac{\hbar\,\omega}{e^{\hbar\omega/kT} - 1} = \frac{\hbar}{2}\,\omega\, \coth\left(\frac{\hbar\omega}{2kT}\right) = \langle \text{energy} \rangle \tag{20.46}$$

is obtained which is the average energy of a single oscillator in thermal equilibrium at temperature T (k = Boltzmann's constant). The first term, corresponding to the particular solution of the respective Riccati equation and is the ground state energy of the HO; the second term corresponding to the solution of the Bernoulli equation is the energy distribution obtained by Planck for the black body radiation. For the choice $2\tilde{\mathscr{C}}\,\kappa_0 = -2$, this second term attains the form of a Fermi–Dirac distribution.

So far, the examples only consider real Riccati equations, but further complex cases can also be found. The Gross–Pitaevskii equation is a (complex) SE with cubic nonlinearity,

$$i\hbar\frac{\partial}{\partial t}\Psi = \left\{-\frac{\hbar^2}{2m}\Delta + V(\mathbf{r},t) + g|\Psi|^2\right\}\Psi\,, \tag{20.47}$$

which is used in a mean field approximation to describe the macroscopic WP of a BEC where $V(\mathbf{r},t)$ can be given by $V(\mathbf{r},t) = \frac{m}{2}\omega^2(t)r^2$, i.e., a HO with TD frequency; g parametrizes the strength of the atomic interaction.

Although Eq. (20.47) cannot be solved analytically, the dynamics of the BEC characterized by this equation can be described in terms of so-called moments M_n ($n = 1 - 4$) (for details, see e.g., [59]), where M_1 represents the norm, M_2 the width, M_3 the radial momentum and M_4 the energy of the WP. It can be shown that these moments satisfy a set of coupled first-order differential equations (where $\frac{d}{dt}M_1$ corresponds to the conservation of probability or particle number). This set can be reduced to a single equation for M_2 which can be expressed, using a new variable $X = \sqrt{M_2}$, in the form of an Ermakov equation,

$$\ddot{X} + \omega^2(t)X = \frac{k}{X^3}\,, \tag{20.48}$$

which, as shown in Sect. 20.2, is equivalent to a complex Riccati equation.

To include dissipative effects, one could add another NL term like the logarithmic one from Eq. (20.35) to the Gross–Pitaevskii equation which would correspond to adding a linear term to the Riccati equation. So one simply has to solve this modified Riccati equation (or the corresponding Ermakov equation) to obtain all moments M_n for the dissipative BEC [60].

This treatment of the BEC is also interesting for another reason. It has been shown by Lidsey [61] that a correspondence can be established between positively-curved isotropic, perfect fluid cosmologies and the two-dimensional harmonically-trapped BEC by mapping the equations of motion for both systems onto a one-dimensional Ermakov equation. The moments M_n defined above can be identified in the cosmological context with M_2 = scale factor, M_3 = Hubble expansion parameter and M_4 = energy density of the universe. So the expanding universe can be represented as an Ermakov or complex Riccati system.

More examples could be mentioned from fields like electrodynamics, optics, quantum optics, supersymmetry, quantum gravity (see, e.g., [62, 63]) and others, but further details would go beyond the scope of this article.

20.6 Conclusions and Perspectives

In classical physics only real quantities have any physical significance and energy in the form of Hamiltonians or Lagrangians plays the dominant role. In quantum physics however, action, i.e., the product of energy and time (or position and momentum), is essentially the quantized entity. The appearance of $i = \sqrt{-1}$, and hence the use of complex quantities in quantum mechanics, is not just a matter of mathematical convenience but has fundamental physical meaning. A nonlinear version of quantum mechanics should therefore be formulated in terms of a complex nonlinear differential equation that contains the same information about the quantum system as the complex linear SE and is somehow linked to an invariant with the dimension of action.

An example of that kind was found in the TDSE where the time-evolution of the quantum uncertainties obeys a complex NL Riccati equation. The linearized version of this Riccati equation is just a complex Newtonian equation of motion for a quantity $\lambda(t)$ where the coupling of phase and amplitude of this quantity corresponds to the conservation of angular momentum for the motion of λ in the complex plane! The imaginary part of $\lambda(t)$ is up to a constant factor just the classical trajectory containing the information about the mean value of position.

The complex TD Riccati equation (or its transformed version, the real nonlinear Ermakov equation) together with the classical Newtonian equation for the system lead to a dynamical invariant with the dimension of action. When the operator corresponding to this Ermakov invariant is factorized, one obtains generalized creation and annihilation operators that also apply in cases where the Hamiltonian is no longer invariant. Specifically, this is also valid for certain dissipative systems when the Ermakov invariant is expressed in terms of the complex Riccati variable. This has been shown using some effective models for the description of such open systems.

In time-independent quantum mechanics the complex wave function fulfils a linear differential equation. We have seen that this TISE is actually a linearized form of a complex nonlinear Riccati equation. Why should one bother with a more complicated nonlinear equation if there is a simpler linear version at hand for which such nice properties like a superposition principle exist? Because, in the linear form, it is not obvious that real and imaginary part, or phase and amplitude, respectively, of the complex wave function are not independent of each other but uniquely coupled via a kind of conservation law, equivalent to the conservation of angular momentum in the complex plane in the TD case. This coupling can be traced back to the quadratic nonlinearity in the Riccati equation and always occurs in systems that can be described by complex Riccati equations.

Finally, an initial link to nonlinear dynamics was made where properties like scale-invariance, bifurcations as a route to chaos and other similar properties already emerge when real Riccati equations apply. Formal similarities with statistical thermodynamics (also related to the solution of real Riccati equations) as well as a complex Riccati equation (or the equivalent Ermakov equation) in the context os BECs were also mentioned. The relations to our nonlinear version of quantum mechanics, in particular the effect of "complexification" will be further investigated. In addition, formal similarities to fields like SUSY quantum mechanics, quantum optics and cosmology shall be explored.

Acknowledgements The author would like to thank the organizers of "New Challenges in Complex System Physics" for inviting him to this interesting and stimulating NATO Advanced Research Workshop in Samarkand.

References

1. Peitgen H-O, Richter PH (1986) The beauty of fractals. Springer, Heidelberg
2. Yang CN (1987) Square root of -1, complex phases and Schroedinger. In: Kilmister CW (ed) Schrödinger – centenary celebration of a polymath. Cambridge University Press, Cambridge, p 53
3. Mielnik B, Rosas-Ortiz O (2004) Factorization: little or great algorithm? J Phys A Math Gen 37:10007
4. Andrianov A, Cannata F (2004) Nonlinear supersymmetry for spectral design in quantum mechanics. J Phys A Math Gen 37:10297
5. Fernández D (2010) Higher-order supersymmetric quantum mechanics. AIP Conf Proc 1287:3
6. Cooper F, Khare A, Sukhatme U (2001) Supersymmetry in quantum mechanics. World Scientific, Singapore
7. Khare A, Sukhatme U (1989) Phase-equivalent potentials obtained from supersymmetry. J Phys A Math Gen 22:2847; Keung W-Y, Sukhatme UP, Wang Q, Imbo TD (1989) Families of strictly isospectral potentials. J Phys A Math Gen 22:L987
8. Steen A (1874) Om Formen for Integralet af den lineaere Differentialligning af an den Orden. Overs over d K Danske Vidensk Selsk Forh 1
9. Redheffer R (1999) Steen's equation and its generalizations. Aequtiones Math 58:60
10. Redheffer R, Redheffer I (2001) Steen's 1874 paper: historical survey and translation. Aequationes Math 61:131
11. Ermakov VP (1880) Second-order differential equations. Conditions of complete integrability. Univ Jzv Kiev Ser III 20(9):1–25
12. Milne WE (1930) The numerical determination of characteristic numbers. Phys Rev 35:863
13. Pinney E (1950) The nonlinear differential equation $y''+p(x)y+cy^{-3}=0$. Proc Am Math Soc 1:681
14. Lewis HR (1967) Classical and quantum systems with time-dependent harmonic-oscillator-type Hamiltonians. Phys Rev Lett 18:510
15. Schrödinger E (1930) Zum Heisenbergschen Unschärfeprinzip. Sitzungsber Preuss Akad Wiss (Berlin) 296
16. Robertson HP (1929) The uncertainty principle. Phys Rev 34:163; Robertson HP (1930)
17. Schrödinger E (1940) An indeterminacy relation for several observables and its classical interpretation. Proc R Irish Acad 46:A9
18. Dirac PAM (1935) The principles of quantum mechanics. Oxford University Press, Oxford, p 34
19. Schrödinger E (1926) Quantisierung als Eigenwertproblem. (Erste Mitteilung.). Ann d Phys 79:361
20. Castaños O, Schuch D, Rosas-Ortiz O (2013) Generalized coherent states for time-dependent and nonlinear Hamiltonian operators via complex Riccati equations. J Phys A Math Theor 46:075304

21. Hartley JG, Ray JR (1982) Coherent states for the time-dependent harmonic oscillator. Phys Rev D 25:382
22. Hartley JG, Ray JR (1981) Ermakov systems and quantum-mechanical superposition laws. Phys Rev A 24:2873
23. Sarlet W, Cantrijn F (1982) A generalization of the nonlinear superposition idea for Ermakov systems. Phys Lett A 88:383
24. Kevrekidis PG, Drossinos Y (2007) Nonlinearity from linearity: the Ermakov-Pinney equation revisited. Math Comp Sim 74:196
25. Schuch D, Moshinsky M (2006) Connection between quantum-mechanical and classical time evolution via a dynamical invariant. Phys Rev A 73:062111
26. Madelung E (1926) Quantentheorie in hydrodynamischer Form. Z Physik 40:322
27. Reinisch G (1994) Nonlinear quantum mechanics. Physica A 206:229
28. Reinisch G (1997) Classical position probability distribution in stationary and separable quantum systems. Phys Rev A 56:3409
29. Caldeira AO, Leggett AJ (1981) Influence of dissipation on quantum tunneling in macroscopic systems. Phys Rev Lett 46:211
30. Caldeira AO, Leggett AJ (1983) Quantum tunnelling in a dissipative system. Ann Phys (NY) 149:374; Caldeira AO, Leggett AJ (1983) Erratum Ibid. Ann Phys (NY) 153:445(E)
31. Bateman H (1931) On dissipative systems and related variational principles. Phys Rev 38:815
32. Schuch D (1997) Nonunitary connection between explicitly time-dependent and nonlinear approachesfor the description of dissipative quantum systems. Phys Rev A 55:935
33. Schuch D (2012) Complex Riccati equations as a link between different approaches for the description of dissipative and irreversible systems. J Phys Conf Ser 380:012009
34. Caldirola P (1941) Forze non conservative nella meccanica quantistica. Nuovo Cimento 18:393
35. Kanai E (1948) On the quantization of the dissipative systems. Progr Theor Phys 3:440
36. Schuch D (1999) Effective description of the dissipative interaction between simple model-systems and their environment. Int J Quantum Chem 72:537–547
37. Yu LH, Sun CP (1994) Evolution of the wave function in a dissipative system. Phys Rev A 49:592
38. Sun CP, Yu LH (1994) Exact dynamics of a quantum dissipative system in a constant external field. Phys Rev A 51:1845
39. Schuch D, Guerrero J, López-Ruiz FF, Aldaya V (submitted) Interrelations between different canonical descriptions of dissipative systems. Phys Lett A
40. Gisin N (1981) A simple nonlinear dissipative quantum evolution equation. J Phys A 14:2259
41. Gisin N (1982) Microscopic derivation of a class of non-linear dissipative Schrödinger-like equations. Physica A 111:364
42. Beretta GP (1987) Quantum thermodynamics of nonequilibrium. Onsager reciprocity and dispersion-dissipation relations. Found Phys 17:365
43. Beretta GP (2006) Nonlinear model dynamics for closed-system, constrained, maximal-entropy-generation relaxation by energy redistribution. Phys Rev E 73:026113
44. Beretta GP (2010) Maximum entropy production rate in quantum thermodynamics. J Phys Conf Ser 237:012004
45. Kostin MD (1972) On the Schrödinger-Langevin equation. J Chem Phys 57:3589
46. Cho BR (1980) Suhak Gwa Mulri (Korean) (Math Phys) 3:37
47. Albrecht K (1975) A new class of Schrödinger operators for quantized friction. Phys Lett B 56:127
48. Hasse RW (1975) On the quantum mechanical treatment of dissipative systems. J Math Phys 16:2005
49. Schuch D, Chung K-M, Hartmann H (1983) Nonlinear Schrödinger-type field equation for the description of dissipative systems. I. Derivation of the nonlinear field equation and one-dimensional example. J Math Phys 24:1652
50. Doebner H-D, Goldin GA (1992) On a general nonlinear Schrödinger equation admitting diffusion currents. Phys Lett A 162:397

51. Nassar AB (1986) Ermakov and non-Ermakov systems in quantum dissipative models. J Math Phys 27:755
52. Nassar AB (1986) Time-dependent invariant associated to nonlinear Schrödinger-Langevin equations. J Math Phys 27:2949
53. Madelung E (1950) Die Mathematischen Hilfsmittel des Physikers. Springer, Berlin, p 432
54. Mrowka B (1951) Zur Darstellung der Quantenmechanik I. Z Phys 130:164
55. Schuch D, Chung K-M (1986) From macroscopic irreversibility to microscopic reversibility via a nonlinear Schrödinger-type field equation. Int J Quantum Chem 29:1561
56. Schuster HG (1984) Deterministic chaos; an introduction. Physik Verlag, Weinheim, p 112
57. Großmann S (1989) Selbstaehnlichkeit, Das Strukturgesetz im und vor dem Chaos. In: Gerok W (ed) Ordnung und Chaos. Wissenschaftliche Verlagsgesellschaft, Stuttgart, p 101
58. Schuch D (2002) New energetic and dynamic quantum effects originating from the breaking of time-reversal symmetry. J Phys A Math Gen 35:8615
59. García-Ripoll JJ, Pérez-García VM, Torres P (1999) Extended parametric resonances in nonlinear Schrödinger systems. Phys Rev Lett 83:1715
60. Schuch D, Kaushal RS (2011) Some remarks on dissipative Ermakov systems and damping in Bose-Einstein condensates. J Phys Conf Ser 306:012032
61. Lidsey JE (2004) Cosmic dynamics of Bose-Einstein condensates. Class Quantum Grav 21:777
62. Kiefer C (1992) Decoherence in quantum electrodynamics and quantum gravity. Phys Rev D 46:1658
63. Kiefer C (2012) Quantum gravitational contributions to the cosmic microwave background anisotropy spectrum. Phys Rev Lett 108:021301

Index

D. Matrasulov and H.E. Stanley (eds.), *Nonlinear Phenomena in Complex Systems: From Nano to Macro Scale*, NATO Science for Peace and Security Series C: Environmental Security, DOI 10.1007/978-94-017-8704-8,